Liquid Phase
High Pressure Chemistry

Liquid Phase
High Pressure Chemistry

Neil S. Isaacs

Department of Chemistry,
University of Reading

A Wiley–Interscience Publication

JOHN WILEY & SONS

Chichester · New York · Brisbane · Toronto

British Library Cataloguing in Publication Data:

Isaacs, Neil Stewart
   Liquid phase high pressure chemistry.
   1. Chemistry, Physical and theoretical
   2. High pressure (Chemistry)
   I. Title
   541′.34   QD453.2   80-40844

   ISBN 0 471 27849 1

   Typeset in Great Britain by John Wright & Sons Ltd., at the Stonebridge Press, Bristol and printed at The Pitman Press, Bath

Contents

Preface

High pressure chemistry like many another area of science is rapidly expanding both in terms of the numbers of its practitioners and their rates of publication. LeNoble's review of the literature up to 1966 recorded some 470 volumes of activation. The succeeding decade saw the publication of more than 1400 volumes of activation and reaction. While copious information of this type is now available, many chemists are still unaware of the value of high pressure techniques, which may be adapted to almost any physical measurement. The justification of this volume, which embodies a wide range of subject matter at a level below that of a specialist publication, lies in increasing the awareness of the practising chemist of techniques which could prove advantageous, and in providing a useful summary of information to those engaged in high pressure research. It is a pleasure to acknowledge the assistance of the following authorities in compiling this volume: Professor LeNoble (University of New York, Stoneybrook), Dr. E. Whalley (National Research Council of Canada), Dr. K. A. H. Heremans (Catholic University of Leuven), and Dr. T. M. Hardman (University of Reading).

Reading, January 1980   N. S. ISAACS

Chapter 1

Apparatus for High Pressure Studies

Neither the chemicals nor the physical properties of condensed systems—liquids and solids—may be said to be highly susceptible to pressure. Consequently, in order to observe the many interesting phenomena brought about by pressure it is necessary to work in the range measured in kilobars or even hundreds of kilobars, possibly also at extremes of temperature. Chemical kinetic studies are relatively undemanding (say up to 5 kbar, and $-100°$ to $+100°$) while preparative chemistry may require the highest pressures attainable,

2

together with the greatest sample space (for example, 20 kbar for some organic syntheses (section 4.4), up to 100 kbar for diamond synthesis). Spectroscopic properties and phase studies may need pursuing to the highest possible limits though samples may often be extremely small. Perhaps the most demanding applications are in the realm of geochemistry where, in seeking to duplicate conditions within the earth's interior, it may be desirable to carry out work at > 100 kbar, 2000 °C. Such conditions call for the toughest materials possible and also great sophistication in design. The following account of apparatus emphasizes the equipment useful in chemical studies with examples from the many hundreds of designs available in the literature. It is usual for the worker in high pressures to design and build his own apparatus, though increasingly high pressure apparatus is becoming available commercially (Appendix 3).

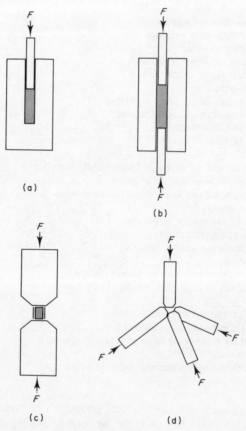

Fig. 1.1. Arrangements for the containment of high pressures.
(a) Simple piston and cylinder.
(b) Cylinder with opposed pistons.
(c) Bridgman anvil.
(d) Tetrahedral multi-anvil device

1.1   THE PRODUCTION OF HIGH PRESSURES

Four principal methods for imposing high pressures upon samples are:
(a) the piston and cylinder (fig. 1.1(a, b))
(b) anvil presses (e.g. fig. 1.1(c, d))
(c) pressurization by phase change
(d) shock waves
Of these, the first is by far the most important from a chemical point of view, though (b) includes techniques necessary for pressures above about 50 kbar. Phase changes are of limited importance and shock wave methods, producing very transient surges of extreme pressure tend to be a highly specialized tool.

1.1.1   The Piston and Cylinder

For liquid and gaseous samples working up to about 50 kbar, some sort of piston and cylinder apparatus is appropriate. In principle, the apparatus consists of a strong steel vessel with uniform bore extending either partly or wholly through the piece, fig. 1.1(a, b) and either one or two close fitting pistons and a means of exerting force on them. In its simplest form, a steel cylinder may withstand about 25 kbar though this figure may be raised considerably by more elaborate design.

1.1.2   The Strength of a Cylinder[178]

This engineering problem was first attacked by Lamé and Clapeyron (1833) and subsequently by many workers on account of the importance of the theory for industrial processes.[1-13] A closed cylinder under pressure ($P_{ext} < P_{int}$) must

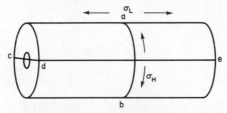

Fig. 1.2. Stresses in an internally pressurized
hollow cylinder

withstand tension in the walls both longitudinally and circumferentially (fig. 1.2), the latter being the greater:

$$\text{Longitudinal stress (N m}^{-2}\text{),} \quad \sigma_L = \frac{\text{Force}}{\text{Area}} = \frac{\pi r^2 P}{2\pi r t} = \frac{Pr}{2t} \qquad (1.1)$$

where $P = P_{int} - P_{ext}$, r = radius of the vessel, t = wall thickness

$$\text{Circumferential (hoop) stress,} \quad \sigma_H = \frac{2rP}{2t} = \frac{Pr}{t} \qquad (1.2)$$

Hence $\sigma_H \approx 2\sigma_L$ and the cylinder must be designed against hoop stress.

4

These equations refer to thin-walled cylinders but must be modified for the thick-walled vessels which are inevitably used, thus

$$\sigma_H = P\frac{(R^2+r^2)}{(R^2-r^2)} \tag{1.3}$$

where R, r are external and internal radii, respectively. This formula is often given in the form

$$\sigma_H = P\frac{(K^2+1)}{(K^2-1)} \tag{1.4}$$

where $K = R/r$ or in the form

$$\sigma_H = \frac{P}{K^2-1}\left(1+\frac{R^2}{r'^2}\right)$$

where r' refers to the radius at any point in the wall. The analysis and testing of this problem is discussed at length in references 14–22. Here, the stress σ_H refers to the inner surface of the vessel and in practice falls off rapidly through the wall (fig. 1.4). Equation (1.3) predicts that a thicker wall will reduce the inside stress but

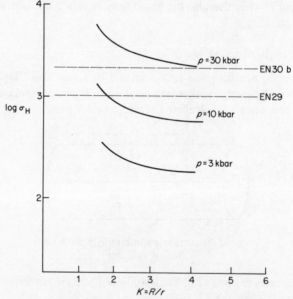

Fig. 1.3. Curves showing stress at the inner surface of a pressurized cylinder as a function of relative wall thickness. The (tensional) strengths of two steels are indicated in relation

that this benefit will rapidly diminish with increasing K, (fig. 1.3). In fact, K values of 3–4 are commonly used on vessels for moderate pressure work.

The ultimate strength of the vessel may be obtained by equating the stress calculated with the breaking stress of the material. This may be expressed as tensile strength, σ_T, a property measured by determining the force required to break in tension a specimen of prescribed dimensions. However, the strain (deformation) in a pressurized cylinder may be considered to take the form of a radial expansion (fig. 1.4). Thus, successive layers of the wall tend to slide past one

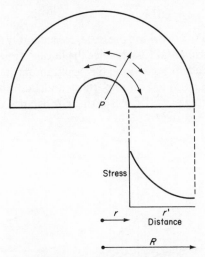

Fig. 1.4. Distribution of stress through the wall of a pressurized cylinder. The strain is greater the nearer to the centre resulting in shear stress being generated

another since the innermost layers expand more. This means there is a considerable shear component in the stress. Steels are considerably weaker in shear than in pure tension (about 60% or $1/\sqrt{3}$) so that the maximum pressure, P_{max}, which can be contained will be

$$P_{max} = 0.6\sigma_T \frac{(R^2 - r^2)}{(R^2 + r^2)} \tag{1.5}$$

or

$$P_{max} = 0.6\sigma_T \frac{(K^2 - 1)}{(K^2 + 1)} \tag{1.6}$$

This is a greatly simplified approach and applies only to elastic behaviour of the metal. In fact, as the yield stress is approached, metal near the inner wall enters a plastic region and relatively facile flow, to relieve stress, will occur. With a fairly ductile metal the plastic region will extend further into the wall as stress increases. The mathematical modelling of this aspect becomes complex.[23] Furthermore, the quality of the steel sample may be variable and contain flaws, gas bubbles, and other inhomogeneities which renders the value of tensile tests somewhat doubtful. In practice, the loading of pressure vessels is determined by a largely empirical code of practice which embodies considerable conservatism in the interest of safety.[24-26] If, for example, a safety factor of 3 is built into the formula above, a safe working pressure for a cylinder, $K = 4$, $\sigma_T = 15{,}000\,\mathrm{kg\,cm}^{-2}$ would be

$$P_{max} = 15{,}000 \times (9.81 \times 10^{-4}) \times \frac{16-1}{16+1} \times \frac{0.6}{3}$$

$$= 2.6\,\mathrm{kbar}$$

While this would be a safe working pressure, considerably higher pressures might be used in small experimental vessels since the tensile strength of the steel may be in excess of the manufacturer's value (i.e. a minimum is often quoted), fatigue phenomena (section 1.1.4) are unimportant and additional strength may be imparted by autofrettage.[1,27,178] Also, shear resistance of steels may increase under hydrostatic pressure.[26a]

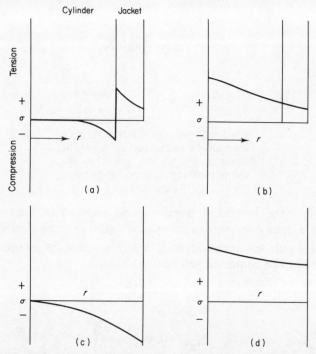

Fig. 1.5. Distribution of stress in the wall of a supported cylinder.
(a) Cylinder with shrink-fit jacket unpressurized.
(b) Cylinder with shrink-fit jacket under pressure.
(c) Autofrettaged cylinder unpressurized.
(d) Autofrettaged cylinder under pressure

Autofrettage is the name given to the process of stressing a pressure vessel such that the inner wall enters the plastic region whereupon stress is more evenly distributed throughout the wall. On releasing the pressure, the outer parts of the wall are left in permanent tensile stress. This residual stress may have the property of raising the bursting pressure by a factor of 2 or so. The correct pressure to employ in order to obtain the desired amount of autofrettage is a matter of experience aided by numerical data such as that due to Manning,[29,30] but is a procedure usually left to specialists. The radial stress curve of an autofrettaged vessel will have the appearance of fig. 1.5. Even if residual stress is not purposely created, the strain produced in cycling a vessel to high pressure will often increase its strength due to 'work-hardening'.

1.1.3 Supported Vessels

The yielding of a pressure vessel occurs when the inner wall is carrying the highest stress and the outer parts almost nothing. In order to make better use of the strength of the steel it would be better if the stress were more evenly distributed. In general this involves creating tensional stress in the outer skin, fig. 1.5. One method of achieving this is autofrettaging, but the same end may be achieved by other means. Permanent compression may be given by the careful fitting of an outer jacket so designed that it is in tension when fitted.[31,32] In order to do this, the outer jacket is heated and the inner vessel cooled so that the one will go inside the other but after coming to the same temperature the jacket is 'shrunk-fit' and extremely tight. An autofrettaged high strength steel liner with lower strength jacket of high elasticity is recommended.[215] This process requires careful calculations and precise machining. Industrial vessels are often made on this principle by winding wire or metal tape at a prescribed tension around the vessel, this being economicaly more favourable than shrink-fitting.[6] A similar technique has been applied to gun-barrels. The disadvantage of permanently stressing the outer wall lies in the fact that at atmospheric pressure the vessel may yield by cracks in the outer wall spreading; in other words, one can only design for

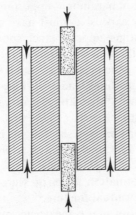

Fig. 1.6. Cylinder support by hydraulic pressure

uniform stress at one pressure. A more elegant, though more elaborate, method of supporting the vessel is to apply compressional stress as the pressure is increased. This may be done by using hydraulic pressure from a subsidiary source acting in an outer jacket (fig. 1.6) or by using a conical outer profile to the vessel which fits into a similar jacket. Stress is increased by pressing the vessel into the jacket by an

8

independent mechanism as pressure is increased (fig. 1.7). Examples of these designs are shown below. With modern steels it is now possible to design monoblock vessels to withstand 25 kbar, which under test conditions are found not to fail in excess of 30 or even 40 kbar.[163]

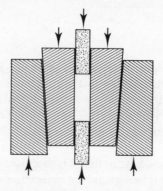

Fig. 1.7. Cylinder support by means of a conical jacket. The force pushing cylinder and jacket together is increased as the cylinder pressure rises

1.1.4 Fatigue and Chemical Attack

Continual cycling of a vessel to high pressures which includes pressure fluctuation due to pump action gradually causes a weakening of the walls which will eventually yield at a pressure previously considered safe.[33] This phenomenon, known as fatigue, is of paramount importance in industrial design since pressure vessels, being very expensive, must have a long life and the onset of fatigue is unpredictable. Fatigue appears much sooner, the higher the pressure used. However, one normally thinks in terms of thousands or even millions of cycles of raising and lowering the pressure so that it is hardly significant in small experimental vessels unless, perhaps, one is working repeatedly near the limit of the cylinder. It has been shown[34] that fatigue is very much affected by the nature of the material in contact with the surface of the vessel. Steel surfaces coated with rubber or polyethylene apparently resist fatigue much better than those in contact with pressurized oil. Presumably, the oil is forced into microscopic cracks thereby widening them and causing large local stresses.[35] At higher temperatures particularly permanent deformation under pressure may occur slowly ('creep') leading to weakening and eventual fracture.[36,37]

More serious for the experimenter is the deleterious effect of certain materials upon steel. Leaving aside substances such as strong acids which are naturally corrosive, mercury will gradually penetrate steel, possibly amalgamating with the iron and will cause serious weakening.[38] Gallium has a similar effect. Hydrogen under pressure is also a serious cause of embrittlement of steel where it apparently reacts with carbon in the steel producing methane and disrupting the lattice.[6] Oxygen under pressure is hazardous on account of actual combustion of the steel which may occur.Neutron irradiation will cause embrittlement, possibly due to release of α-particles forming helium.

Effect of temperature   The strength of steels falls off as temperature is increased, usually drastically above 350°, fig. 1.8. Torsional strengths of selected steels at elevated temperatures have been reported by Crosland[39] and typically show yield pressures reduced by some 20% at 370°. This problems can be countered partly by a suitable choice of steel and partly by design. Further information on this aspect has been produced by Faupel.[40] Low temperatures may also cause problems for, whereas strengths of steels by tensile tests increase somewhat, brittleness may increase greatly. Careful choice of steels which resist cryogenic embrittlement is necessary.[41]

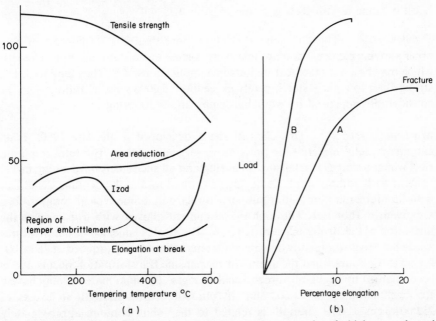

Fig. 1.8. Typical effects of tempering temperature on the properties of a high strength steel (EN30)

1.1.5   Materials for Pressure Vessels

With rare exceptions, pressure vessels are constructed from a type of steel, chosen to have suitable properties[42] for the intended purpose. A large variety of steels are available, each manufacturing country usually having its own system of designation and standard.[43–51] The tensile test indicates the behaviour of a given sample under tension by measuring elongation against load. Fig. 1.8(b) shows some typical results. Steel A at first behaves elastically, with strain $\propto$ stress but eventually deforms plastically with considerable elongation occurring before fracture. Steel B fractures very soon after its elastic limit which is much higher than that of A. It is much stronger than A, though more brittle. An ideal steel for pressure work should have a high tensile strength (shear strength) combined with at least 10% elongation before rupture: that is, strength combined with toughness.[52-55] Tendency to brittle fracture is highly undesirable.[56] Low alloy steels are the commonest materials for pressure vessels; samples with strengths in excess of 15,000 kg cm^{-2} are standard while special steels up to 25,000 kg cm^{-2} have been made.[46-49] These are essentially carbon steels containing 5–6% of various transition metals such as nickel, chromium, manganese, molybdenum, vanadium, tungsten, or niobium. A typical formulation (British Standard En30B) is:

C	Si	Ni	Cr	Mn	Mo
0.26–0.34	0.1–0.35	3.9–4.3	1.1–1.4	0.4–0.6	0.2–0.4

A list of some useful steels is given in Table 1.1.

Stainless steels These in general contain more than 20% of chromium with other alloying elements and are relatively corrosion-resistant and non-magnetic which may be a desirable feature for some applications.[59,60] They tend to be less strong than low alloy steels, rarely exceeding 15,000 kg cm^{-2}, though may be considerably improved by work-hardening and autofrettage.

Maraging steels[61,62] This class of steels, developed in the late 1950s, is an extremely useful material for pressure vessel construction. The term implies a steel which becomes martensitic on ageing and all such steels have a high nickel content with additional elements such as cobalt and molybdenum, sometimes with titanium and aluminium but very low carbon. Some typical compositions are given in Table 1.1. The high nickel content alloyed with iron ensures the formation of the strong martensite phase. The steel is annealed at temperatures above 800° to give a readily machined material: this is then hardened at 450–500° for up to 12 hours when the ageing or martensitic transformation occurs and is accompanied by negligible dimensional changes. The final material may have a Rockwell hardness of >60, and an ultimate tensile strength in excess of 25,000 kg cm^{-2}. The strength is related to the cobalt content, approximately 12.5% Co being optimal while the other alloying elements confer ductility and hardness. Stainless maraging steels are also made which contain 10–15% chromium, of lower strength but greater corrosion resistance. The metallurgy of these materials is exceedingly complex; further information may be found in references within reference 48.

Tool steels These contain often large proportions of tungsten, molybdenum, and chromium and are extremely hard though brittle. Their strength is in compression rather than tension so that these materials may be very suitable for pistons. Samples may be obtained in accurately ground cylindrical form.

Non-ferrous metals The titanium alloys mentioned in Table 1.1 are of high strength, corrosion resistant, and non-magnetic and have been used for certain

Table 1.1 Representative high strength steels and their characteristics

Type	Designation					Tensile strength (3 in sample)/$kg\,cm^{-2}$	Quench temp/°C	Anneal temp/°C	Elongation %	Hardness R_c
	UK	France	USA	Germany	USSR					
3% NiCr	En23	30NC11		22NiCr14	30KhN3A	9,000	840	600	22	53
1½% NiCrMo	En24	35NCD6	4,340	34CrNiMo6		8,000	840	675	24	58
2¼% NiCrMo	En25				38KhN3MA	11,600	840	300	12	58
2½% NiCrMo	En26					11,700	850	580	19	58
3% NiCrMo	En27	30NCD12				12,900	830	590	18	53
						12,300	830	400	14	
3% CrMoNi	En28	35NCD14		32CrMo12		10,700	830	580	18	48–50
3% CrMo	En29	30CD12		35NiCr18		12,600	910	525	20	50
1¼% NiCrMo	En30	35NC15		35NiCrMo16		17,700	820	200	15	50+
	En30B	35NCD16						100		65
1% C–Cr	En31	10006	51,100 / 52,100	100Cr6	KhShKh15	—				
5% NiCrMo	KE355			(1.2601)		17,000	1000	820	10	60
12% Cr(MoV)	KE180			(1.6351)			1000	230		52–56
18% Ni maraging			AMS6512 Böhler ISODISC			20,000	830	485	8	30
CrMoVB						15,800		550		53

Nickel steels	Ni	Cr	Fe	Co	Mo	Ti	Al	Si	C	Mn	W
Incoloy	32	21	47								
Inconel	85	15									
Hastalloy C	78	16	6	5							
Rene 41	47–53	18–20	5	10–12	9–10	3	1.5	0.5	0.1	0.1	
Stellite 31	9.5–12	27–31	0–3	53–69	0–1.5	0	0	0–1.5	1	0–1	4

special applications such as high pressure NMR tubes. Beryllium–copper can be made almost as tough as alloy steel while beryllium itself is quite strong and has the advantage of being transparent to X-rays. Proprietary nickel or cobalt alloys of a very complex nature such as the Rene or Stellite series are extremely tough and very suitable for high pressure work.

Composite materials   Diamond, tungsten carbide, and boron carbide form three of the strongest crystal lattices known. These substances are not however readily available as large single crystals but are manufactured in fine crystalline form. Tungsten carbide may be sintered into a coherent mass with metallic cobalt (5–15%) and forms an extremely hard composite known as carboloy. This may be formed into the desired shapes during manufacture and finally polished. Carboloy is one of the strongest materials known in compression so that it finds applications in pistons, anvils, and similar components. Sintered diamond and other carbides show great potential for the preparation of very strong materials.

1.1.6   Heat Treatment

A billet of steel for pressure vessel construction must be free of defects, cracks, and gas bubbles which may only be achieved in manufacture, by hot working, vacuum melting, or sometimes scavenging by aluminium. The material as obtained by slow cooling will be relatively soft and ductile and in need of heat treatment to bring it to its maximum potential hardness.[6,57,58]

The stable state of iron at ambient temperature is the body-centred cubic α-phase containing dissolved carbon (ferrite) with cementite (Fe_3C) or an intimate mixture of these phases known as pearlite. In this form, the steel is soft and is said to be annealed. At temperatures above 750° the stable phase is austenite (face-centred cubic γ-Fe containing 0.8% C). On very rapid cooling this phase is preserved as a metastable phase (martensite) which is very hard and brittle.[56] It appears that small amounts of transition metals assist the preservation of this martensitic structure. The rapid cooling of the specimen results in stresses being set up which are relieved by tempering at a moderate temperature so that the resulting steel loses much of its brittleness. The exact temperature regime needed to bring a given steel to maximum toughness should be available from the manufacturer but typically it would consist of heating for a prolonged period at 950°, usually in a molten salt bath, quenching by plunging into oil or water, and finally tempering at 250°. There is a maximum size of sample which can be successfully hardened by this method since the cooling rate in larger specimens will be insufficiently fast. Tempering temperatures may be quite critical and if too high may increase brittleness by precipitation of carbon—fig. 1.8. Furthermore, it is often impossible to harden uniformly a specimen with a closed bore since gas pockets form excluding the cooling liquid. A fairly small, open-ended cylinder is the more favourable subject for hardening. The final properties of the sample may be checked by a hardness test (Rockwell, Brinell, or Vickers) in which an indentation is made under standard conditions and its dimension measured. For example, the Rockwell C test uses a diamond cone and 150 kg force. The diameter of the pit produced gives a hardness number, R_c, which may be related to tensile strength. Such a test, of course, gives information only concerning the surface. The Izod impact test is a measure of the force needed to break a notched specimen and hence the tendency to fracture.

Heat treatment, necessary for the development of strength, introduces further problems. Some distortion of the sample inevitably occurs during hardening due both to the phase change occurring and to the formation of scale on the surface. Machining to fine tolerances must be finished after heat treatment and this is rendered extremely difficult by the very hardness of the steel. Finishing is usually accomplished by grinding and lapping.

1.1.7 Components of Pressure Equipment[1−5,65−69]

Machining The containment of liquids and particularly gases at high pressures is ever subject to leakage and a high order of precision in the machining of components is absolutely necessary. Pistons and cylinders usually need to be matched to a clearance of about 2×10^{-4} cm and polished to a mirror finish to give best results although greater clearance further up the piston is acceptable and even desirable. We have found it advantageous to have low-pressure cylinders bright chromium plated, which gives an excellent seal against rubber and reduces problems of wear and corrosion. Shrink-fit jackets need to be made to this tolerance also and conical vessels and supports are even more demanding if even support is to be achieved. The difficulties are accentuated by the hardness of the material so that grinding, honing, and lapping are often resorted to.

Pistons The piston needs to be capable of withstanding compressive stress without crumbling or becoming deformed and so binding in the cylinder. Metals of the greatest hardness ($R_c \geqslant 50$) may be used even though they are brittle or relatively poor in tension. In the 5 kbar region there are no special difficulties and silver steel is quite satisfactory. At higher pressures tool steel might be considered or a ball-bearing steel and the surface may be case-hardened. For ultra-high pressures it is necessary to use carboloy or a similar composite. It is important that the high stress be carried by no greater length of exposed piston than necessary hence the diameter of the piston is often increased about twofold outside the cylinder. Truly uniaxial stress should be applied, otherwise a bending moment may be imparted to the piston causing fracture. A guide bushing on the cylinder, often of brass, may be used to aid this aim. It is essential to avoid ingress of particles of grit which may cause scoring or binding of a tightly fitting piston.

Seals The junction between two moving parts (e.g. a piston and cylinder) or between two demountable parts, such as a closure, requires a seal to prevent leakage. In addition to the requirement that it should perform this function seals may also need to permit movement with minimal friction and to withstand high temperatures or corrosive liquids at least for a period, while being simple and readily replaced since they invariably have a limited lifetime.

The Bridgman unsupported area seal is often used at a piston tip. This consists (fig. 1.9a) of a hard steel head of the same cross section as the piston with a stem of approximately 0.5–0.6 the piston diameter, recessed into the end of the piston. Partial threading may be used to facilitate removal, and a disc of soft material

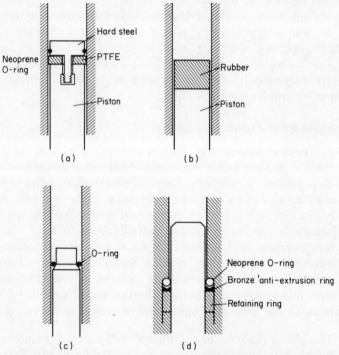

Fig. 1.9. Some types of piston seals. (a) Bridgman unsupported area seal. (b) Poulter seal. (c) Wedge ring seal. (d) Patterson seal

such as PTFE is placed between the two. The high pressure at the tip is transmitted to the annular area of the packing which, since the force is constant, experiences a higher pressure than the fluid contained. Deformation of the packing occurs causing effective sealing since a solid material such as PTFE is much less prone to flow than a liquid. The Bridgman seal is simple, robust, and very effective. Its disadvantages include relatively high friction and, at very high pressures, the tendency of the packing to pinch off the piston tip. Extrusion between piston and cylinder may also occur with a badly fitting apparatus and may cause jamming.

Other packing materials such as polyethylene, leather, Kel-F (polychlorotri-fluoroethylene), or hard rubber are also effective to about 12 kbar. A simple deformable piston tip (Poulter seal) may be effective at lower pressures such as fig. 1.9(b). The PTFE tip with O-rings seems to hold to 3–4 kbar. O-rings, normally of neoprene or viton, will readily deform and extrude under high pressure. While providing a good seal, provision against extrusion usually must be made by the use of triangular section anti-extrusion rings ('delta rings') placed in contact such as in the design of figs. 1.9(d) and 1.11. Metal O-rings of copper, lead, or gold, solid or hollow may be used and do not need anti-extrusion rings. In part, these seals remain effective at high pressures since the liquids they are containing may become very viscous. Dynamic seals require careful design.[69]

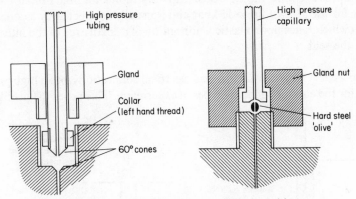

Fig. 1.10. High pressure connections to tubing

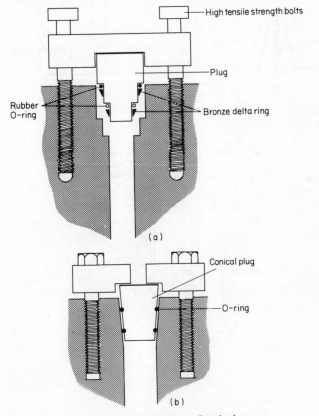

Fig. 1.11. Some designs of end-plugs

Plugs and tubing connections   Closure plugs may be threaded (in which case the threads should be hardened and ground to a close tolerance) or may be constructed to be held in place by a steel plate fitted with high tension bolts (fig. 1.10a, b). In either case, one or more O-ring seals backed by delta rings may be used to improve sealing, fig. 1.11. Steel capillary tubing for high pressure work is available for use at least up to 15 kbar and is usually connected by a cone fitting (fig. 1.10) which relies upon elastic deformation of a small area of the interface to produce the seal.

Electrical leads   It is necessary to lead electrical connections into a high pressure cylinder for the conduct of many types of experiment. This poses the problems of

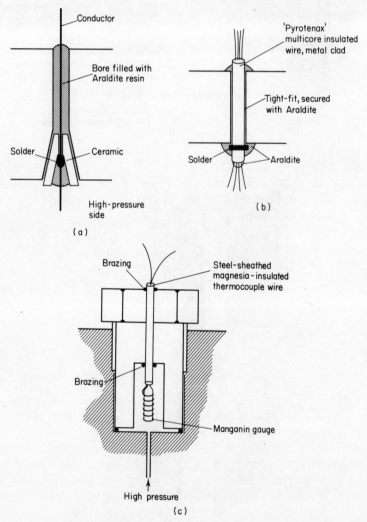

Fig. 1.12. Electrical leads in pressure vessels

sealing against leaks, preventing the extrusion of the wires under high pressure, and maintaining electrical insulation. The wire may be soldered into a ceramic cone (fig. 1.12a) which is fitted into a plug, the exposed portion being both sealed in and insulated by epoxy resin. A very useful and effective connection may be made using commercial multi-core magnesia-insulated steel-jacketed wire such as 'Pyrotenax'. This may be tightly inserted into a plug and epoxy resin forced into the space by gas pressure (fig. 1.12b) which will also seal the exposed magnesia. Dental cement also makes a good electrically insulating seal. Numerous variants are mentioned in the literature,[1-5,63,64,69] including a device for up to ten leads, and are available commercially (Appendix 3).

Windows   For optical work it is necessary to build one or more windows into the high pressure cylinder. Window materials which have been used include plate glass, quartz, sapphire (alumina), and diamond. Transmission characteristics are shown in fig. 1.13. While of these materials diamond is the strongest, cost rules this out for any but very small windows. Sapphire is preferred for most

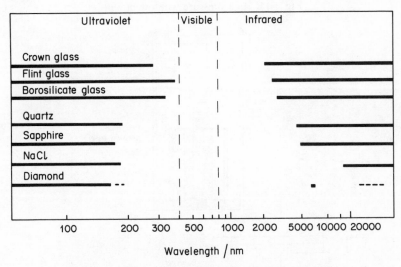

Fig. 1.13. Transmission regions for window materials

applications as it is readily available, inexpensive, and may be polished to 'optical' flatness. In practice, a thick disc of window material is mounted against a retaining ring, both surfaces being very flat, and may have an outer steel rings or an O-ring to locate them. A small area in the centre is unsupported and constitutes the window. Sealing is thus by the 'unsupported area' principle, the pressure between the window and retaining ring being higher than in the cylinder so making a good seal. Pressures of 6–7 kbar may thus be contained with little trouble,[70-72] fig. 1.14, though this basic type of design has been used successfully

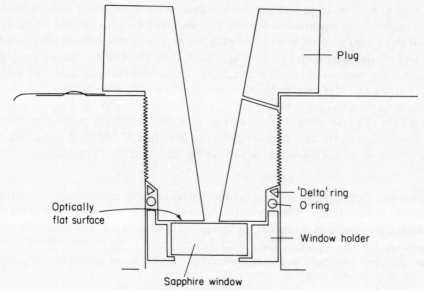

Plug

'Delta' ring
O ring

Optically
flat surface

Window holder

Sapphire window

Fig. 1.14. High pressure window mount

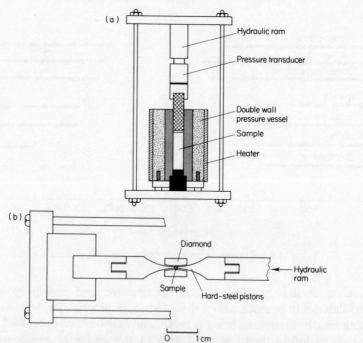

(a)

Hydraulic ram

Pressure transducer

Double wall
pressure vessel

Sample

Heater

(b)

Diamond

Hydraulic
ram

Sample   Hard-steel pistons

O   1 cm

Fig.1 15. (a) High pressure preparative unit working to 15 kbar,
200 °C.
(b) Diamond cylinder cell for high pressure spectroscopy (after
Wong and Whalley[75]). (Reproduced by permission of the
American Institute of Physics)

with glass windows at 25 kbar.[73]* Various window designs are discussed by Besson.[74] CaF_2 has been used to 2 kbar, 200° for infrared studies.[245]

1.1.8 The Generation of High Pressure

The high pressures aimed at are achieved by application of the lever principle often through several stages. It is quite feasible to reach more than 15 kbar by direct man-power though for much higher pressures, electrically powered hydraulic pumps may constitute the primary pressure source.

The simplest approach is to place a load of a few tons directly on the piston which may most conveniently be managed by the use of a commercial hydraulic press, particularly if this has the facility for maintaining a constant pre-set load. In this way, a simple cylinder may be pressurized to 15 kbar with a bore of about 2.5 cm and a working volume of 50 ml. Heating can be maintained by winding with resistance tape providing an inexpensive preparative unit working at up to 200°, fig. 1.15(a). Care must be taken to ensure that no great length of the piston protrudes when at pressure and that the load is as near as possible uniaxial. If a hydraulic press is not available, one may improvise with a heavy hydraulic hand-operated truck jack compressing the piston and cylinder in a strong framework of steel I-beams. A simple apparatus of this type is principally of use for preparative chemistry where the precise conditions are not too critical. A similar principle is used but on a very small scale in the construction of single crystal diamond spectroscopic cells, the pressure for the very small area piston now being applied by an intensifier,[75] fig. 1.15(b).

1.1.9 Pressure Intensifiers

If hydraulic pressure is used to displace a relatively large-bore piston which bears on a small bore piston, the pressure generated in the small bore cylinder is greater than that applied by the ratio of areas of the two pistons (fig. 1.16). This is the principle of the pressure intensifier which constitutes the most common approach to the production of high pressures in a cylinder:

$$P = p\frac{R^2}{r^2} + F \qquad (1.7)$$

where p is the primary and P the final pressure, R and r the radii of the two cylinders, and F a frictional constant. If it is desired to use the manometric pressure of the hydraulic fluid in the primary cylinder, p, as a measure of the high pressure produced, the intensifier must be calibrated in order to determine F using a suitable gauge in the high pressure cylinder. An intensifier is capable of only a single stroke after which the piston must be returned to the extended position. Since the displacements experienced with liquid samples are small (no more than about 10%) this suffices providing neither pistons leak appreciably. In order to return the pistons for a further intensification stroke, it is convenient to

---

* Polycarbonate or lucite windows may be used to 2 kbar as, for example, in the stopped-flow systems, *Rev. Sci. Instr.* **51**, 806, 896 (1980).

20

insert a valve connected to a hand-pump. The low pressure cylinder may often be made from ordinary mild steel but needs to be rigidly attached to the high pressure cylinder either by a screw thread or a massive frame. The large bore piston may conveniently make use of commercial hydraulic seals and, indeed, may be constructed from a commercial hydraulic ram. The high pressure microreactor unit (operating at 4 kbar, 600°) has an inbuilt intensifier, fig. 1.17.[164] A single stage intensifier for use at 10 or 15 kbar is readily constructed but greater sophistication is needed to apply this principle to the limit, about 100 kbar. Details of representative designs are described below.

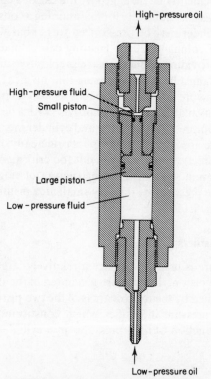

Fig. 1.16. A pressure intensifier.
(Reproduced by permission of the
National Physical Laboratory)

Primary pumps for actuating an intensifier may be hand- or power-operated.[76] Hand pumps include lever and valve types and screw types, both of which are commercially available. More convenient is an electrically operated diaphragm pump which is supplied with air or nitrogen at a pre-determined pressure and delivers hydraulic fluid at a proportionally (e.g. 100 ×) higher pressure from a reservoir.[79] Commercial diaphragm pumps may be upgraded to run at much higher pressures than originally intended.[77,78]

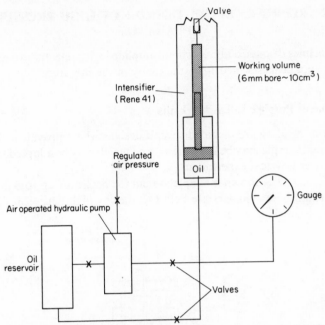

Fig. 1.17. Microreactor for preparative chemistry at up to 600 °C,
4 kbar. (After Hagen[164])

1.1.10 Safety Factors*

The main hazard in high pressure work comes from a sudden failure of the pressure vessel or other component under pressure. In fact there is little danger from a small research vessel with liquid under pressure since there is so little energy stored in the system. One mole of water at 1 kbar contains a stored energy of only 72 J. If a crack should develop, the release of only 4% of the fluid restores atmospheric pressure (Table 2.1). Since the degree of compression falls off with pressure, these quantities increase more steeply at higher pressures, thus at 5 kbar the energy is 10^3 J and 13% leakage is required. Solids under pressure offer even less hazard since their compressibilities are, on the whole, less than those of liquids. Gases on the other hand have a much greater potential for causing damage. One mole of an ideal gas at 1 kbar (which would be compressed to 2.2% of its volume at 1 bar) would release 10^5 J on expansion and 97.8% of the gas would pass through a crack before reaching atmospheric pressure. There is much more danger in damage from flying fragments or ejected valve spindles and similar components carrying considerable force. Thus, there is little danger from compressed liquids or solids at, say, 4 kbar but beyond this and when using gases, suitable shielding should be placed between apparatus and operator. Several discussions of the fracturing of vessels under pressure are available, including the effect of fatigue (not a serious problem at moderate pressures).[81]

* In the United Kingdom, the Safety Code of the High Pressure Technology Association is a recommended guide.[243]

1.2 REPRESENTATIVE DESIGNS OF HIGH PRESSURE APPARATUS

The principles discussed above will be amplified with reference to examples of published apparatus selected from the many in the literature.

1.2.1 General Purpose Intensifier Units

Under this heading are included several designs which provide a chamber in which a liquid sample may be pressurized and which may be adapted to a variety of chemical or physical experiments.

In figs 1.16, 1.18 is shown a simple system for pressures up to 5 kbar.[82] The large and small bore cylinders are held together by high tension bolts. A hand

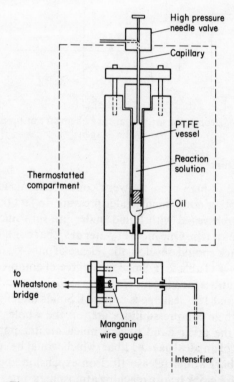

Fig. 1.18. Intensifier connected to reaction vessel with sampling valve.[82] (Reproduced by permission of the Chemical Society)

hydraulic pump working to 100 bar supplies oil to the primary cylinder which is intensified at a 50 : 1 ratio in the upper cylinder. This may be used for experimental purposes directly or may supply the oil through a high pressure capillary tube to an experimental vessels, from which samples may be withdrawn while under pressure if desired.

A vessel due to the National Physical Laboratory and working to 25 + kbar is shown in fig. 1.19, which is similar in principle but illustrates several additional features.[83] The structure is generally more massive than that of fig. 1.18 and is held in a strong tension frame to withstand the thrust of the small bore piston. The piston is sealed with a Patterson type O-ring seal and anti-extrusion ring, and a guide insures against the piston breaking through bending.

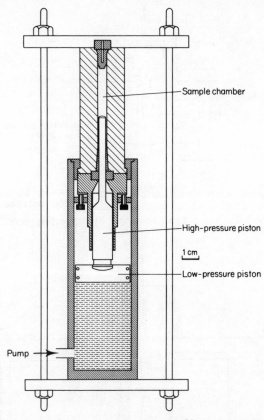

Sample chamber

High-pressure piston

1 cm

Low-pressure piston

Pump

Fig. 1.19. NPL 25 kbar intensifier unit.[83] (Reproduced by permission of the National Physical Laboratory)

Several commercially available intensifiers are on the market actuated by a hydraulic pump[84] or a screw device which makes it unnecessary for the experimenter to construct this item. A Russian-built version of this type of press shows[86] additional features (fig. 1.20). The high-pressure cylinder is tapered externally to fit into an outer compression jacket. Compression of the cylinder is achieved by a separate hydraulic cylinder at the bottom. The high pressure carboloy piston is driven by the intensifier above and cylinder support is steadily increased as the internal pressure rises. Oil inlets are provided for pumping back into position both the large bore pistons after use. Again, a massive support frame is necessary. Access to the experimental cylinder for electrical connections is

24

through the lower plug. The apparatus of Gonikberg,[87] designed for 40 kbar, is similar in principle but more massive. Double conical support for the cylinder is provided and maintained by the lower hydraulic piston. An even higher intensifier ratio is used to power the high pressure piston. Other similar designs have been used.[1-5, 53]

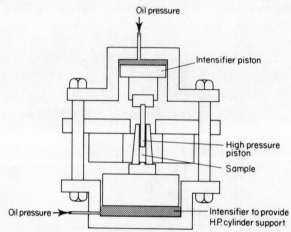

Fig. 1.20. Conical pressure vessel with supporting jacket[86]

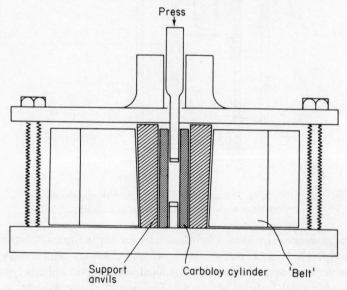

Fig. 1.21. Supported cylindrical apparatus for containment of hydrogen on corrosive materials[244]

The containment of hydrogen and of corrosive materials at high pressures can be accomplished by using carboloy cylinders as well as pistons. Since this material is not strong in tension it must be supported laterally, as in the Los Alamos design, fig. 1.22, in which a set of tapered steel segments are pressed into the annular space between the cylinder and a massive steel 'belt' by means of bolted flanges.[244]

The highest hydraulic pressures produced in a piston and cylinder are probably those in the stepped piston due to Hall,[80] fig. 1.22. The conical cylinder

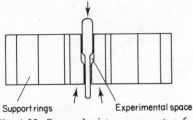

Support rings   Experimental space

Fig. 1.22. Stepped piston apparatus for
200 kbar. (After Hall[80])

is supported by several tension rings (similar to the 'Belt' apparatus, fig. 1.47) and the chamber is the annular space between the two large sections of piston and cylinder which acts in effect as a further intensifier in that the force applied to the piston at A is transmitted to a much smaller area in the sample chamber. In this way, pressures in excess of 100 kbar have been reached. Compression of the piston itself to prevent 'pinching off' permitted 200 kbar being reached.

External support to the cylinder has also been achieved by the use of a rubber packing which is progressively compressed by the intensifier fluid as the pressure is raised (fig. 1.23). Perhaps the ultimate in sophistication of supported cylinders is the controlled clearance principle whereby hydraulic support to the cylinder is

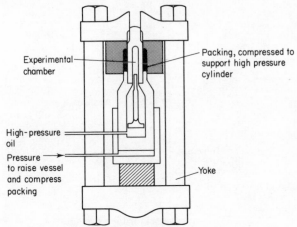

Fig. 1.23. High pressure vessel with cylinder supported by
ancillary compression. (After Tsiklis[53])

provided from an external source of fluid and regulated so as to maintain a constant desired clearance between piston and cylinder. This is particularly important for absolute pressure measurements (section 1.3.1) using a pressure balance.

1.2.2 Apparatus for Use at Elevated Temperatures

Apparatus for high temperature studies needs special consideration since the strength of low alloy steels falls off drastically above about 400° (fig. 1.8).[88, 89] If these are to be used for high temperature experiments, the walls need to be insulated from the furnace which may be located inside and electrical leads conducted out. Water-cooling if necessary will then maintain the strength of the vessel. In this way, devices have been constructed in which samples have been maintained at 50 kbar and 5000° or 100 kbar and 3000°,[80] fig. 1.27. Some high alloy steels maintain high strength to temperatures above 500°, much depending

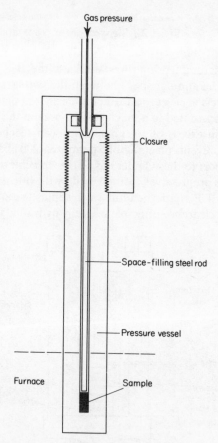

Fig. 1.24. Sample vessel for high pressure
geochemical applications

upon the length of the experiment. Thus, Inconel vessels may withstand 4 kbar at 600° while Stellite may be used for 4 kbar at 700° or 1 kbar at 880° for long periods and 940° for a short exposure. Some nickel-based alloys such as Nimonic 105 and Rene 41 are even better in this respect. The vessel for geochemical investigations, fig. 1.24, may be used at 10 kbar, 750° and heating is applied only to the lower part of the vessel in which the sample capsule is located, kept in place by the backup rod, pressure being applied by the admission of argon.[90, 92]

A precise apparatus for measuring $P–V–T$ relationships in liquids[93, 94] is shown in fig. 1.25. The intensifier uses a bellows of steel rather than a piston to pressurize the upper cylinder in which is located the piezometer. The temperature is maintained at up to 250° and 15 kbar by the heater located in the high pressure cylinder around the sample which is placed in a PTFE capsule while volume measurements are made by observing the piston movement. A simple device for

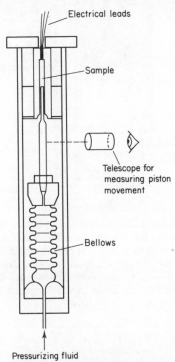

Fig. 1.25. Bellows intensifier for compressibility measurement. (After Millet and Jenner[93])

reaching high temperatures (2000°, 20 kbar) uses a piston and cylinder compressed by an external hydraulic press (as in fig. 1.26) while the sample is placed in a three-layer capsule, the middle layer being conducting (graphite or graphite/silicon carbide) while the outer ones are insulating.[95] Heating is achieved by passing a heavy current through the conducting layer in much the same way as in the heated capsule devised by Eyring,[96] and by Brouha and Rijnbeek.[97] The latter workers used a PTFE inner container equipped with a conical plug which could carry up to ten electrical leads. A cylinder supported by several conical rings was used with a tungsten carbide piston and external press to obtain up to 40 kbar, 300°. For compression of solids, a relatively shallow cylinder may be used since leakage problems are small. In the apparatus, fig. 1.27, the sample is contained in

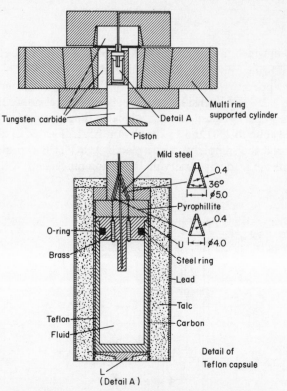

Fig. 1.26. Method of heating sample while insulating from the cylinder (M. Brouha and A. G. Rijnbeek[97]). Schematic view of the Teflon cell with manganin coil and sample in the supported pressure vessel and detailed drawing of the Teflon cell assembly. Dimensions in millimetres. (Reproduced by permission of the American Institute of Physics)

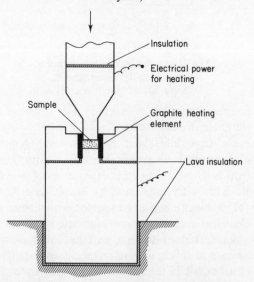

Fig. 1.27. High temperature piston apparatus for solid samples. (After Brouha and Rijnbeck[97])

three interleaved containers of lava, graphite, and lava respectively.[98] The cylinder is electrically insulated by a lava seating and an electrical current may be passed (40 volt, 300 amp) through piston and cylinder, heating occurring in the graphite resistance.

1.2.3 Optical Vessels

It is often necessary to pass radiation into a vessel under pressure for spectrometric purposes, either for monitoring concentrations, for observing the effects of pressure on the spectra themselves or for effecting photochemical transformations under pressure. Window materials are without exception much weaker in tension than steel; nevertheless, it is possible to build windows into

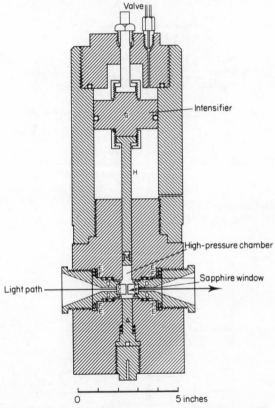

Fig. 1.28. NPL optical high pressure cell. (Reproduced by permission of the National Physical Laboratory)

pressure vessels which will withstand 4–6 kbar or with rather more trouble, > 20 kbar.[102] Current practice in this area has been reviewed.[99] A typical design is that of fig. 1.28 used at the National Physical Laboratory.[100] An intensifier located at the top pressurizes the chamber containing a pair of thick cylindrical windows of quartz or sapphire, optically flat and bearing directly on the optically

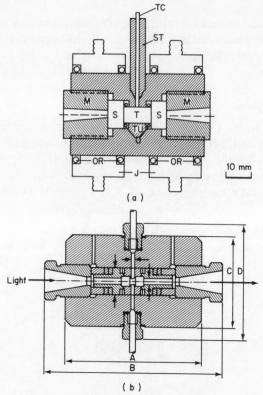

Fig. 1.29. High pressure optical cells.
(a) Reference 105.
(b) Nova Swiss. (Reproduced by permission of Nova Swiss)

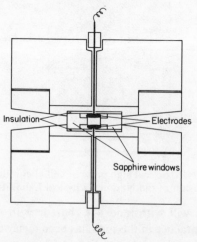

Fig. 1.30. High pressure optical cell for temperature-jump kinetic studies using resistive heating. (After Grieger et al.[101])

flat faces of the rings, D, each with double O-ring seals and held in place by the threaded retaining rings. The pressure holds the windows against the rings and the fit ensures an even distribution of pressure and sealing at this interface. The optical space may be altered by the choice of window thickness. A closure at the end is available for filling the cell. This device may be operated at 3–4 kbar and up to 100°. When used in conjunction with a spectrophotometer, it is necessary to use a mirror system (fig. 1.30) or fibre optics since the unit is too bulky to be incorporated into the cell compartment. Much more compact are the commercial cells shown in fig. 1.29 which can give a 45 mm cell space in a device 120 mm wide for use at 4 kbar or a 10 mm space for use at 7 kbar,[85] and also the apparatus of Lentz and Oh[105] (for T-jump measurements), among many designs which have been published.

A similar design used by Grieger (fig. 1.30) has sapphire windows recessed into a PTFE body containing two large electrodes with external leads. Window faces are sealed to the retaining rings by Canada balsam–epoxy resin. The cell is used for temperature-jump kinetic measurements, an electrical pulse being administered to effect the heating. It can be used to 2 kbar. For Raman spectroscopy a prism has been incorporated into one window (fig. 1.31) permitting scattered light to be collected through an off-axis exit hole.[103] The

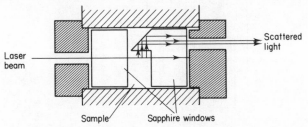

Fig. 1.31. Raman cell for use at high pressure embodying a built-in prism. (After Figuière et al.[103])

single crystal diamond piston cell for optical work has been mentioned and the diamond anvil cell is described in section 1.2.7. A versatile optical cell for the range from ultraviolet to infrared has been developed.[104]

It is usually desirable to exclude the liquid under investigation from contact with the steel vessel. This may be done by using an internal tight-fitting jacket of PTFE[105] or Hastaloy-C.[106] The liquid is admitted through a PTFE capillary to fill the optical cell and the pressurizing fluid is directly applied to the capillary which, being of narrow bore, does not permit mixing with the sample. Alternatively, a fixed path cell with mercury seal may be enclosed in the optical cell, the dimensions remaining fixed under pressure which is equalized inside and out, fig. 1.32(a). An elegant approach is the cell designed by LeNoble,[107] fig. 1.32(b), consisting of two closely fitting quartz cylinders with window ends, filled by hypodermic syringe when the hole and slot are aligned then sealed by twisting the two components. With this type of cell, the optical path length will change with pressure, but the number of molecules in the beam will not.

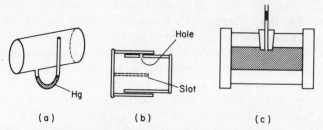

(a)   (b)   (c)

Fig. 1.32. Types of pressure-equalized optical cells

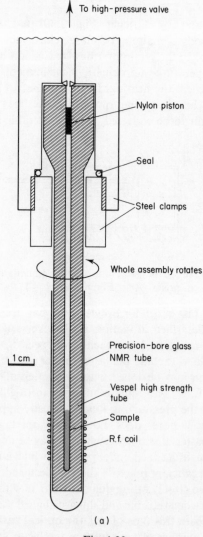

To high-pressure valve

Nylon piston

Seal

Steel clamps

Whole assembly rotates

1 cm

Precision-bore glass
NMR tube

Vespel high strength
tube

Sample

R.f. coil

(a)

Fig. 1.33

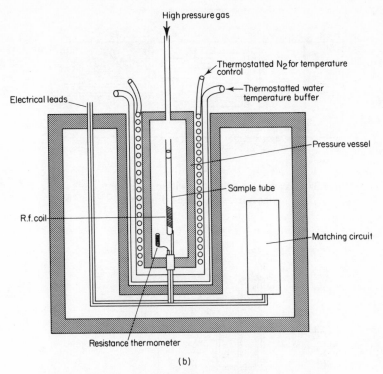

High pressure gas

Thermostatted N_2 for temperature control

Thermostatted water temperature buffer

Electrical leads

Pressure vessel

Sample tube

R.f. coil

Matching circuit

Resistance thermometer

(b)

Fig. 1.33. High-pressure NMR techniques.
(a) Spinning high pressure sample tube. (After Merbach[176])
(b) Static high pressure NMR probe. (After Merbach[176])

1.2.4 Magnetic Resonance Apparatus

The observation of magnetic resonance (NMR, e.s.r.) requires a thermostatted liquid sample to be contained in a narrow tube of non-magnetic material and to be subjected to a powerful magnetic field and radiofrequency radiation, preferably while spinning rapidly. These are challenging conditions to be met if high pressure is also to be imposed but have been achieved by several groups of workers since 1954.[170–172] Recently, successful high resolution NMR spectra at pressures up to 3 kbar have been recorded.[173–177] The material of the sample tube is normally glass. If this is to withstand the applied pressure, a much stronger material is needed. Quartz can be used to above 1 kbar if care is taken to remove surface scratches,[175] otherwise Cu–Be or Ti alloys or the polyimide Vespel (DuPont) may be used. Pressurization is achieved via a symmetrical valve after which the sample is disconnected and may be permitted to rotate within the magnet, fig. 1.33. An alternative approach is the non-spinning system in which a small loss in line-width is offset by greater convenience. The high pressure sample vessel of Cu–Be alloy is built into the probe and glass sample tubes may be used since the pressure is equalized inside and out, fig. 1.34.[176]

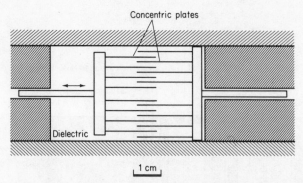

Fig. 1.34. Capacitance cell for use within a pressure vessel; the concentric plates are adjustable

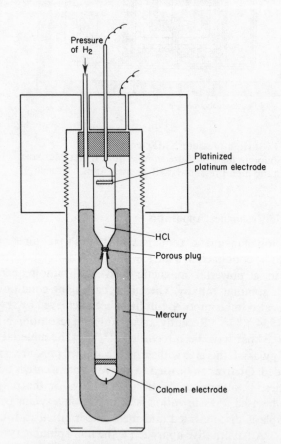

Fig. 1.35. High pressure vessel for the study of the cell Pt, H_2 | HCl | Hg_2Cl_2 | Hg. (After Hainsworth et al.[168])

1.2.5 Electrical Measurements

A pressure vessel may be fitted with electrical leads, so it is straightforward to design apparatus for measurement of conductivity, capacitance, thermo-e.m.f., and so on. Conductivity apparatus has been frequently used[105,123−126] up to 10 kbar.[162] The apparatus shown in fig. 1.38 is typical for low temperature applications.[127] The cell has a mercury seal which permits equalization of pressure. Careful pressurization is necessary to prevent cracking of glass-to-metal seals of the electrodes. A detailed high-precision design has been previously

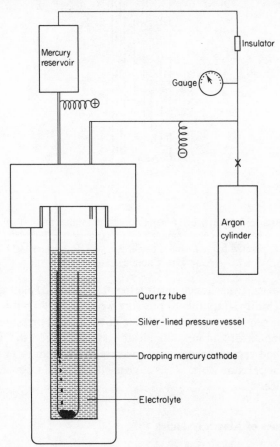

Fig. 1.36. A miniature dropping mercury electrode for use
under pressure. (After Hills and Ovenden[167])

described.[162] Many useful designs have also been depicted by Hills[167] including apparatus for use at the high temperatures necessary for working with molten salts (fig. 1.37). Cell e.m.f.'s may be measured, for example in the apparatus fig. 1.35. The glass electrode may be used under pressure[168] provided pressure equalization is ensured, and even the dropping mercury electrode has been adapted to permit polarographic measurements (fig. 1.36). Dielectric constant is

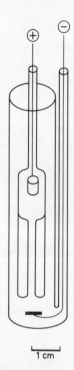

1 cm

Fig. 1.37. Quartz cell for high pressure–high temperature
conductivity measurement. (After Bannard and Hills[128].)
(Reprinted from *J. Amer. Chem. Soc.*, **46**. Published 1924,
American Chemical Society)

determined from a.c. capacitance measurements.[233] Parallel plate and concentric
capacitors are readily adaptable to pressure work (fig. 1.34) but it is a matter of
concern that the geometry of the plates does not change as pressure is applied.[169]
Transport numbers are more difficult to estimate since the usual moving
boundary method requires the ability to sense the boundary position but has
been adapted for pressure work,[171] as have the use of pressure coefficients of cells
with transference.[167]

1.2.6 Techniques of Kinetics Studies

1. *Sampling* The most obvious technique, the removal of samples periodically
for analysis, is often satisfactory. The pressure vessel is provided with a needle
valve which may be opened in a controlled fashion while the vessel is pressurized
and a sample withdrawn. A typical apparatus is shown in fig. 1.18 in which
contact between solution and the steel vessel is minimized by placing the former
in a PTFE cylinder with free piston[82] or hypodermic syringe with piston sawn
off.[108] There may be a certain amount of 'dead space' between syringe tip and
valve so the initial volume should be discarded at each sampling. An apparatus
from which 10 ml of sample can be withdrawn, working at 2–3 kbar is readily

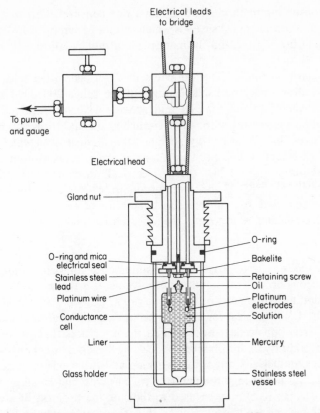

Fig. 1.38. Conductivity cell for kinetic studies.[127] (Reprinted with permission from *J. Phys. Chem.*, **73**. Copyright 1969, the American Chemical Society)

built and is suitable for reactions with a half-life of at least 1 hour since, with faster reactions, the time taken to seal the sample and the temperature rise upon pressurization introduce serious errors.[109−111] These latter disadvantages may be overcome if reagents may be mixed *in situ*. Several workers have described mixing apparatus in which the components are placed in separate chambers, pressurized and temperature equilibrated.[112,165] They are then mixed rapidly by an externally-controlled electromagnet and sampling or spectroscopic measurement may commence immediately.[113,114] There is inevitably a temperature drop as each sample is taken and a rise as the required pressure is re-established. To minimize the effect on the reaction the pressure may be maintained automatically by, for example, the use of a diaphragm pump with controlled output.

If the reaction mixture is corrosive to metals, containing for example acids or halogens, it may be necessary to use a tantalum[115] or Hastalloy[116] capillary through the valve. Corrosion problems usually become worse under pressure. Alternatively, the solution may be placed in a PTFE capsule (fig. 1.35a), pressurizing each for a definite time before removing for analysis. This, however, is a time-consuming method of obtaining a rate constant. This technique may also be used for geochemical systems which cannot be sampled through a valve. High temperature capsules may be made of metal (gold, platinum,[117] fig. 1.35(b)).

2. *Spectrometric methods* The optical cells described, fitted with quartz or sapphire windows, may be used throughout the range 200–3300 nm, i.e. UV through the visible to the near IR and provide a versatile analytical system since, for almost every reaction, some change in the spectrum will occur. Fast reactions may be followed by stopped-flow apparatus which may be designed for immersion in a pressure vessel[122, 166] or relaxation methods[238]— temperature-jump, fig. 1.28, using electrical discharge or laser pulse heating,[101, 118 – 120, 239 – 241] or pressure-jump.[114, 242] For the latter, the sample is pressurized and the pressure then allowed to drop by a succession of small decrements, monitoring the absorbance and displaying it on an oscilloscope.

3. *Other methods* Conductivity is very suitable for following ionogenic reactions in aqueous or, at least, conducting solvents. It is the method of choice for high-precision measurements of solvolytic rates.[105, 123 – 126, 237] The apparatus shown in fig. 1.38 is typical for ambient temperature operation. Since only the progress of a reaction needs to be measured, it is not crucial that the cell retains its dimensional stability under pressure, so long as this does not change during the experiment. Conductivity apparatus working at 10 kbar has been described[162] and also that for use at high temperature such as with molten alkali halides.[128]

Pressure-change itself may be used to indicate the progress of a reaction in which a volume change is occurring.[234, 113] This would only be feasible if the change observed were small compared to the total pressure and a high precision pressure gauge is needed. However, good results have been recorded for some acid- and base-catalysed reactions.

1.2.7 Anvil Devices

The other major principle whereby high pressures may be produced is the use of a set of hydraulically compressed anvils. The limitation to pressures produced in a cylindrical vessel is ultimately the tensile strength of the steel, which is capable of withstanding much greater stress in compression. This applies to an even greater degree for hard, brittle materials such as tungsten carbide, which are immensely strong in compression and even more so if diamond tipped. In order to take advantage of the properties of these materials, Bridgman, and many other workers since, have developed devices in which sets of hard members (anvils) are forced together by powerful hydraulic rams. In the simplest arrangement, one ram set in a tension frame forces two anvils together, fig. 1.39, with pressures which may achieve 50 kbar. The anvil faces are normally plane though may be recessed slightly but the sample volume is essentially small and of very thin section. It is clear that fluids cannot be contained except in a capsule so the apparatus is primarily intended for compression of solids. Even so, many solids are ductile and it is common to contain the sample within a ring of pyrophyllite, a porous (and therefore compressible) volcanic rock which has a high resistance to plastic flow.[130] The area behind the faces of the anvils increases rapidly to disperse the stress but failure of the anvils is ultimately due to shear forces causing cracking parallel to these shoulders. In order to reach higher pressures up to the megabar region[131], lateral support is required which is provided by various types of concentric support rings.

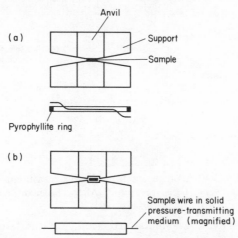

Fig. 1.39. Anvil configurations with samples encased in a solid pressure-transmitting medium (such as $AgNO_3$); (b) permits a larger sample volume than (a)

Modifications of this basic design include 4-, 6-, and even 8-anvil systems arranged in symmetric arrays (tetrahedral, octahedral, etc.) and capable of reaching the highest pressures. Sliding anvils have been proposed[129] though, as yet, little developed. Heating of the very thin sample under pressure may be achieved by electrical resistance[132] or by laser radiation.[134] One severe disadvantage of the anvil method lies in the non-uniformity of pressure and the shear components which may be present.[135] The use of multi-anvil arrays improves this aspect.[133] The commercial synthesis of diamond and boron carbide employs large anvil presses to attain the necessary conditions.

1- and 2-Anvil devices

The design in fig. 1.40 due to Balchan and Drickamer,[136−138] uses two sintered tungsten carbide anvils with truncated conical tips, the lower resting on a stationary platen and the upper powered by an external press. The pistons are mounted in hard steel jackets and the working space is surrounded by a carboloy ring backed by a steel support ring. The sample, in the original purpose, a very thin metal strip (0.0005 in) for resistance measurement, is encased in silver chloride, an insulating solid pressure-transmitting medium and then pyrophyllite discs and retaining ring, the space between the anvils being only of the order 0.1 in diam. During use, the anvils became slightly deformed at the tips but the pressure was found to have the effect of work-hardening the carboloy immediately underneath. After regrinding, the anvils were capable of reaching 425 kbar and many metallic phase transitions were observed.

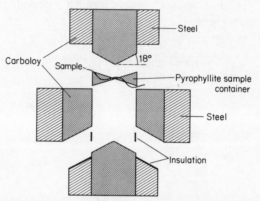

Fig. 1.40. Arrangement for resistance measurement at very high pressure. (After Balchan and Drickamer[136])

The apparatus in fig. 1.41 is similar in concept, but was designed for Hall effect measurements. The anvils were constructed of sintered carbide (iron–titanium carbide) and the support rings were required also to be of non-magnetic material. In a 1000-ton press, the apparatus was used at 50 kbar and a temperature range from 25° to −100°.[139, 140]

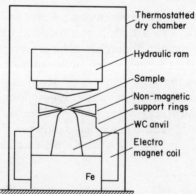

Fig. 1.41. Anvil apparatus for magnetic measurements at 50 kbar. (After Pitt[139])

Double anvil presses are suitable for X-ray or neutron diffraction studies[141] and may be made very small but provision must be made for the incident radiation (a very narrow collimated beam) and the diffracted rays which emerge within a cone of angle up to 30°, fig. 1.42.[142] Tungsten carbide anvils will absorb

X-radiation and limit the collection of diffracted rays to the plane of the sample. This may be alleviated by the use of anvils made from light elements such as diamond or boron carbide[143] with encapsulation in boron/epoxy resin[150] (fig. 1.44). A most versatile piece of equipment is the diamond anvil, fig. 1.43, in which

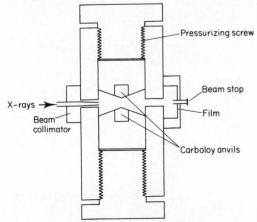

Fig. 1.42. High pressure anvil apparatus for X-ray diffraction. (After Jamieson and Lowson[142])

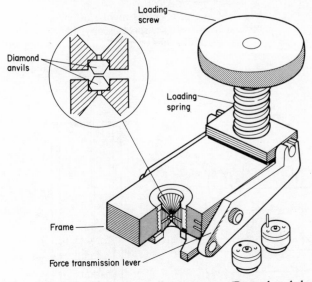

Fig. 1.43. The diamond anvil apparatus. (Reproduced by permission of de Beers Industrial Diamond Division)

pressures over 1 Mbar can be generated between the accurately lapped faces of two small diamonds by a lever system operated by a hand screw. The combination of hardness and transparency to radiation makes this suitable for high pressure visible, infrared, and Raman as well as X-ray spectroscopic work[144–147] and may be adapted to very high temperature work also[148, 149] 'in pile' neutron irradiation[152] (fig. 1.46). The use of screw clamps to reach moderately high pressures is convenient for many solid state experiments; for example, the apparatus fig. 1.45 may be made small enough to fit into a Dewar vessel for cryostatic measurements at several hundred kbar.[151] Ultra-high

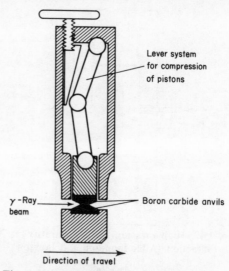

Fig. 1.44. Lever-operated anvils for γ-ray resonance (Mossbauer spectroscopy). (After Holtzapfel et al.[217])

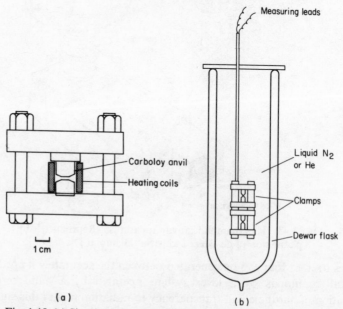

Fig. 1.45. (a) Simple clamped anvil apparatus.
(b) Several clamped anvils arranged for experiment in Dewar flask

pressure systems may be built along these principles but using massive lateral support such as in the two stage apparatus of Lorent[154] (700 kbar) and the well-known 'Belt' apparatus constructed by Hall[155] in which the first syntheses of diamond were achieved, fig. 1.47. A variable laterally-supported vessel for the

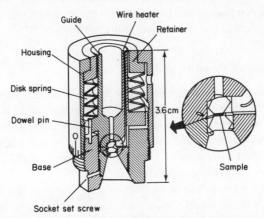

Fig. 1.46. WC anvil apparatus for 'in-pile' neutron irradiation of materials under pressure.[152] (Reproduced by permission of the American Institute of Physics)

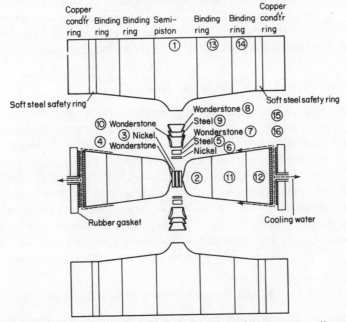

Fig. 1.47. 'The Belt' ultra-high-pressure high-temperature anvil apparatus[155]. 'Exploded' assembly. (Reproduced by permission of the American Institute of Physics)

300 kbar region has been recently described.[156] The advantages and disadvantages of different configurations of anvil apparatus have been discussed by Kumazawa.[129] Clamps of Be–Cu alloy may be used to contain apparatus for magnetic measurements.[235] A multi-anvil device has been reported in which pressurization is achieved by immersion in a hydrostatic fluid. The pressure arises from the large force placed on the (spherical) exterior and transmitted to a very small region at the centre thus effecting intensification.[164]

1.2.8 Generation of Pressure by Phase Change

If a closed vessel were completely filled with a liquid which expanded on freezing, very high pressures could in principle be generated by allowing it to cool until freezing occurred. Three such liquids are water, bismuth, and germanium.[158] Allowing for the effect of compression on the freezing point, it has been calculated that a pressure of 180 kbar could theoretically be obtained from the solidification of liquid germanium (section 2.18).

Similarly, high pressures may be obtained by allowing the melting of solids which form a less dense liquid phase, the usual situation, or by heating of liquids or the use of other suitable phase changes. There is perhaps some difficulty in controlling the pressures obtained in this way but potentially the method might have utility for preparative work on account of the simplicity of the apparatus required.

Kan has demonstrated the feasibility of obtaining pressures around 6 kbar by allowing a sealed vessel of frozen ethanol to expand on warming to room temperature.[159] If the ethanol is frozen under a pressure of 5 kbar, before sealing the vessel, an eventual pressure of 11 kbar is obtained on warming.

1.2.9 Generation of High Pressure by Shock Wave

The highest pressures which may be generated in the laboratory result from the passage of shock waves, produced from an explosive source, through a condensed material. In this way pressures as high as 10 Mbar may be produced and quite accurately measured though the duration of this stress is only transient, usually of the order 10 ns. The technique has been reviewed by Duvall and Fowles[160] and is used mainly for the measurement of PVT data at extreme conditions thus lying outside the scope of this monograph. A nuclear explosion can generate transient pressures in the 20 Mbar region within surrounding samples and, at the same time, produce secondary neutrons which may act as the probe for the effects produced,[157] perturbations of the nuclei at these extreme conditions.

1.3. MEASUREMENT OF HIGH PRESSURE

The ideal, of a measuring device which is accurate, reproducible, easy to construct and use, and suitable for a wide range of pressures, is a demanding one. In practice, a range of very different measuring devices are used, each of which possess some advantage for the particular application.[178-182] The problem of accuracy is an acute one. Ultimately any pressure measurement must be referred to a primary standard. According to the definition of pressure (section 1.1.2) absolute acceleration should be used, though this is a dynamic property and we normally wish to measure static pressures. Thus, the acceleration of the earth's gravitational field is usually the ultimate reference although, in principle we could use the PVT properties of some form of matter, a gas or crystal for instance, for which theory is adequate to enable pressures to be deduced from volume changes. Primary standards may then be used to calibrate secondary pressure transducers which may in principle use any convenient pressure-dependent property although a linear dependence is most desirable. Such devices may then have the convenience of simplicity, small bulk, or direct-reading capability and are the manometers normally in use.

1.3.1 Absolute Pressure Measurement

The pressure balance; this consists basically of a piston and cylinder designed so that the piston may be loaded with weights which are added until their force

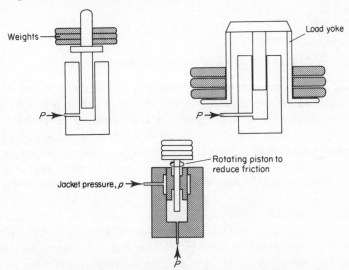

Fig. 1.48. Schematic layout of some pressure balance designs: (a) simple pile of weights; (b) lowered centre of mass using a load yoke; (c) controlled piston clearance using subsidiary pressure system

balances the force on the piston due to the pressure to be measured, fig. 1.48. At this point the piston neither ascends nor descends and the pressure is given by

$$P = m/A$$

where m = mass of piston + weights, A = piston area. Various configurations have been used in order to avoid using a high pile of weights and instability resulting,[183, 184] such as the yoke principle with lowered centre of gravity and the

46

steelyard type, fig. 149, to produce an equivalent force from a smaller loading.[185] In order to eliminate friction, it is necessary to rotate the piston slowly during measurement and also to use such a carefully machined apparatus that a film of oil remains between piston and cylinder and there is a very slight leakage of lubricant. At higher pressures this may become excessive on account of

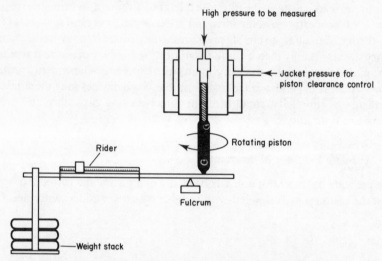

Fig. 1.49. Pressure balance using the steelyard principle. (After Yasumani[185])

deformation of the cylinder. The problem may be overcome by the 're-entrant cylinder' approach of Bridgeman,[186] a self-sealing device or, better, by applying at the correct level, a subsidiary oil pressure to the cylinder (controlled clearance piston gauge)[187, 188] (fig. 1.48c). The controlled clearance piston gauge is highly accurate and usable up to at least 25 kbar, though in order to obtain the highest accuracy a complicated series of corrections need to be made to allow for distortion, leakage, etc.[189] Instruments measuring up to 7 kbar are available commercially.[190]

The mercury column; a hydrostatic head of mercury is one of the most obvious manometers and widely used for low pressures. The column length becomes unwieldy above about 40 bar (30 m) but can be used as a differential manometer in conjunction with some other gauge. Due to simplicity of design and high precision, the mercury column has been used as a standard, though various precautions must be observed. The pressure existing between two levels x_1 and x_2 is given by

$$P = \int_{x_1}^{x_2} \rho g \, dx \qquad (1.8)$$

where ρ is the density of mercury. The highest precision demands a knowledge of the variation of density with height, which changes due to compression, and also strict temperature control. Taking these factors into account, a precision of 1 in 10^4 may be achieved up to 3 kbar.[191] To avoid using one very tall column, a series of connected U-tubes with arms filled with mercury in the lower halves and water in the upper has been proposed.[53]

P–V–T behaviour   An adequate equation of state for some form of matter, applicable to the high pressure region, would permit the measurement of pressure from changes in volume or density. Gases all show large deviations from ideality and equations of state contain too many arbitrary constants. The theory of even the simplest types of crystal is not adequate for pressures to be calculated from measurements of v/v_0 alone; i.e. one cannot calculate compressibilities from theory. If, however, the volume and compressibility of a sample are both known pressure may be determined from the relationships,

$$\text{compressibility } \beta = -\frac{1}{v}\frac{\mathrm{d}v}{\mathrm{d}P}$$

$$P = -\int_{v_0}^{v}\frac{1}{\beta v_0}\mathrm{d}v \tag{1.9}$$

A gauge due to Ruoff using this principle determines the length of a silicon rod (using ultrasonic methods) as a function of pressure.[192] The method may also be applied by determining the lattice parameter of a cubic crystal such as NaCl or CsCl (by X-ray diffraction) as a function of pressure.[193, 194] The compressibility of these salts is established by previous experiment (fig. 1.50). While non-linear, such a method may be used up to the highest pressures.

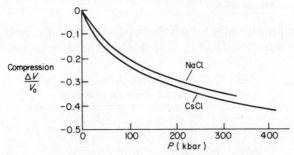

Fig. 1.50. Changes in lattice parameters of some cubic crystals under pressure. (After Decker[193, 194])

1.3.2   Secondary Measuring Devices (Transfer Gauges)

(a) *Resistance measurement*

The resistance of metals increases with pressure. Bridgman[195] (around 1911) measured many metals and found manganin (approx $80:15:5\%$ Cu:Mn:Ni) to have the largest pressure coefficient and for the past 50 years the manganin gauge has been one of the most popular measuring devices. Typically, a coil of fine (e.g. 40 gauge) manganin wire, cotton-insulated and of about $100\,\Omega$ resistance (though greater sensitivity is claimed with a much greater length, some $5\,k\Omega$), is wound non-inductively on a ceramic or PTFE former and immersed in a highly insulating medium. Electrical connections are led out to a sensitive Wheatstone bridge which may contain an identical manganin coil in the opposite arm for temperature compensation.[196] The gauge must be conditioned by repeated heating and cooling and pressure cycling when its resistance becomes very nearly linear to 100 kbar. The coefficient of resistance is about 2.4×10^{-6} bar^{-1}, the value differing slightly between samples so that each gauge must be calibrated and calibration repeated at intervals, to obtain α and β,

$$P = \alpha(R_p - R_0) + \beta(R_p - R_0)^2 \tag{1.10}$$

although the quadratic term is usually ignored. The manganin gauge is capable of $1:10^4$ precision at up to 10 kbar. Calibration is usually carried out at a 'fixed point'[197] and, since the gauge is linear, one such measurement suffices for most purposes. A gold–chromium alloy has been used but is less linear than manganin.[198]

Semiconductors might be expected to show large pressure coefficients of resistance though little information is available. The resistance of carbon and of zener diodes, both quite non-linear, have been suggested as pressure transducers. The breakdown voltage of the latter falls linearly with pressure to at least 20 kbar.[199] Gallium arsenide single crystals may be useful, their resistivity (ca. 10% increase for 8 kbar) varying linearly.[200,201,212–214]

(b) *Capacitance transducers*

Hydrostatic pressure affects both the dimensions and dielectric properties of non-isotropic materials. On this principle, a capacitive transducer has been

Table 1.2. Temperature and pressure characteristics of some dielectric materials

	$\dfrac{10^6}{C_0}\dfrac{dC}{dT}$ (K)	$\dfrac{10^{12}}{C_0}\dfrac{dC}{dP}$ (Pa)
CaF_2	263	−37.8
$CaCO_3 \parallel$	331.5	12.0
$CaCO_3 \perp$	335	71.0
As_2S_3	76.7	110.6
$Bi_{12}GeO_{20}$	73.7	−102.8

developed by the National Bureau of Standards based on changes in the capacitance of a crystal of parallel-cut calcite, for which

$$C = C_0(1 + AP + BP^2) \tag{1.11}$$

Other crystals for which such a quadratic law holds have also been examined and their characteristics given in Table 1.2. This type of transducer is potentially very useful though as yet largely untried.[202]

(c) *Spectroscopic methods*

Spectral shifts with pressure are on the whole rather small both for vibrational and electronic spectra and may be non-linear.[203] Spectral shifts have, however, been used extensively in high pressure measurement in apparatus which is accessible to light such as the diamond anvil cell. A scale based on the pressure-induced shift of the wavelength maximum of ruby fluorescence has been developed, $\Delta\lambda = 0.036\,\text{nm}\,\text{kbar}^{-1}$, fig. 1.51. Measurements may be carried out *in situ* using very small ruby crystals within the sample being examined. Non-hydrostatic pressure causes broadening as well as a band-shift. This scale appears to be linear to about 100 kbar and a pressure of 1 Mbar on this scale has been claimed.[205]

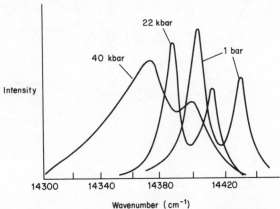

Fig. 1.51. The effect of pressure on the fluorescence
emission spectrum of ruby (Cr^{+++}-doped Al_2O_3)

For measurement of lower pressures, spectral shifts of charge–transfer complexes in solution, such as tetracyanoethylene-hexamethylbezene have been suggested since the sensitivity to pressure of these transitions is unusually large.[204]

Bourdon gauges The commonest type of direct reading gauge for moderate pressures depends upon the mechanical deformation by pressure of a hollow element, usually a flattened and curved tube. This deformation is amplified mechanically and made to rotate a pointer, fig. 1.52. Such gauges must be calibrated but may be made with a large and finely divided dial so that a reading of about 0.5% of the full scale deflection may be made. Bourdon gauges may be obtained in a variety of ranges, the maximum being 0–10 kbar. Their advantage lies in extreme simplicity, ease of reading, and ruggedness. Disadvantages include the need to recalibrate periodically for the highest accuracy, and hysteresis between a rising and falling pointer.

'Strain' gauges A very useful and commercially available pressure transducer is the so-called strain gauge in which a hollow element of strengthened steel

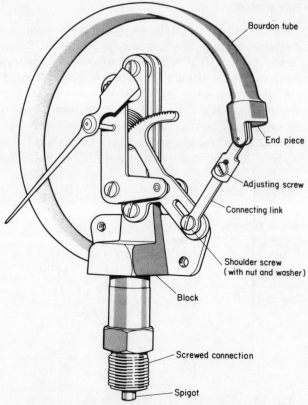

Fig. 1.52. Bourdon gauge mechanism. (Reproduced by permission of Budenberg Gauge Co. Ltd.)

undergoes expansion when pressurized from the inside (fig. 1.53). This strain brings about a change in resistance of elements on the outer surface, which is measured. Such devices may be used up to 15 kbar at least, are linear, and with suitable circuitry may be made direct reading. Unlike the manganin gauge, no electrical connections within the high pressure system have to be made.

Thermal conductivity gauge The thermal conductivity of a gas in particular rises steeply with pressure. This can be detected by the change in temperature and hence of electrical resistance of a wire filament carrying a constant current. A simple transducer based on this principle has been proposed and accuracy of ca. 2% up to several kbar claimed.[206] The device is non-linear and dependent upon the composition of the gas whose pressure is being measured.

1.3.3 Fixed Points

Phase changes, liquid–solid or solid–solid in pure substances, usually occur at a definite combination of pressure and temperature. A typical phase diagram is

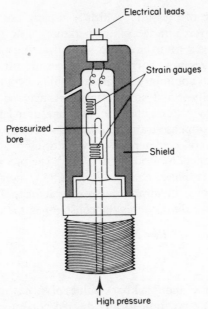

Electrical leads

Strain gauges

Pressurized bore

Shield

High pressure

Fig. 1.53. Layout of a typical strain gauge
pressure transducer

shown in fig. 1.54. At a given temperature, the phase change will occur at a fixed pressure and, once the transition curve has been determined, a single pressure measurement may be obtained at any given temperature with reference to it by observing the phase change. A series of suitable substances may be used to

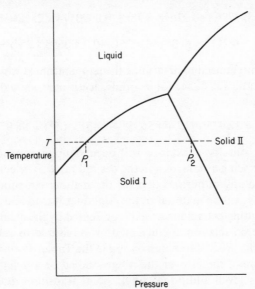

Liquid

T

Temperature

P_1

Solid II

P_2

Solid I

Pressure

Fig. 1.54. Typical phase diagram for use in pressure
calibration

calibrate over a range of pressures, which is usually more convenient than following the curve over a range of temperatures. For the lower range of pressures, phase changes from liquid to solid (freezing pressures) are often employed. At higher pressures, solid–solid changes are usually used. Freezing may be detected by a discontinuity in compressibility, evolution of heat, or the onset of optical opacity. Solid–solid transitions may be evident from discontinuities in conductivity (for metals) or of volume, or in latent heat evolution. An enormous volume of phase transition data has been accumulated over the years, much of which is collected in reviews by Pistorius[207] and by Merrill,[208] as well as in numerous papers by Bridgman.[209–211] The pressure–dependence of melting points or other phase changes is invariably non-linear but over a moderate range may fit a quadratic curve satisfactorily. It is convenient to tabulate data for use in calibration as the constants in the equations,

$$P = X + YT + ZT^2 \tag{1.12}$$

and

$$T = A + BP + CP^2$$

(where T is expressed in °C and P in kbar). Values of X, Y, Z and A, B, C are given in Table 1.3 so that interpolation to any desired conditions may be quickly carried out.

Suppose it is desired to calibrate a manganin gauge operating up to 4 kbar, 25°. Since the latter is linear in its response, one calibration point suffices though two would give additional reliability. Inspection of Table 1.3 reveals that carbon tetrachloride and nitrobenzene would be suitable substances, their melting pressure at 25° being given by

CCl_4: $P = 0.5796 + (0.02678 \times 25) + (0.0000574 \times 25^2) = 1.28$ kbar

$PhNO_2$: $P = -0.2173 + (0.04148 \times 25) + (0.0000695 \times 25^2) = 0.863$ kbar

Conversely one might need to determine the temperature at which one needed to work in order that a substance for example, acetic acid, should not freeze when 5 kbar is applied:

$$T = 22.975 + (17.4485 \times 5) - (0.4513 \times 5^2) = 98.9 \,°C$$

The choice of a suitable substance will depend upon a variety of factors in addition to there being a transition in the desired range. The compound must be highly pure since any impurities depress the melting temperature. It must be stable with respect to the material of the container it is in and a check must be made that the melting behaviour is well represented by the quadratic function as indicated by the correlation coefficient. It is advisable to select a substance showing a value of $r \geqslant 0.9999$ (expressed as g in the Table). Poor correlations may be due to a failure of the fit over the range studied or unreliable data.

In Table 1.4 is given additional single point transition data; the following references contain phase data:[219] (many inorganic and organic compounds),[220] (Bi halides),[221] (silver halides),[222] (sulphates and other oxy-salts),[223] (Sn salts),[224]

Table 1.3. P–T correlations for solid–liquid transitions

[T/°C; P/kbar] Material (liquid–solid transitions unless otherwise stated)	P limits/ kbar	T limits/ °C	$T = A + BP + CP^2$			$P = X + YT + ZT^2$			No. of points	*r
			A	B	C	X	Y	Z		
Yellow P	1–6	44–200	44.4394	30.1327	0.7730	−1.358	0.02867	0.0000473	7	g
Sodium	1–12	97.5–180	97.500	8.5116	−0.1543	−6.1071	1.7197	−0.002849	8	g
Potassium	0–12	62–180	63.65	14.4226	−0.3998	7.555	0.1964	0.0000811	8	0.9997
Mercury	0–12	−40 25	−40.338	5.4024	−0.04096				8	g
Bismuth	0–12	220–270	271.3	−3.4352	−0.08499	3.8358	0.2533	−0.0009863	13	g
SiCl$_4$	0–11	−70 213	−69.1	30.6459	−0.4524	2.3072	0.03576	0.0000270	11	g
Carbon disulphide L–S	0–35	−115 209	−93.66	9.914	−0.0339	10.055	0.1097	0.0000583	11	g
Benzene	0–11	86–200	8.0971	25.5776	−0.6652	−0.1852	0.03220	0.0001054	9	g
Cyclohexene	25–175	25–175	−172.2	29.1787	−0.1908	55.5662	1.7892	0.01285	5	0.96
Chlorobenzene	2–12	−11 110	−25.33	9.4961	0.1450	3.0814	0.08027	0.0000158	9	0.9568
Bromobenzene	0–12	−30 130	−30.82	18.3038	−0.4007	1.6390	0.05821	0.0001639	7	g
Dichloromethane	10–30	0–150	−88.60	9.3335	−0.03520	10.12	0.1179	0.0000720	5	g
Dibromomethane						−0.3903	0.04285	0.0000395	5	0.9942
Chloroform	3–18	−10 175	−49.61	14.1202	0.07097	16.6566	−0.9539	0.01008	16	0.9940
Carbon tetrachloride	0–9	−22 −220	−22.6	36.0766	−1.0726	0.5796	0.02678	0.0000574	10	g
Bromoethane	15–28	5–75	−95.62	7.6955	0.05911	14.3334	0.1629	0.0004537	7	0.9996

Table 1.3 (cont.)

[T/°C; P/kbar] Material (liquid–solid transitions unless otherwise stated)	P limits/ kbar	T limits/ °C	$T = A + BP + CP^2$			$P = X + YT + ZT^2$			No. of points	*r
			A	B	C	X	Y	Z		
1-Bromopropane	14–40	25–200	−81.66	7.6803	0.01046	11.979	0.1160	0.000129	9	0.9904
Bromoform	0–11	7.8–215	7.780	24.002	−0.04808	−0.2958	0.03774	0.0000733	12	g
Ethanol	15–29	−5 110	−100.38	6.7206	−0.01909	16.035	0.1670	0.0001014	8	g
1-Butanol	11–35	25–150	−73.58	9.4478	0.08202	8.9705	0.1069	0.0003986	10	0.9991
Tert-butanol			24.80	34.4936	−3.7081	−0.6012	0.02035	0.0001620	6	g
Tert-pentanol	0–2.5	−8 −30	−8.45	23.5281	−3.0973	0.3749	0.04685	0.0008105	6	0.9997
Acetic acid L–SI	0–2	16–55	16.60	24.3988	−2.4487	−0.9267	0.05503	−0.0000899	5	g
Acetic acid L–SII	2–11	55–160	22.975	17.4485	−0.4513	−0.1806	0.02466	0.0002888	8	g
ClCH$_2$COOH			62.74	14.200	−0.2677	−3.5646	0.04605	0.0001767	6	g
Ethyl acetate	12–30	25–125	−4.1636	1.5341	0.0814	5.0500	0.2993	−0.000648	4	0.9992
Dimethyl oxalate	0–8	54–208	54.24	20.8259	−0.2088	−2.6183	0.04562	0.0000299	6	0.9997
Urethane L–SI			47.94	11.8316	−0.6925	1.5122	−0.1447	0.0023626	5	g
L–SIII			21.80	14.6477	−0.2509	−0.4797	0.0436	0.0002202	6	g
L–SII			44.75	11.6620	−0.9240	16.4702	−0.5464	0.005023	4	g
Aniline			−5.239	19.578	−0.4057	0.3068	0.04812	0.0001262	6	g
p-Toluidine	0–8	44–210	43.7	16.9803	0.7385	0.9983	−0.007828	0.000206	6	0.9494
Diphenylamine			54.32	26.7856	−0.7858	−1.6624	0.0255	0.0000893	5	g
Phenol L–II			25.60	20.5724	−0.3979	−0.6755	0.0345	0.0001194	6	g
o-Cresol L–I			31.79	16.5600	−0.6920	−0.9026	0.01489	0.0004996	6	g

Compound										n	r
o-Cresol L–II			12.48	17.2733	0.2631	1.4561	0.0268	0.0000779		5	g
Benzophenone			48.34	27.6635	−0.7809	1.5892	−0.1044	0.001046		6	0.9997
Acetamide (S–S)	−48		81.56	12.709	−0.7565					6	0.9997
Furan IV–I	−21	8–9	−255.1	26.6217	−0.03992						
Furan L → I	−63	1–9	−84.706	12.7621	−0.5844	13.8125	0.3174	0.002066		9	0.9992
	−16										
Nitrobenzene			5.970	22.7429	0.4171	−0.21447	0.04148	0.0000695		5	g
4-Nitrophenol			112.64	26.68	−1.0818	−2.9137	0.01483	0.0000983		5	g

*r = multiple correlation coefficient; g signifies $r \geqslant 0.9999$.

(S),[225] (benzene and nitrobenzene),[226] (many inorganic compounds),[227] (Na halides),[228] (many salts),[229,230] (alkali nitrates),[231] (alkali nitrites),[232] (Sb halides). The effect of pressure on the melting point depends upon the volume change with which the phase change is associated, which is further discussed in section 2.3.

Table 1.4. Some fixed points for pressure calibration

Substance	Transform-ation	$T/°C$	P/bar
Hg	liq–I	−20	3,648
		−10	5,602
		0	7,569
		10	9,549
		20	11,541
		25	12,542
		50	17,590
		100	27,891
H_2O	liq–VI	25	9,743
CO_2	liq–II	0	3,372
		8.5	4,000
CCl_4	liq–I	75.8	3,000
		149.5	6,000
	I–II	20	3,310
C_6H_6	liq–I	96.6	4,000
C_6H_5Cl	liq–I	16.7	4,000
Ga	II–liq	25	21,290
Bi	II–III	25	26,965
NH_4I	I–II	25	536
PhOH	I–II	5	1,598
NH_4NCO	I–II	25	2,151
Camphor	I–II	25	2,897
KNO_3	I–II	25	2,944
High pressure transition			P (kbar)
Bi	I–II	25	25.4
	III–V		88
Pb	I–II		161
Ba			59 and 141
Rb			330

1.3.4 Temperature Measurement Under Pressure[236]

Accurate measurement of temperature within a high pressure enclosure will usually involve the use of a thermocouple. The e.m.f. of most thermojunctions is somewhat pressure dependent so that, for the highest accuracy, a correction needs to be made.[4,218] A typical set of correction curves for the Pt–Pt/Rh couple is shown in fig. 1.55.

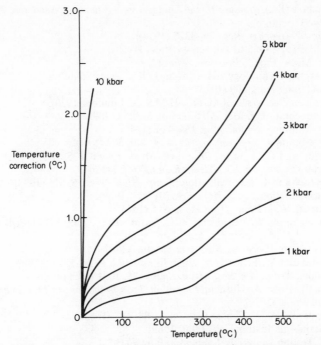

Fig. 1.55. Typical correction curves for the Pt–Pt/10%Rh thermocouple as a function of pressure. (After Bell et al.[218])

REFERENCES

1. S. E. Babb, *Technique of Inorganic Chemistry*, Vol. VI, 83 (1966).
2. D. C. Munro, *High Pressure Physics and Chemistry*, **1**, 11 (1963).
3. K. E. Weale, *Chemical Reactions at High Pressures*, Spon, London (1967).
4. P. W. Bridgman, *Physics of High Pressures*, Bell, London (1949).
5. R. S. Bradley (Ed.), *High Pressure Physics and Chemistry*, 2 vols., Academic Press, New York (1963).
6. E. W. Comings, *High Pressure Technology*, McGraw-Hill, New York (1956).
7. W. R. D. Manning, *High Temp. High Pres.*, **1**, 123 (1969).
8. H. Tongue, *Design and Construction of High Pressure Plant*, Chapman and Hall, London (1959).
9. E. E. Sechler, *Elasticity in Engineering*, Wiley, New York (1952).
10. W. R. D. Manning, *Ind. Eng. Chem.*, **49**, 1969 (1957).
11. J. L. M. Morrison, *Proc. Inst. Mech. Engrs.*, **159**, 81 (1948).
12. British Standards Institute, **BS 1500** (1969).
13. V. I. Sokolov, *Khim. Machinostroenie*, 14, (1938).
14. B. Crosland and J. A. Bones, *Proc. Inst. Mech. Engrs.*, **172**, 777 (1958).
15. B. Crosland, S. H. Jorgensen, and J. A. Bones, *Trans. Am. Soc. Mech. Engrs.*, **81**, 95 (1958).
16. B. Crosland and B. A. Austin, *Proc. Inst. Mech. Engrs.*, **180**, 118 (1965).
17. B. Crosland, *Proc. Inst. Mech. Engrs.*, **180**, 243 (1965).
18. B. Crosland, *Welding Research Council Bulletin*, No. 94 (1964).
19. J. H. Faupel, *Trans. Am. Soc. Mech. Engrs.*, **78**, 1031 (1956).
20. J. H. Faupel and A. R. Furbeck, *Trans. Am. Soc. Mech. Engrs.*, **75**, 345 (1953).

58

21. R. W. Nichols (Ed.), *Pressure Vessel Engineering Technology*, Academic Press, New York (1969).
22. K. T. Kim, *J. Geophys. Res.*, **79**, 3325 (1974).
23. B. Hill, *Plasticity*, Oxford University Press (1950).
24. Am. Soc. Mech. Engrs., *Pressure Vessel Code*.
25. High Pressure Technology Association (U.K.), *Safety Code*.
26. *J. Iron and Steel Inst.*, 600 (1973).
27. A. E. Macrae, *Overstrain of Metals*, H.M.S.O., London (1930).
28. V. A. Shapochkin, *Fiz. Met. Metallorev.*, **2**, 303 (1960).
29. W. R. D. Manning, *Engineering*, **159**, 101 (1945).
30. W. R. D. Manning, *Engineering*, **169**, 479, 509, 562 (1950).
31. B. Crosland and D. J. Burns, *Proc. Inst. Mech. Engrs.*, **175**, 1083 (1961).
32. R. A. Strub, *Trans. Am. Soc. Mech. Engrs.*, **75**, 73 (1963).
33. D. C. Harvey and B. Langyal, *High. Temp. High Press.*, **5**, 515 (1973).
34. G. H. Haslam, *High Temp. High Pres.*, **1**, 705 (1969).
35. C. E. Turner, *High Temp. High Pres.*, **6**, 1 (1974).
36. V. S. Robinson, A. W. Pense, and R. D. Stout, *Weld J.*, **43**, 5315 (1964).
37. J. Rogan and J. M. Alexander, *Met. Rev.*, **1**, 60 (1966).
38. P. W. Bridgman, *Proc. Am. Soc. Arts Sc.*, **46**, 325 (1911).
39. B. Crosland and W. F. K. Kerr, *High Temp. High Pres.*, **1**, 133 (1969).
40. J. H. Foupel, *Trans. Am. Soc. Mech. Engrs.*, **78**, 1031 (1956).
41. J. L. Christian and A. Hurlich, *Theory of Transformations in Metals and Alloys*, Pergamon Press, (1977).
42. R. D. Stout and A. W. Pense, *Met. Prog.*, **88**, 147 (1965).
43. British Standards Institute, **BS 1500** (1969).
44. Am. Soc. Testing Materials, *Standards, Part 4* (1973).
45. C. M. vonMeysenburg, P. Gramberg, E. Gaube, and M. Dickhaenser, *Tech. Uberwach.*, **16**, 7 (1975).
46. A. M. Rall, *Met. Progr.*, **88**, 178 (1965).
47. J. Glan, *Brit. Power Eng.*, **1**, 47 (1960).
48. *J. Iron and Steel Inst.*, 815a (1972).
49. *J. Iron and Steel Inst.*, 1043b (1970).
50. L. Vereshchagin, *Doklady Akad. Nauk.*, **132**, 1059 (1960).
51. H. M. Priest and J. A. Gilligan, *Design and Manufacture of High Strength Steels*, U.S. Steel Corp., Pittsburg, Penn. (1959).
52. H. Spaehn and I. Class, *Zeit Werkstofftechnik*, **4**, 401 (1973).
53. D. S. Tsiklis, *Handbook of Techniques in High Pressure Chemistry, Research and Engineering* (Eng. trans.), Plenum, New York (1968).
54. G. T. Williams, 'What Steel shall I Use?', *Am. Soc. for Metals*, Cleveland, Ohio (1941).
55. *J. Iron and Steel Inst.*, **741** (1973).
56. R. W. Nichols, *Proc. Roy. Soc.*, **285**, 104 (1965).
57. G. M. Enos and W. E. Fontaine, *Elements of Heat Treatment*, Wiley, New York (1953).
58. S. Nicholson, *Heat Treatment of Steels*, D.S.I.R., Wellington, N.Z. (1967).
59. British Standards Institute, **BS 1500** (1969).
60. C. P. Sullivan, M. J. Donachie, and P. R. Morral, *Cobalt Base Superalloys*, Cobalt Information Centre (1970).
61. A. Magnee, J. M. Drapier, J. Dumont, D. Coutsouradis, and L. Hasraken, *Cobalt-containing High Strength Steels*, Centre d'Information du Cobalt, Brussels (1074).
62. *18% Nickel Maraging Steels*, International Nickel Company.
63. I. Simon, *Rev. Sci. Instr.*, **28**, 963 (1957).
64. M. Brouha and A. G. Rijnbeck, *Rev. Sci. Instr.*, **44**, 852 (1973).
65. C. C. Bradley, *High Pressure Methods in Solid State Research*, Butterworth, London (1969).

66. F. W. Wilson, *Chem. Eng.*, **69**, 159 (1962).
67. R. Kiyama, *Rev. Phys. Chem. Jap.*, **19**, 1 (1945).
68. D. M. Waschener and W. Paule, *Rev. Sci. Instr.*, **29**, 675 (1958).
69. P. W. Bridgman, *Rev. Mod. Phys.*, **18**, 1 (1946).
70. D. C. Harvey and B. Lengyel, *High Temp. High Pres.*, **3**, 631 (1971).
71. E. Whalley, A. Lavergne, and P. T. T. Wong, *Rev. Sci. Instr.*, **47**, 845 (1976).
72. W. B. Holtzapfel, *Chem. Abs*, **82**, 60702 (1975); S. Claesson *et al.*, *Trans. Farad. Soc.*, **74**, 3048 (1970).
73. E. Whalley, A. Lavergne, and P. T. T. Wong, *Rev. Sci. Instr.*, **47**, 845 (1976).
74. J. M. Besson, J. P. Pinceaux, and R. Piotrzkowski, *High Temp. High Pres.*, **6**, 23 (1974).
75. P. T. T. Wong and E. Whalley, *Rev. Sci. Instr.*, **45**, 904 (1974).
76. H. J. Dittmers, *Chem. Zeit. Chem. App.*, **94**, 940 (1970).
77. E. Whalley and A. Lavergne, *Rev. Sci. Instr.*, **32**, 1062 (1961).
78. L. F. Vereshchagin, Yu. S. Konyaev, and E. V. Polyakov, *High Temp. High Pres.*, **3**, 355 (1971).
79. J. B. Hyne, H. S. Golinkin, and W. G. Laidlaw, *J. Amer. Chem. Soc.*, **88**, 2104 (1966).
80. H. T. Hall, *Rev. Sci. Instr.*, **29**, 267 (1958).
81. *J. Iron and Steel Inst.*, 601 (1973).
82. N. S. Isaacs and E. Rannala, *J. Chem. Soc.*, *Perkin II*, 709 (1978).
83. S. L. S. Thomas, H. S. Turner, and W. F. Wall, *Proc. Inst. Mech. Engrs.*, **182**, 271 (1967).
84. See appendix.
85. See appendix.
86. L. F. Vereshchagin, L. Kh. Freindlin, A. M. Rubinshtein, and N. K. Numanov, *Izv. Akad. Nauk SSSR Otdel Khim Nauk*, 809 (1951).
87. V. P. Butazov, G. P. Shakovskoi, and M. G. Gonikberg, *Tr. Inst. Kristallog. Akad. Nauk SSSR*, **11**, 233 (1955).
88. B. Crosland and W. F. K. Kerr, *High Temp. High Pres.*, **1**, 133 (1969).
89. British Standards Institute, **BS 1500** (1969).
90. O. F. Tuttle, *Geol. Soc. Am. Bull.*, **60**, 1727 (1949).
91. G. W. Morey and J. M. Hesselgesser, *Econ. Geol.*, **46**, 821 (1951).
92. G. W. Morey and J. M. Hesselgesser, *J. Am. Ceram. Soc.*, **36**, 279 (1953).
93. M. Millet and G. Jenner, *J. Chim. Phys.*, **67**, 1766 (1970).
94. M. Millet and G. Jenner, *J. Chim Phys.*, **67**, 1667 (1970).
95. L. B. Robinson, F. W. Vohldiek, and C. T. Lynch, *Metall. Soc. Conf.*, Vol. 22, 180, Gordon and Breach, New York (1964).
96. M. Tamayama and H. Eyring, *Rev. Sci. Instr.*, **38**, 1666 (1967).
97. M. Brouha and A. Rijnbeck, *Rev. Sci. Instr.*, **44**, 853 (1973).
98. M. E. Muhle and R. G. Bautista, *High Temp. High Pres.*, **4**, 213 (1972).
99. J. R. Ferraro and L. J. Basile, *Appl. Spectr.*, **28**, 505 (1974).
100. N. S. Isaacs and E. Rannala, *J. Chem. Soc.*, *Perkin II*, 1555 (1975).
101. A. D. Yu, M. D. Wansbluth, and R. A. Grieger, *Rev. Sci. Instr.*, **44**, 1390 (1973).
102. J. M. Besson, J. P. Pinceaux, and R. Piotrkowski, *High Temp. High Pres.*, **6**, 101 (1974).
103. P. Figuiere, M. Ghelfenstein, and H. Szwarc, *High Temp. High Pres.*, **6**, 61 (1974).
104. H. B. Tinker and D. E. Morris, *Rev. Sci. Instr.*, **43**, 1024 (1972).
105. H. Lentz and S. O. Oh, *High Temp. High Pres.*, **7**, 91 (1975).
106. T. W. Swaddle and P. C. Kong, *Can. J. Chem.*, **48**, 3223 (1970).
107. W. J. LeNoble and R. Schlott, *Rev. Sci. Instr.*, **47**, 770 (1976).
108. Y. Kondo, H. Tojima, and N. Tokuru, *Bull. Chem. Soc. Jap.*, **40**, 1408 (1967).
109. W. J. LeNoble and R. Mukhtor, *J. Amer. Chem. Soc.*, **96**, 6191 (1974); **97**, 5938 (1975).
110. D. R. Strauhe and T. W. Swaddle, *J. Amer. Chem. Soc.*, **93**, 2783 (1971).
111. D. R. Strauhe and T. W. Swaddle, *J. Amer. Chem. Soc.*, **94**, 8357 (1972).

60

112. R. A. Grieger and C. A. Eckert, *J. Amer. Chem. Soc.*, **92**, 7149 (1970).
113. R. J. Withey and E. Whalley, *Trans. Farad. Soc.*, **59**, 895 (1963).
114. K. R. Brower, *J. Amer. Chem. Soc.*, **90**, 5401 (1968).
115. W. E. Jones, L. R. Carey, and T. W. Swaddle, *Can. J. Chem.*, **50**, 2739 (1972).
116. D. L. Carle and T. W. Swaddle, *Can. J. Chem.*, **51**, 3795 (1973).
117. A. D. Edgar and R. G. Platt, *High Temp. High Pres.*, **3**, 1 (1971).
118. C. D. Hubbard, C. J. Wilson, and E. F. Caldin, *J. Amer. Chem. Soc.*, **98**, 187 (1976).
119. K. Heremans, *High Pressure Chemistry*, H. Kelm (Ed.), 311, Riedel (1978).
120. B. B. Hasinoff, *Can. J. Chem.*, **54**, 1820 (1976).
121. K. R. Brower, *J. Amer. Chem. Soc.*, **90**, 5401 (1968).
122. K. Heremans, F. Ceuterick, J. Snauwaert, and J. Wauters, *Techniques and Applications of Fast Reactions in Solution*, W. J. Gettins and E. Wyn-Jones (Eds.), Reidel (1979).
123. D. D. McDonald and J. B. Hyne, *Can. J. Chem.*, **48**, 2494 (1970).
124. A. B. Lateef and J. B. Hyne, *Can. J. Chem.*, **47**, 1369 (1969).
125. W. J. LeNoble and A. Shirpik, *J. Org. Chem.*, **35**, 3588 (1970).
126. M. L. Tonnet and E. Whalley, *Can. J. Chem.*, **53**, 3414 (1975).
127. B. T. Baliga and E. Whalley, *J. Phys. Chem.*, **73**, 654 (1969).
128. J. E. Bannard and G. J. Hills, *High Temp. High Pres.*, **1**, 571 (1969).
129. M. Kumazawa, *High Temp. High Pres.*, **3**, 243 (1971).
130. B. Okai and J. Yoshimoto, *High Temp. High Pres.*, **5**, 675 (1973).
131. L. F. Vereshchagin, E. N. Yakovlev, B. V. Vinogradov, V. P. Sakun, and G. P. Stepanov, *High Temp. High Pres.*, **6**, 505, (1974).
132. M. Nishikowa and S. Akimoto, *High Temp. High Pres.*, **3**, 161 (1971).
133. H. Mazaki, *J. Sci. Instru.*, **6**, 1672 (1973).
134. L. Ming and W. A. Barrett, *Rev. Sci. Instr.*, **45**, 1115 (1974).
135. A. I. Prikhova, *Sinteticheskaya Almazy*, 5 (1975).
136. A. S. Balchan and H. G. Drickamer, *Rev. Sci. Instr.*, **32**, 308 (1961).
137. H. G. Drickamer, *Rev. Sci. Instr.*, **32**, 212 (1961).
138. H. G. Drickamer and A. S. Balchan, *Modern Very High Pressure Techniques*, R. H. Wentorf (Ed.), Butterworth, London (1962).
139. G. D. Pitt, *High Temp. High Pres.*, **1**, 111 (1969).
140. W. B. Holtzapfel and D. Severin, *Chem. Abs.*, **82**, 60702 (1975).
141. M. D. Banus, *High Temp. High Pres.*, **1**, 483 (1969).
142. J. C. Jamieson and A. W. Lowson, *J. Appl. Phys.*, **33**, 776 (1962).
143. T. N. Kolobyanina, S. S. Kabalkina, L. F. Vereshchagin, M. F. Kachan, and V. G. Losev, *High Temp. High Pres.*, **4**, 203 (1972).
144. J. Walker, *Nature*, **265**, 498 (1977).
145. W. B. Holtzapfel, *High Temp. High Pres.*, **2**, 241 (1970).
146. Industrial Diamond Research Bureau Publications, Sunninghill, Ascot, U.K.
147. L. Merrill and W. A. Bassett, *Rev. Sci. Instr.*, **45**, 290 (1974).
148. G. J. Piermavona and S. Block, *Rev. Sci. Instr.*, **46**, 973 (1975).
149. H. K. Mao and P. M. Bell, *Science*, **191**, 851 (1976).
150. D. B. McWhon and W. L. Bond, *Rev. Sci. Instr.*, **35**, 626 (1964).
151. G. Fujii, Y. Oda, and H. Nagano, *Jap. J. Appl. Phys.*, **11**, 591 (1972).
152. G. L. Kulchinski, C. W. Maynard, and R. C. Walsh, *Rev. Sci. Instr.*, **37**, 871 (1966).
153. S. D. Christan, *Appl. Spectr.*, **30**, 227 (1976).
154. R. E. Lorent, *Rev. Sci. Instr.*, **44**, 1691 (1973).
155. H. T. Hall, *Rev. Sci. Instr.*, **31**, 125 (1960).
156. D. P. Kendall, P. V. Dembowski, and T. E. Davidson, *Rev. Sci. Instr.*, **46**, 629 (1975).
157. C. E. Ragan, M. G. Silvert, and B. C. Diven, *Proc. 6th AIRAPT Conf.*, Plenum, 993 (1979).
158. J. Lees, *High Temp. High Press.*, **1**, 601 (1969).
159. Ya S. Kan, *Zhur. Tekh. Fiz.*, **18**, 1156 (1948).

160. G. E. Duvall and G. R. Fowles, *High Pressure Physics and Chemistry*, R. S. Bradley (Ed.), Vol. 2, Academic Press, London (1963).
161. S. Yamamoto, *Proc. 4th AIRAPT Conf., Kyoto, 810* (1974).
162. F. Grønlund and B. Andersen, *Acta. Chem. Scand.*, **23**, 2452 (1969).
163. M. Linton, *Proc. 4th AIRAPT Conf.*, Kyoto, 778 (1974).
164. J. Wanagel and A. L. Ruoff, *Proc. 6th AIRAPT Conf.* (Boulder), Plenum, 840 (1979).
165. D. A. Palmer, H. Schmidt, R. van Eldik, and H. Kelm, *Inorg. Chem. Acta* **29**, 261 (1978).
166. K. Heremans, J. Snauwaert, and J. Rijkenburg, *High Pres. Science and Technology*, Vol. 1, K. D. Timmerhaus and M. S. Barber (Eds.), Plenum, New York, 646 (1979).
167. G. J. Hills and P. J. Ovenden, *Adv. Electrochem. Electrochem. Eng.*, **4**, 185 (1966).
168. W. R. Hainsworth, H. J. Rowley, and D. A. McInnes, *J. Amer. Chem. Soc.*, **46**, 1437 (1924).
169. G. P. Johari and W. Dannhauser, *J. Chem. Phys.*, **48**, 5114 (1968).
170. G. B. Benedek and E. M. Purcell, *J. Chem. Phys.*, **22**, 2008 (1954).
171. F. T. Walls and S. J. Gill, *J. Phys. Chem.*, **59**, 278 (1955).
172. J. Jonas, *Adv. Mag. Res.*, **6**, 73 (1973); *Ann. Rev. Phys. Chem.*, **26**, 167 (1975).
173. J. Jonas, *Rev. Sci. Instr.*, **43**, 643 (1972).
174. J. Van Jouanne and J. Heidberg, *J. Magnet. Res.*, **7**, 1 (1972).
175. H. Yamada, *Rev. Sci. Instr.*, **45**, 540, (1974).
176. H. Vanni, W. L. Earl, and A. E. Merbach, *J. Magnet Res.*, **29**, 11 (1978).
177. W. L. Earl, H. Vanni, and A. E. Merbach, *J. Magnet. Res.*, **30**, 571 (1978).
178. T. E. Davidson and D. P. Kemall, *Mechanical Behaviour of Materials Under Pressure*, H. Ll. Pugh (Ed.), Elsevier, Amsterdam (1970).
179. C. Y. Liu, K. Ishizaki, J. Paauwe, and I. L. Spain, *High Temp. High Pres.*, **5**, 359 (1972).
180. R. Leclercq, *Exp. Thermod.*, **2**, 213 (1975).
181. W. G. Brombacher, *Instrum. Control Syst.*, **40**, 87 (1976).
182. A. A. Giordini and E. C. Lloyd (Eds.), *High Pressure Measurements*, Butterworth (1963).
183. A. Michels, *Ann. Phys.*, **73**, 577 (1924).
184. A. Michels, *Ann. Phys. N.Y.*, **72**, 285 (1923).
185. K. Yasumani, *Proc. Japan. Acad.*, **43**, 310 (1967); *Rev. Phys. Chem. Jap.*, **37**, 1 (1967).
186. P. W. Bridgman, *Proc. Am. Acad. Arts. Sc.*, **44**, 201 (1909).
187. D. P. Johnson and D. M. Newhall, *Trans Am. Soc. Mech. Engrs.*, **75**, 301 (1953).
188. D. P. Johnson and P. L. M. Heydemann, *Rev. Sci. Instr.*, **38**, 1094 (1967).
189. *Nat. Bur. Standards Monograph No. 65*, U.S. Govt. Printing Office, Washington, D.C.
190. See Appendix.
191. K. E. Bett, P. F. Hayes, and D. M. Newitt, *Phil. Trans. Roy. Soc.*, **59A**, 247 (1954).
192. A. L. Ruoff, R. C. Lincoln, and Y. C. Chen, *J. Phys. D, Appl. Phys.*, **6**, 1295 (1973).
193. D. L. Decker, *J. Appl. Phys.*, **42**, 3239 (1971).
194. D. L. Decker and T. G. Walton, *J. Appl. Phys.*, **43**, 4749 (1972).
195. P. W. Bridgman, *Proc. Am. Acad. Arts Sc.*, **47**, 321 (1911).
196. C. A. Gall and A. W. Birks, *Strain*, **3**, no. 4, 27 (1967).
197. R. S. Zeto and H. B. Vanfleet, *J. Appl. Phys.*, **40**, 2227 (1969).
198. H. E. Darling and D. H. Newhall, *Trans. Am. Soc. Mech. Engrs.*, **75**, 311 (1953).
199. W. Włodarki, *High Temp. High Pres.*, **6**, 115 (1974).
200. M. K. R. Vyas, *High Temp. High Pres.*, **6**, 237 (1974).
201. G. Connell, *High Temp. Pres.*, **6**, 310 (1974).
202. J. H. Colwell, *A Solid Dielectric Capacitive Pressure Transducer*, Nat. Bur. Stand., Washington, D.C.
203. H. G. Drickamer and C. W. Frank, *Electronic Transitions and the High Pressure Chemistry and Physics of Solids*, Chapman and Hall, London (1973).
204. A. H. Ewald, *Trans. Farad. Soc.*, **64**, 733 (1968).

62

205. H. K. Mao and P. M. Bell, *Science*, **191**, 851 (1976).
206. P. J. Freund and G. M. Rothberg, *Rev. Sci. Instr.*, **44**, 769 (1973).
207. C. W. F. T. Pistorius, *Prog. Solid State Chem.*, **11**, 1 (1976).
208. L. Merrill, *J. Phys. and Chem. Ref. Data*, **6**, 1205 (1977).
209. P. W. Bridgman, *Phys. Rev.*, **3**, 126 (1914).
210. P. W. Bridgman, *Phys. Rev.*, **6**, 1 (1915).
211. P. W. Bridgman, *Proc. Am. Acad. Arts. Sc.*, **74**, 399 (1942).
212. G. Pitt and B. Lees, *Phys. Rev. B.*, **2**, 4144 (1972).
213. G. Connell, *High Temp. High Pres.*, **1**, 77 (1969).
214. G. Pitt and W. Gunn, *High Temp. High Pres.*, **4**, 101 (1972).
215. P. W. Bridgman, *Studies in Large Plastic Flow and Fracture*, McGraw-Hill, New York (1952); *J. Appl. Phys.*, **24**, 560 (1953).
216. *Characteristics of En Steels*, British Iron and Steel Corp.
217. L. D. Roberts *et al.*, *Phys. Rev.*, **179**, 656 (1969); W. B. Holtzapfel, *High Temp. High Pres.*, **2**, 241 (1970).
218. P. M. Bell and J. L. England, and F. R. Boyd, *Year Book of the Carnegie Inst. of Washington*, **66**, 545 (1968).
219. Landolt-Bornstein, *Tables of Physical Constants*, Springer, Berlin (1970).
220. A. J. Darnell and W. A. McCollum, *J. Phys. Chem.*, **72**, 3032 (1968).
221. B. C. Deaton, *J. Appl. Phys.*, **36**, 1500 (1965).
222. C. W. F. T. Pistorius, *J. Chem. Phys.*, **43**, 2895 (1965); **44**, 4532 (1966); **46**, 2167 (1967).
223. M. Midorikawa, Y. Ishibashi, and Y. Takugi, *J. Phys. Soc. Jap.*, **37**, 1583 (1974).
224. G. C. Vezzoli and F. Dachille, *Inorg. Chem.*, **9**, 1973 (1970).
225. M. K. Zhokhovskii and V. S. Bosdanov, *Zhur. Fiz. Khim.*, **39**, 2520 (1965).
226. C. W. F. T. Pistorius, *Prog. Solid State Chem.*, **11**, Part 1, 1 (1976).
227. C. W. F. T. Pistorius, *J. Chem. Phys.*, **45**, 3513 (1966).
228. J. W. Johnson, W. J. Silva, and D. Cubiciotti, *J. Phys. Chem.* see **70**, 1169, 2985, 2389, 2989 (1966); **71**, 808, 3066, 1958, (1967); **72**, 1272, 1664, 1669 (1968); **73**, 3054 (1969).
229. S. E. Babb, P. E. Chaney, and B. B. Owens, *J. Chem. Phys.*, **41**, 2210 (1964).
230. B. B. Owens, *J. Chem. Phys.*, **42**, 2259 (1965).
231. E. Rappoport and G. C. Kennedy, *J. Phys. Chem.*, **26**, 1995 (1965); **27**, 1349 (1966).
232. B. F. Bowles, G. F. Scott, and S. E. Babb, *J. Chem. Phys.*, **39**, 831 (1963).
233. A. Gerschel, *Organic Liquids*; A. D. Buckingham, E. Lippert, and S. Bratos (Eds.), John Wiley, Chichester (1979).
234. T. Moriyoshi, *Bull. Chem. Soc. Jap.*, **44**, 2582 (1971).
235. R. P. Guertin and S. Fraser, *Rev. Sci. Instr.*, **45**, 863 (1974).
236. F. K. Meyer and A. E. Merbach, *J. Phys. E*, **12**, 185 (1979).
237. W. Foag and H. Tiltscher, *High Temp. High Pres.*, **10**, 465 (1978).
238. K. Heremans, *High Pressure Chemistry*, H. Kelm (Ed.), Reidel (1978).
239. B. B. Hasinoff, *Can. J. Chem.*, **57**, 77 (1979).
240. A. Jost, *Ber. Bunsenges. Phys. Chem.*, **79**, 850 (1975).
241. E. F. Caldin, M. W. Grant, and B. B. Hasinoff, *J. Chem. Soc.*, *Faraday I*, **68**, 2247 (1972).
242. M. W. Grant, *J. Chem. Soc.*, *Faraday I*, **69**, 560 (1973).
243. G. Saville, *Proc. 6th AIRAPT Conf. (Boulder)*, Plenum, London, 830 (1979).
244. D. H. Liebenberg, R. L. Mills, and J. C. Bronson, *Proc. 6th AIRAPT Conf. (Boulder)*, Plenum, London, 396 (1979).
245. M. Buback, E. U. Franck and H. Lendle, *Z. Naturforsch.* **34A**, 1489 (1979).

Chapter 2

Effects of Pressure on Physical Properties of Matter

The conjugate variable to pressure is volume as summed up in Boyle's Law for ideal gases (i.e. 'dilute' matter): $PV =$ constant, or the general relationship,

$V = (\partial G/\partial P)_T$ where G is the Gibbs free energy. All matter suffers a diminution of volume upon the application of pressure although the effect is very much smaller for condensed matter than for gases. In the liquid phase, chemical studies are principally carried out in a range not exceeding 20–30% compression while the physicists domain may exceed this 2–3 times. The ultimate in compressed matter is the neutron star in which a compression of matter to approximately $10^{-12}\%$ of its 'normal' value is achieved by enormous gravitational forces.

2.1 DENSITY PROPERTIES OF LIQUIDS

Many physical properties depend upon the density of matter and these include refractive index, dielectric constant, heat capacity, sound velocity. These properties will be, in general, affected by pressure in the same sense as a lowering of temperature which also brings about an increase in density. Other properties depend upon intermolecular forces and these include the kinetic properties viscosity and diffusion, electrical transport, and also solvation. Compression will decrease the average intermolecular distance and will tend to reduce rotational and translational motion but in a way which is structure and solvent dependent. At very high pressures, more deep-seated properties such as molecular orbital energies and symmetries may be affected. It is predicted that even hydrogen should show metallic properties at pressures in excess of 1 Mbar. These factors rarely enter into discussions of pressure effects below 10 kbar. The theory of liquids[1] is understood only imperfectly although there are now a succession of models on which to base liquid behaviour, necessarily of a statistical type incorporating interaction potentials, attractive and repulsive both between nearest neighbours and more distant pairs of molecules. The simplest molecular model is the 'smooth, hard sphere' but theory may be modified to incorporate non-spherical shapes, 'rough, hard spheres' in which a frictional element is introduced, or polarizable hard spheres in which multipole interactions contribute. Each model may have merit in a particular situation. Water is inevitably a special case and many of its properties change with pressure in accordance with an initial breaking of the hydrogen-bonded structure. A liquid is considered to contain molecules which occupy space in excess of that needed for close packing. This excess is referred to as the 'free volume' and it is this which is reduced in initial compression. At higher pressures when free volume has largely disappeared reduction in van der Waals dimensions may occur and the compressibility is greatly diminished.

2.1.1 The Compression of Liquids

The compressibility of liquids, the relative volume change per unit change of pressure, is of the order of 0.01% per bar. That liquids are compressible was apparently first noted by John Canton in 1762 although reliable compressibility data did not appear until the improved technology of the times permitted Amagat[2] and then Bridgman[3–7] and others to attack the problem. Very many workers have undertaken measurements of the compressibilities of common liquids such as water and mercury. Gibson,[8-11] Weale,[12,13] and coworkers have examined many organic liquids and authoritative reviews on the thermodynamics of volume change by Whalley[248,249] comprehensively covers the subject.

Apart from being an intrinsic physical property of a liquid, compressibility is important in testing theories of the liquid state since it is determined by the balance between attractive and repulsive potentials. It enters into many pressure-dependent thermodynamic expressions, is essential knowledge for the design and use of high pressure equipment, and may be required for correction of concentration terms, etc.

2.1.2 The Measurement of Compressibility

A pressure–volume plot for a liquid, unlike an ideal gas, is not linear. The compressibility β, where

$$\beta = -\frac{1}{V} \cdot \left(\frac{dV}{dP}\right) \text{(units, pressure}^{-1}) \tag{2.1}$$

decreases with pressure and increases with temperature. Compressibility may be reported at a particular pressure from the slope at that point, and is usually extrapolated to standard pressure (1 bar, β^0), although the P–V behaviour of the substance may be investigated to very high pressures. Other related quantities used by some authors are the tangent bulk modulus, K and the secant bulk modulus, $\bar{K}$, where

$$K = 1/\beta \tag{2.2}$$

$$\bar{K} = \frac{V_0 P}{V_0 - V_p} \tag{2.3}$$

i.e. $\bar{K}$ is an average value of bulk modulus over a finite pressure range.[14]

A further complication in defining compressibility is that, on compression of a liquid, as for a gas, heat is evolved due to the work of compression against repulsive intermolecular forces. The temperature rise produced will increase the volume and will affect the value of β, which is temperature dependent. Two situations are therefore defined; either the experiment is performed at constant temperature to obtain the *isothermal compressibility*, $\beta_T(K_T, \bar{K}_T)$, or the volume change produced by a sudden compression or expansion is measured allowing no heat to escape from sample to surroundings (i.e. at constant entropy) and the *adiabatic compressibility*, $\beta_s(K_s, \bar{K}_s)$ obtained. Situations between these two can, of course, be made to occur but are not easily treated. Some values of compressibilities are shown in Table 2.1. Measurement of isothermal compressibility requires the knowledge of volume and volume change with pressure at constant temperature. The pitfalls of such measurements have been detailed by Hayward[14] who notes that discrepancies of 10% between different workers' results are common and, indeed, may be as large as 30%.

Table 2.1. Isothermal compressibilities, β_T, of some liquids at 1 bar, 20° (unless otherwise indicated)

	β_T $(10^2\,\mathrm{kbar}^{-1})$		β_T $(10^2\,\mathrm{kbar}^{-1})$
Hexane	15.40	Methanol	12.14
2,3-Dimethylbutane	17.97	Ethanol	11.19
Heptane	13.4	1-Propanol (0°)	8.43
Octane	11.3	1-Butanol (0°)	8.10
Cyclopentane	13.31	1-Pentanol (0°)	7.71
Cyclohexane	11.30	Ethanediol	3.64
Cycloheptane	9.22	1-Octanol (0°)	6.82
Cyclooctane	8.03		
Dodecane (37.8°)	9.9	Acetic acid	9.08
Pentadecane (37.8°)	9.1	Ethyl acetate	11.32
Octadecane (60°)	9.4	Acetone	12.62
Benzene	9.44	Diethyl ether	18.65
Toluene	8.96		
m-Xylene	8.46	Water	4.58
		Mercury	0.40
Chlorobenzene	7.45		
Bromobenzene	6.46		
Anisole	6.60		
Aniline	4.53		
Nitrobenzene	4.93		
Phenol (60°)	6.05	Molten salts:	
		$NaNO_3$, 623 K	1.99
Carbon disulphide	9.38	KNO_3, 657 K	2.19
Carbon tétrachloride	10.50	$RbNO_3$, 629 K	2.26
Chloroform	9.96	$CsNO_3$, 709 K	3.09
Bromoethane	12.94		
Iodoethane	9.82		
1,1-Dichloroethane	7.97		
Tetrachloroethylene (25°)	7.56		
Trichloroethylene (25°)	8.57		

At low pressures (say to 100 bar), glass apparatus may be used, employing gas pressure and measuring ΔV by the movement of the meniscus in a capillary,[9,15-20] fig. 2.1. At higher pressures, the movement of a piston in a cylinder[21-25] or the change in length of a metal bellows containing the liquid[26-29] may be used to define the volume change, e.g. fig. 1.25. In all cases, allowance must be made for distortion of the vessel which may not be isotropic and therefore cannot be estimated from elasticity theory. It is recommended that this correction be obtained experimentally by observing the behaviour of mercury under the same conditions in the vessel, this liquid having the best P–V data available.[30,31] Thermostatting must be very good. For the low pressure range, at least 0.01° constancy is required and a long equilibration time may be necessary. Other sources of error include air bubbles trapped in the apparatus, errors in absolute volume measurement (V), and compression of rubber or plastic components (which are best absent). It is possible to reach a precision of the order of 0.4% with the observation of these caveats. A comprehensive review of techniques has been written by Whalley.[248]

Measurement of piston displacement or meniscus movement may be made with a cathetometer. Piston movement may be amplified by actuating changes in

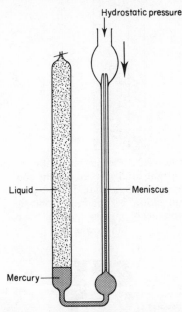

Hydrostatic pressure

Liquid   Meniscus

Mercury

Fig. 2.1. A piezometer for use with organic liquids; the whole apparatus is immersed in a water-filled pressure chamber with a window for observation of the meniscus[19]

a potentiometer slide wire to give a voltage reading[7,32] or the sensitive differential transformer principle in which piston movement is signalled by the movement of a small magnet into a coil.[33] To avoid contact between experimental liquid and the vessel, one may resort to encapsulation in a deformable container of a material such as lead or indium, as used by Bridgman,[31] a bellows of inert material may also be a method of achieving this end.

A piezometer originally due to Aimé (1843) has a reentrant capillary (fig. 2.2). It is filled with liquid and inverted under mercury. Upon pressurization, mercury enters the vessel via the capillary and drops inside. After equilibration, the vessel is removed and the trapped mercury weighed. This represents, after appropriate corrections, the compression of the liquid sample.[34] Each experiment, however, only yields one data point.[35] Another approach, used by Whalley[5] and by Weale,[36] gives more data in each experiment. The piezometer (fig. 2.3) is equipped with a capillary stem fitted with a series of contacts, each defining a successively smaller volume. After filling with liquid and inverting over mercury, the

68

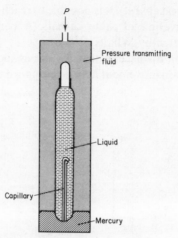

Fig. 2.2. Piezometer with re-entrant capillary; the volume contraction of the liquid upon pressurization is obtained from the volume of mercury which enters the vessel

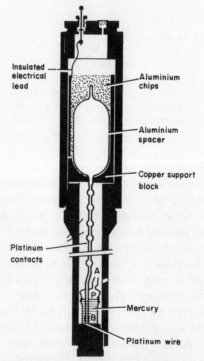

Fig. 2.3. Electrically-indicating piezometer.[54] (Reproduced by permission of the Chemical Society)

piezometer is pressurized and pressures needed to raise the mercury level to each contact successively recorded. The detection of each step is made by the electrical short circuit which results. Other precise piezometers have been described by Millero[171,172] and Kell.[173,174]

The change in density with pressure equally gives a value of β_T with the advantage that corrections for the vessel are unnecessary. The measurements of Razumikhin employ this principle but it has been little used, since it is usually less precise to make measurements of density change than of volume change. Use of the hydrometer principle requires that the change in volume of the float with pressure be known.

Adiabatic compressibility measurements may be made by sudden release of pressure and rapid reading of the new level in a capillary piezometer,[16,17] and the values of β_s can then be converted to β_T by the relationship,

$$\beta_T = \beta_s + \frac{\left(\dfrac{d\bar{v}}{dT}\right)_p^2 \rho T}{C_p} \quad \text{or} \quad \beta_T - \beta_s = \frac{\alpha^2 \rho T}{C_p} \tag{2.4}$$

where $\bar{v}$ is the specific volume ($m^3\,kg^{-1}$) and so $(d\bar{v}/dT) = \alpha$ represents thermal expansivity; $T =$ temperature (K), $\rho =$ density ($kg\,m^{-3}$) and C_p the specific heat at constant pressure ($J\,kg^{-1}\,K^{-1}$) whence β_T, β_s will be in Pa^{-1}.

The preferred method of determining β_s is by a measurement of the velocity of sound , $\mathbf{v}$, through the fluid. The passage of a sonic wave brings about alternate compression and rarefaction of the medium under adiabatic conditions and its velocity is determined by compressibility thus, 37–40,[222] accurate to 0.02% and needing a sample only some $10\,cm^3$ volume.

$$\mathbf{v} = (1/\beta_s \rho)^{1/2} \tag{2.5*}$$

Sonic measurements are easily made by tuning a resonant cavity containing the sample with an audio-oscillator. Ultrasonic beams of short wavelength, generated piezoelectrically, permit the use of small samples. Much data have been reviewed by Coates.[18]

2.1.3 Structural Additivity Relationships to Compressibility

A number of empirical relationships have been proposed in which compressibility arises from contributions of structural elements of the sample. While none of these relationships give results of great accuracy,[41] such an estimation may be better than no value at all in the absence of experimentally determined compressibility.

(a) Wada's equation[42]

$$\beta_T = (M/\rho \Sigma B)^7 \quad (0-2\%) \tag{2.6}$$

$$\beta_s = (M/\rho \Sigma A)^7 \quad (2-5\%) \tag{2.7}$$

* It follows also that $\beta_T/\beta_s = C_p/C_v = \gamma$, the ratio of principal specific heats, and $\mathbf{v} = (\gamma/\beta_T \rho)^{1/2}$. This relationship is principally of use in obtaining the otherwise inaccessible C_v.

where M = molecular weight (g), ρ = density (g ml^{-1}), and B, A are constants pertaining to covalent bonds in the sample (Table 2.7).

(b) Rao's equations[43-45]

$$\beta_T = [\bar{p}Z(6 \ln Z - 11)(1.01325 \times 10^6)] \tag{2.8}$$

where

$$Z = \frac{82.06T}{\bar{p}v_M} \quad \begin{array}{l}\text{(a factor expressing departure} \\ \text{from ideality of the vapour)}\end{array} \quad (12\text{--}30\%)$$

Table 2.2. Effect of temperature and homology on isothermal compressibilities

Compound	298.15 K	308.15 K	318.15 K	333.15 K
	$\beta_T (10^{10}\,N^{-1}\,m^{-2})$			
Benzene	9.66	10.44	11.28	12.77
n-Hexane	16.69	18.31	20.27	23.85
n-Heptane	14.38	15.70	17.12	19.69
n-Octane	12.82	13.85	15.06	17.05
n-Nonane	11.75	12.66	13.59	15.36
n-Decane	10.94	11.78	12.62	14.15
n-Undecane	10.31	11.08	11.87	13.22
n-Dodecane	9.88	10.53	11.32	12.54
n-Tridecane	9.48	10.06	10.75	11.93
n-Tetradecane	9.10	9.69	10.33	11.49
n-Pentadecane	8.82	9.40	10.03	11.03
n-Hexadecane	8.57	9.17	9.78	10.66

$\bar{p}$ is the vapour pressure (atmospheres) of the sample at $T(K)$

$$\left(\frac{1}{\rho\beta_s}\right)^{1/2} = C(T_c - T) \quad (2\text{--}4\%) \tag{2.9}$$

C is a constant for the compound which must be evaluated from one experimental measurement of β_s and ρ, and T_c is the critical temperature. Sound velocities may be calculated by an additivity rule due to Rao;

$$v = 0.032808\left(\frac{\Sigma D \cdot \rho}{M}\right)^3 \quad (1\text{--}4\%) \tag{2.10}$$

where D is a structural constant (see Table 2.7).

The percentage figure in brackets after each equation summarizes the statistical results of Gold and Ogle[41] and is a measure of the reliability of each equation for predicting the properties of organic liquids at the 95% confidence level.

Table 2.3. Compressibility data summarized by the Tait equation (eq. (2.14))

Liquid	$T(°C)$	B(bar)	C	Reference
Dichloromethane	25	1028	0.2360	86
	25	1066	0.2391	30a
Chloroform	25	1052	0.2391	
Carbon tetrachloride	25	798	0.2001	34
	25	867	0.2129	9
	25	854	0.2126	30a
Hexane	0	713.9	0.2172	13
	25	579.1	—	
	40	508.6	—	
	60	424.5	—	
Heptane	0	788.6	—	
	25	653.1	—	
	40	582.9	—	
	60	498.8	—	
Octane	0	931.4	—	
	25	776.5	—	
	40	696.8	—	
	60	598.1	—	
Benzene	25	970	0.2159	
	45	829	—	
	65	701	—	
	25	1042	0.2264	34
Pentadecane	37.8	983	0.2721	87
Bromoethane	0	1245	0.1014	27
	25	965	0.1012	
1-Bromobutane	−1	1172	0.0921	
	20.5	1094	0.0968	
Chlorobenzene	25	1249	0.2159	9
	45	1097	—	
	65	960	—	
	85	835	—	
	25	1191	—	
Bromobenzene	25	1404	—	
	45	1247	—	
	65	1103	—	
	85	972	—	
Nitrobenzene	25	1865	—	9
	45	1678	—	
	65	1504	—	
	85	1429	—	
Aniline	25	2006	—	
	45	1798	—	
	65	1606	—	
	85	1429	—	
Acetone	25	815	0.2356	30a
1,1-Dichloroethane	25	879.3	0.2296	
1,2-Dichloroethane	25	1231	0.2233	
1,1,2,2-Tetrachloroethane	25	1498	0.2156	
1,2-Dichloroethylene	25	1036	0.2518	
Trichloroethylene	25	1231	0.2233	
Tetrachloroethylene	25	1213	0.2425	

Table 2.4. Constants of the Hudleston equation (eq. (2.20)) for some liquids

Liquid	$T(°C)$	J	K	Reference
Bromoethane	−50	11, 118	9137	28
	0	10, 525	12, 435	
	25	10, 248	12, 998	
1-Bromobutane	−50	10, 974	13, 031	
	0	10, 515	15. 108	
	49	10, 229	13, 039	

Table 2.5. Compressibility data summarized by eq. (2.12)[a] $V/V_0 = 1 - \theta P + \phi P^2 + \psi P^3$

Liquid	$T(°C)$	$10^7\,\theta\,(\text{bar})$	$10^9\,\phi\,(\text{bar}^2)$	$10^{13}\,\psi\,(\text{bar}^3)$	Reference
Carbon disulphide	20	619	6.372	2.529	88
	40	686	7.444	3.044	
	60	761	8.556	3.540	
	80	847	9.861	4.128	
Isopentane	30	2111	900	10,945	
	50	3420	800	13,660	
Carbon tetrachloride	25	1070	69.4	710	47b
	37.5	1198	82.6	510	
	50.29	1329	100	750	
	62.5	1484	137	2000	
	75	1672	192	4000	
Benzene	25	968	42.9	216	
	39.6	1090	62.7	340	
	50	1191	75.4	420	
	60	1295	89.8	620	
	75.8	1494	122	1000	
Cyclohexane	25	1140	81	810	
	37.6	1267	95	850	
	50.1	1415	116	1030	
	62.4	1576	134	750	
	75	1784	195	3100	
Methanol	20	676	7.252	2.920	88
	60	787	8.48	3.627	
	80	851	9.808	4.058	
Ethanol	20	666	7.119	2.888	
	40	725	8.051	3.332	
	60	781	8.846	3.667	
	80	843	9.752	4.071	
n-Propanol	20	605	6.691	2.755	88
	40	662	7.553	3.155	
Isobutanol	20	646	7.135	2.958	
1-Pentanol	20	608	6.655	2.744	

Table 2.5. (*cont.*)

Liquid	$T(°C)$	$19^7 \theta$ (bar)	$10^9 \phi$ (bar^2)	$10^{13} \psi$ (bar^3)	Reference
Allyl alcohol	9.6	778	21.06	30.6	
	35	892	25.3	35.6	
Diethyl ether	20	855	10.09	4.288	
	40	998	12.65	5.568	
Acetone	20	790	11.50	6.686	
	40	769	8.596	3.547	
Bromoethane	20	722	8.087	3.336	
Iodoethane	20	659	7.135	2.920	

[a] For compilations of compressibility data, see references 8–9.

Table 2.6. Bond contributions for isothermal and adiabatic compressibilities for use with Wada's equations, eqs. (2.6), (2.7)

Bond	Constant A	Constant B
C—C	−1.10	1.07
C—O	2.05	2.78
C—S	5.43	—
C—N	0.40	0.24
C—H	5.10	4.16
C—F	—	6.57
C—Cl	12.91	12.55
C—Br	15.54	15.33
C—I	19.65	—
O—H	4.64	5.07
N—H	5.57	5.00
C=C	5.68	6.36
C=O	9.93	9.08
C=S	16.83	—
C=N	7.60	—
N=O	8.17	8.28
C≡N	14.13	—
Ring	4.80	−0.43

Table 2.7. Structural contributions for calculating velocity of sound by Rao's method

Types of compounds	Constant D
Basic structure:	
Methane	1850
Benzene	4534
Cyclohexane	5363
Naphthalene	6566

Table 2.7. (*cont.*)

Types of compounds	Constant D
Substituted radicals:	
$-\overset{\vert}{\underset{\vert}{C}}-, -\overset{\vert}{CH}, -CH_2, -CH_3$	872
$-COO-$	1220
$-\overset{O}{\overset{\Vert}{C}}-H$	449
$-\overset{O}{\overset{\Vert}{\underset{\vert}{C}}}-$	872
$-NH$	638
$-NH_2$	478
$-COOH$	942
$-C\equiv N$	819
$-O-$	273
$-OH$	137
$-Cl$	610
$-Br$	692
$-I$	893
$-NO_2$	893
$-S$	550
$=S$	550
Double bonds	-254
Triple bonds	-507
Position contributions:	
Ortho	0
Meta	59
Para	117

2.1.4 Work of Compression

The energy or work necessary to compress a liquid from P_0 to P is given by, eq. (2.1),

$$E = \int_{P_0}^{P} P\left(\frac{\partial V}{\partial P}\right)_{\mathrm{T}} \cdot \mathrm{d}P$$

When the *P–V* relationship is expressed as a polynomial,

$$V_p/V_0 = 1 + \theta P + \phi P^2 + \psi P^3$$

this becomes, eq. (2.11),

$$E = -V_0[\tfrac{1}{2}\theta(P^2 - P_0^2) + \tfrac{2}{3}\phi(P^3 - P_0^3) + \tfrac{3}{4}\psi(P^4 - P_0^4)^4] \tag{2.11}$$

Suppose we wish to calculate the energy of compression of 1 litre of ethanol at 1 kbar; $V_0 = 1$ (litre), $P_0 = 1$, $P = 1000$ bar and, taking values of θ, ϕ, and ψ from Table 2.6,

$$E = 1[\tfrac{1}{2} \times 666 \times 10^{-7} \times 10^6] + \tfrac{2}{3}[7.12 \times 10^{-9} \times 10^9]$$
$$- \tfrac{3}{4}[2.8 \times 10^{-13} \times 10^{12}]$$

$$= 38.26\, 1.\,\text{bar}$$

$$11.\,\text{bar} = 101.78\,\text{J} \quad \text{therefore } E = 3894\,\text{J per litre compressed}$$

$$= 230\,\text{J mol}^{-1}\,\text{kbar}^{-1}$$

Table 2.8. Energies of compression of some liquids
$(P_0 = 1, P = 1000\,\text{bar})$

	$E/\text{J mol}^{-1}\,\text{kbar}^{-1}$
Carbon disulphide	188
Carbon tetrachloride	460
Benzene	548
Cyclohexane	552
Acetone	350
Diethyl ether	518
Methanol	158
Ethanol	230

The entropy change on compression has been little studied. From the expression,

$$dS = C_p \frac{dT}{T} - \alpha v_m\, dP$$

it follows that the isothermal entropy of compression is given by

$$\int_{P_1}^{P_2} dS = - \int_{P_1}^{P_2} \alpha v_m\, dP$$

and demands a knowledge of the variation of thermal expansivity, α $(= -1/v \cdot dv/dT)$ as a function of pressure. This is known to decrease markedly as pressure increases, for example:[253]

carbon tetrachloride, 48 °C

P/kbar	0.001	0.11	0.7	1.9	2.36	3.89
State	l	l	l	l	s	s
$10^4\,\alpha/\text{K}^{-1}$	12.5	10.9	8.5	5.86	4.93	3.72

benzene, 48°C

P/kbar	0.001	0.10	0.86	1.55	2.15	3.78
State	l	l	l	l	s	s
$10^4\,\alpha/\text{K}^{-1}$	12.5	11.7	8.42	7.23	5.84	3.75

The variation of molar volume, v_m, with pressure follows from the isothermal compressibility so that a graphical evaluation of the integral can in principle be carried out.

2.1.5 Equations of State for Liquids

The liquid state may be viewed as an aggregation of molecules close-packed in some regions but with 'holes' or regions of free space, supposed to be of molecular dimensions, the molecules and holes being in a state of 'random' motion. There is approximately 3% of free space in a liquid which is the average increase in volume on melting of a close-packed solid and initially (up to approximately 1 kbar) compression of the liquid serves to reduce this free space. After this, further increases of pressure tend to force the molecules into closer proximity than their normal van der Waals radii and so the compressibility falls off sharply at higher pressures. The $P-V-T$ properties of a liquid, as of a gas, may be represented by a virial equation,

$$P\bar{v} = A + BP + BP + CP^2 + DP^3 + \ldots$$

with coefficients up to the seventh or eighth being required owing to the high density of matter and interactions between molecules beyond nearest neighbours being significant. The theoretical treatment of such system is at present beyond the capabilities of the largest computers owing additionally to the non-spherical symmetries of potential functions and to the non-random motion of the molecules, i.e. there is some residual 'structuring' which must be taken into account.

The van der Waals equation of state   In general terms the pressure is given by the resultant of attractive and repulsive terms for an aggregate of hard particles (molar volume $\bar{v}$) occupying a volume V,

$$P = RTf(\bar{v}, V) - \frac{q}{V^2} \tag{2.12}$$

where q is a constant and from which compressibility may be calculated. The volume function to be used depends upon the model assumed, Thus Frisch gave, for pure liquids with assumed spherical molecules,[46]

$$f(\bar{v}, V) = (V^2 - \bar{v}V + \bar{v}^2)/(V - \bar{v})^3 \tag{2.13}$$

which may be modified for mixtures of spherical molecules[47] or for spheroidal molecules.[48] The agreement between experiment and theory for various equations of state has been explored by Ewing and Marsh,[49] who conclude that the most successful equation for cycloalkanes is that which allows for a non-spherical shape. Such equations serve to test our understanding of the liquid state but for the storage of compression data, it is more convenient to use empirical equations which accurately reproduce the data by the incorporation of as many constants as necessary. Several sophisticated semi-theoretical expressions have been successful in reproducing $P-V$ data for liquids; that due to Keane[256] is an example, eq. (2.13a), but there are obvious difficulties in its application which is not justified unless data of the highest precision is available:

$$P = \frac{a \cdot x_0}{N^2}\left\{\left(\frac{\rho}{\rho_0}\right)^N - 1\right\} - \frac{(a-N)x_0}{N} \cdot \ln\frac{\rho}{\rho_0} \tag{2.13a}$$

(a, N are constants; $x = \rho \cdot (\partial\rho/\partial P)$).

The Tait equation[50-52]   This is the oldest and perhaps still the most popular equation, originally devised (although in a different form*) to represent the behaviour of sea water and which accurately reproduces compressibility data of most liquids to about 12 kbar in terms of two empirical constants B and C (in the form due to Tammann (1895) and Wohl[53]), thus:

$$\frac{V_0 - V_p}{V_0} = C \log \frac{(B+P)}{(B+P_0)} \tag{2.14}$$

where V_0, V_p are volumes at P_0, P pressures; normally $P_0 = 1$ bar. The constant C is temperature independent and varies only slightly from liquid to liquid; its value has been taken as 0.2058 for hydrocarbons[100] and *ca.* 0.215 for other organic liquids, 0.3150 for water. B is dependent upon temperature and the nature of the liquid, Table 2.3 above. It may be shown[54] that the equation (2.14) may be written as the power series, eq. (2.15)

$$\frac{V_p}{V_0} = 1 - \theta P - \phi P^2 - \psi P^3 \tag{2.15}$$

where

$$\theta = C \log \frac{e}{B}$$

$$\phi = -C \log \frac{e}{2B^2}$$

$$\psi = C \log \frac{e}{3B^3}$$

whence

$$\theta = \beta_T^0 \quad \text{and} \quad \theta/\phi = 2B, \quad C = \frac{\theta \cdot B}{\log e}$$

In this form, a least squares fit is readily obtained by computer. Ginell[55] has attempted to justify the form of the Tait equation in terms of the cluster theory of liquids. Values of the Tait constants, B and C, are included in Table 2.3. From (2.14) it follows by differentiation that

$$-\frac{1}{V_0}\left(\frac{\partial V}{\partial P}\right)_T = \beta_T = \frac{C}{B+P} \tag{2.16}$$

or the average compressibility β' between P_0 and P is given by

$$\beta' = C/(B+P) \cdot \frac{V_0}{V_P} \tag{2.17}$$

Alternative forms of the Tait equation are

$$\beta_T = C/(B+P) \tag{2.18}$$

---

*$\dfrac{\Delta V}{V_0 P} = \dfrac{1}{a+bP}$ where a, b are constants.

and

$$\frac{V_P}{V_0} = (1 + nP\beta_T)^{-1/n} \tag{2.19}$$

where n is a temperature-independent constant.

Hudleston's equation An expression connecting volume and pressure was derived by Hudleston[56] in 1937 from an expression for the force between two liquid molecules, F, given by eq. (2.20):

$$F = \chi(L - L_0)\exp[-\theta(L - L_0)] \tag{2.20}$$

where L = average intermolecular separation, L_0 = effective range of F (i.e. distance at which $F \to 0$) and χ, θ are constants.
    Thus:

$$\log\left\{\frac{PV'^{2/3}}{1 - V'^{1/3}}\right\} = J + K(1 - V'^{1/3}) \tag{2.21}$$

where $V' = V_p/V_0$. Comparison of the behaviour of some liquids by Weale[57] and by Jenner[58] suggests that this expression better represents their data than does the Tait equation. Some value of J and K are included in Table 2.5. It will be noted that these, obtained by computed fit of the data, do not change smoothly with temperature.

Other empirical equations The following are mentioned[58,59] though will not be further discussed; Hayward[30]

$$\bar{K}_T = K_0 + mP[+nP^2\dots] \tag{2.22}$$

K_0 being the bulk modulus at zero pressure and m a constant. At higher pressures it is necessary to include a quadratic or higher term. A linear relationship between $K(=1/\beta)$ and P is also suggested by other authors.[60-63]
Tumlirz:[64]

$$(P + q)(V - s) = \text{constant}$$

q, s are constants.
Tammann:[65]

$$V = V_\infty + \frac{AK}{(K + P)}$$

V_∞ = volume at infinite pressure, A, K are constants.
Peczalski:[66]

$$V = A^{b/P + cP^2/2\dots}$$

$A, b, c, \dots$ are constants.
Modified Tait equation:[67-69]

$$\frac{V_p^*}{V_0^*} = (1 + nP\beta_T)^{-1/n}$$

where V_0^* is the molar volume at 1 atmos and n is a constant and V_p^* is the volume at P; it follows that

$$\frac{1}{\beta_{T,p}} = \frac{1}{\beta_{T,0_|}} + nP$$

which is essentially the linear bulk modulus equation (2.22). It is claimed that to a few hundred bars this equation is an improvement on the form in eq. (2.14).

2.1.6 The Magnitude of Liquid Compressibility

While the compressibilities of all (ideal) gases are the same, those of liquids are more variable though the values fall within an order of magnitude (4 to 18×10^{-2} kbar^{-1}). Compressibility always falls with increasing pressure, fig. 2.4, and the variation between different liquids becomes less marked. Inspection of Table 2.1 reveals no obvious correlation between compressibility and chemical structure for organic liquids. Along a homologous series of n-alkanes, β_T falls reaching a minimum around C_{10}[70] presumably a consequence of the available free volume. A power series (eq. (2.23))

$$\beta_T = \sum_{i=0}^{i=m} A_i n_i \tag{2.23}$$

where A_i are constants and n the number of carbons has been fitted to these data for alkanes C_6 to C_{16}, needing at least five terms to give a good fit.[71] Chain-branching tends to increase compressibility as also does a rise in temperature, both tending to increase free volume as evident from densities. Cyclic compounds appear to be somewhat less compressible than their acyclic counterparts. Hydrogen-bonded liquids such as water are of moderate to low compressibility and it is inferred that compression is accompanied by the breaking of hydrogen bonds, a reduction of structure[11,72] which is responsible for abnormalities in physical properties at low temperatures. The compressibilities of analogous halides fall in the order Cl > Br > I; it is evidently naive to consider the more polarizable halides as being more compressible. There appears to be no correlation between electronic properties of substituted benzenes and their compressibilities.

Invariably, β_T falls with increasing pressure since the repulsive potential is steeper than the attractive. For most liquids it increases with temperature since thermal expansion increases the internuclear distance. In this, water is exceptional and its compressibility falls with temperature passing through a minimum around $50°$,[20,80] fig. 2.5. It seems likely that temperature affects the nature of structured units present.[73] The lack of simple patterns in compressibility data reflects the complexity of the liquid state which is only now beginning to yield to a theoretical interpretation.[74,75] It has been suggested that compressibilities comprise two components, a geometrical, β_T^g and a structural, β_T^s. The first results from reluctance of the molecules to diminish their equilibrium intermolecular distance while the second arises from change in mutual orientation upon compression. Since we may write,

$$\beta_T = -\frac{1}{V}\left(\frac{\partial V}{\partial P}\right) = \frac{1}{V}\left(\frac{\partial^2 H}{\partial P^2}\right)_T + \frac{T}{V}\left(\frac{\partial^2 S}{\partial P^2}\right)_T \qquad (2.24)$$

$$\equiv \beta_T^g \qquad\qquad \equiv \beta_T^s$$

the enthalpy and entropy derivatives have been equated respectively with the geometrical and structural components of compression.[76] Some values obtained in this way are given in Table 2.9.

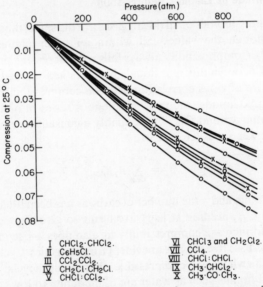

I   CHCl$_2$·CHCl$_2$.
II   C$_6$H$_5$Cl.
III CCl$_2$CCl$_2$.
IV   CH$_2$Cl·CH$_2$Cl.
V   CHCl:CCl$_2$.
VI   CHCl$_3$ and CH$_2$Cl$_2$.
VII CCl$_4$.
VIII CHCl:CHCl.
IX   CH$_3$·CHCl$_2$.
X   CH$_3$·CO·CH$_3$.

Fig. 2.4. Compression curves for some organic liquids.[34] (Reproduced by permission of the Chemical Society)

Table 2.9. Geometrical and structural components of compressibility[76]

	$\beta_T(10^2\,\mathrm{kbar}^{-1})$ (experimental)	β_T^g	β_T^s
		(calculated)	
CCl$_4$	10.5	−12.5	22.2
C$_6$H$_6$	9.44	−12.5	22.0
CS$_2$	9.38	−9.8	18.4
H$_2$O	4.58	8.0	−3.4

Despite a slight lack of agreement, the figures seem to indicate that initial compression brings about greater attractive forces but is disfavoured by the entropy term for normal liquids. Water, on the on the other hand, behaves in the opposite way (Fig. 2.5). Breaking of hydrogen bonds on compression results in an unfavourable enthalpy term although this is partly compensated by the entropy.

The physical interpretation of the constants of the Tait equation, B and C, have long been considered.[8–11,77,78] The term $(B+P)$ has been taken to represent the repulsive 'pressure' exerted by molecules of the liquid internally. While B varies widely according to the chemical structure, C is rather constant (ca. 0.2 for organic liquids). The attractive 'pressure' due to molecular interactions has been evaluated by Gibson[79] as the quantity $(T.(\partial P/\partial T)_v + B)$ and is found to increase with pressure.

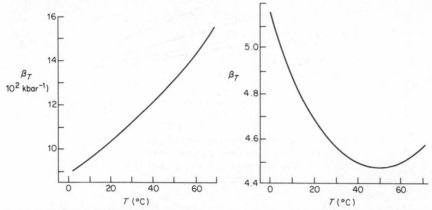

Fig. 2.5. Compressibilities as a function of temperature: (a) carbon tetrachloride, (b) water[20]

A sudden discontinuity in compressibility denotes a phase change and is found to occur at a solid–liquid boundary and, less marked, at a solid–solid one. In solution, this effect can signal the transformation of a surfactant from the dispersed to the micellar state (section 3.2.8). In a micelle, the polar groups are arranged at the surface of the aggregate and hydrocarbon residues within, the whole being more compressible than the free molecules,[217] suggesting that the interior behaves similarly to a liquid hydrocarbon.

2.1.7 Compressibilities of Liquid Mixtures and Solutions

For liquids which mix without volume change, the compressibility is a linear function of composition (mole fraction). This kind of behaviour is found in mixtures of paraffins such as hexane, heptane, and octane.[15] Values of the Tait constants also appear to follow this general principle for such well-behaved solutions. Independence of behaviour of each component is not found for mixtures of less chemically similar substances. Ethanol–water mixtures show specific volumes which deviate from additivity on the positive side for water-rich mixtures and on the negative side for alcohol-rich ones.[16] The effect is of a small amount of water apparently increasing the compressibility of ethanol while a small amount of ethanol reduces the compressibility of water. No doubt the origins of the small effects lies in changes in the hydrogen-bonded structures of the liquids. Fairly irregular behaviour in aqueous solutions may be expected,[17] and even greater deviations should be found in mixtures of donor and acceptor molecules between which there is considerable interaction.

Ionic solutes bring about a marked decrease in compressibility, figs. 2.6, 2.7. It is supposed that solvent surrounding the ions suffers electrostriction, and is

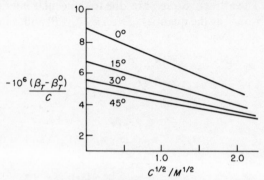

Fig. 2.6. Compressibility function of aqueous sodium chloride;[171] β_T^0, β_T = compressibilities of water, NaCl solution. The slopes give $\bar{\beta}$, the partial molar compressibility of NaCl

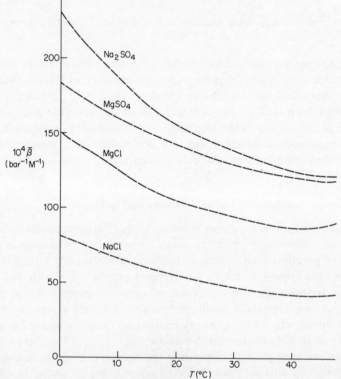

Fig. 2.7. Partial molar compressibilities of some salts in aqueous solution[171]

strongly attracted to the charge centres thereby becoming effectively highly compressed. Thus, the compressibility of electrostricted solvent will be low and as the ionic strength increases this phase constitutes an increasing proportion of the total solvent.

The concentration dependence fits an equation such as eq. (2.25);

$$\beta_{soln} = \beta_0 + A_K c + B_K c^{3/2} \tag{2.25}$$

where β_{soln} and β_0 are compressibilities of solution and solvent respectively; A_K and B_K are constants.[171] Specifically A_K relates to ion–solvent interactions at infinite dilution and B_K to ion–ion interactions. Thus one can define apparent molar compressibilities of salts, $\bar{\beta}$, in solution; as: $\bar{\beta} = -(\partial \bar{V}_m/\partial P)$, units now being $M^{-1} bar^{-1}$ whence

$$10^3 A_K = \bar{\beta} - \beta_0 \bar{V}_m \tag{2.26}$$

where $\bar{V}_m$ is the partial molar volume (Chapter 3); $\bar{\beta}$ then refers to the compressibility of the solute itself in solution and is usually extrapolated to zero concentration. Also

$$10^3 B_K = S_K - \beta_0 S_v$$

in which the constants S_K and S_v are defined in terms of the concentration dependence of partial molar compressibility and of partial molar volume, respectively:

$$\bar{\beta} = \bar{\beta}_{c \to 0} + S_K c^{1/2} \tag{2.27}$$

and

$$\bar{V}_m = \bar{V}_{m_{c \to 0}} + S_v c^{1/2} \tag{2.28}$$

It follows that

$$\bar{\beta} = \frac{1000}{c}(\beta_{soln} - \beta_0) + \beta_0 \bar{V}_m \tag{2.29}$$

Values of $\bar{\beta}_{c \to 0}$ are given in fig. 2.7 for several aqueous salts, from which it appears that they pass through a broad minimum between 40° and 50° which is reminiscent of the compressibility behaviour of water, fig. 2.4. This then supports the thesis that in these salt solutions it is the free water, unaffected by the proximity of the ions, which is undergoing compression.

Partial molar adiabatic compressibilities can be defined similarly using values of β_s rather than β_T in eq. (2.26). The slope of a plot of partial molar compressibility (or volume) against concentration is an index of interactions between solute ions, and is greater for multiply-charged ions than for singly charged. On addition of a crown ether, a macrocyclic polyether having the property of chelating alkali metal cations, the plot becomes invariant with concentration[221] suggesting that the charge is well screened by the crown ether and solute–solute interactions greatly weakened.

Partial molar compressibilities of some organic oxygen compounds in water are given in Table 2.10. The striking feature of these figures is that at low temperature most are negative and the compounds show partial compressibilities which are below that of bulk water. Values rise with temperature, however, and above 25° most of these non-ionic solutes are more compressible than water. The high temperature coefficient of $\bar{\beta}$ must indicate the presence of specific solute–solvent interactions such as hydrogen bonding which are very dependent upon temperature.[224] As expected from this model, there is little difference between homologues; the effect is centred on the oxygen function and additional

Table 2.10. Limiting partial molar compressibilities of alcohols and ethers in water[224]

Solute	$\bar{\beta}_0$ (10^4 cm^3 mol^{-1} bar^{-1})		
	10 °C	25 °C	40 °C
Methanol	8.66	12.6	16.0
Ethanol	1.68	10.0	16.8
Propan-1-ol	−6.51	7.4	17.7
Butan-1-ol	−13.26	5.2	16.8
Pentan-1-ol	−19.17	4.7	21.8
Butan-2-ol	−6.2	5.9	
Pentan-3-ol	−13.5	6.6	
Hexan-3-ol	−18.0	6.4	
Cyclopentanol	−14.0	4.1	
Cyclohexanol	−14.3	4.3	
Cycloheptanol		4.8	
Oxetan		12.1	
Tetrahydrofuran	−5.0	10.6	
Tetrahydropyran	−2.4	14.7	
1,3-Dioxolan	−2.5	10.4	
1,3-Dioxan		13.8	
1,3-Dioxepan	−3.2	11.6	24.8
1,3,5-Trioxan	−5.3	6.0	19.8
Formaldehyde dimethyl acetal	7.3	19.4	
Formaldehyde diethyl acetal	3.0	26.6	
1,2-Dimethoxyethane		15.8	
(water		8.18)	

hydrocarbon constituent makes little contribution. Surfactants show a sharp increase in $\bar{\beta}_0$ when passing from the dispersed to the micellar phase[225] as well as an increase in volume. This is evidently associated with the hydrocarbon 'core' of the micelle into which water does not penetrate.

For mixtures of liquids rather than dilute solutions it is most appropriate to treat departures of various thermodynamic functions from the values expected on the basis of a strictly linear variation with composition. These deviations are known as excess functions.

2.1.8 Excess Functions

The Gibbs free energy change which results from mixing $(x-1)$ moles of one liquid with x moles of another, since the entropy of the system must increase, is given by

$$\Delta G_m^I = RT(1-x)\ln(1-x) + RTx\ln x \tag{2.30}$$

and if a solution obeys this relationship it is said to be ideal. Many properties are found not to be additive in this way and the deviation from ideality is expressed in excess functions such as the excess Gibbs free energy of mixing, ΔG_m^E

$$\Delta G_m^E = \Delta G_m^{expt} - \Delta G_m^I \tag{2.31}$$

The first derivative of this with respect to pressure is the molar excess volume, V_m^E,

$$V_m^E = \left(\frac{\partial G_m^E}{\partial P}\right)_T \tag{2.32}$$

which is given by the volume change on mixing the two components isothermally. The molar excess volume is pressure dependent for non-ideal mixtures which have been examined and tends to diminish at high pressures as shown by the results for acetonitrile–water and 3-methylpyridine–water (fig. 2.8).[175,176] The measurement of V_m^E entails mixing two liquids in prescribed proportions while

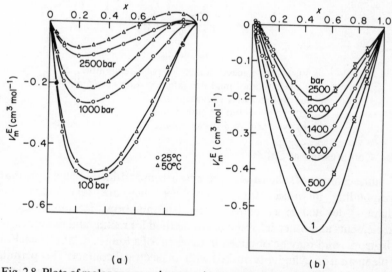

Fig. 2.8. Plots of molar excess volume against composition at different pressures and temperatures for (a) $(1-x)$ water $+x$ acetonitrile[175] and (b) $(1-x)$ 3-methylpyridine $+x$ water[176]

under pressure and determining the volume change (piston movement) necessary to maintain the pressure. In the cases examined, the effect of increasing temperature is also to reduce the excess volume, so both high pressures and temperatures encourage ideal behaviour.

Excess functions may reveal unexpected trends in intermolecular interactions. Binary mixtures of the homologous alcohols, cyclopentanol, cyclohexanol, and cycloheptanol have negative excess compressibilities indicating intermolecular hydrogen bonding as a feature, but this is much larger for cyclopentanol–cycloheptanol mixtures than for any others, fig. 2.9. Evidently, steric considerations permit an especially favourable interaction between these two molecules in the mixture.[222] Many other mixed systems show positive excess compressibilities, e.g. benzene–dichloroethane.[223]

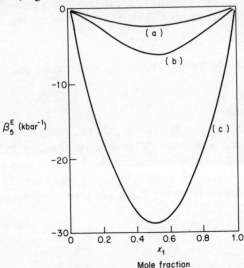

Fig. 2.9. Excess adiabatic compressibilities of cycloalkanol mixtures (after Benson et al.[222]);
(a) cyclohexanol–cycloheptanol;
(b) cyclopentanol–cyclohexanol;
(c) cyclopentanol–cycloheptanol

2.1.9 Compression of Solids

Volume changes of solids due to pressure are more difficult to measure than the corresponding quantities for liquids since they are very much smaller. Bridgman[7] devoted much effort to the determination of compressibilities of solid elements and other substances. His method for metals and materials, which are isotropic and may be obtained in the form of a long rod, was to enclose the sample in an iron cylinder furnished with an electrical contact. This permits, by means of a potentiometer, the accurate measurement of relative movement of the rod with the iron container, and hence the compressibility relative to iron. The absolute compressibility of iron was initially determined with some difficulty. For an isotropic sample, the linear compressibility modulus is the cube root of the bulk modulus. Table 2.11 sets out some data in the form of constants of the quadratic equation (2.33):

$$-\frac{\Delta V}{V_0} = \beta_T P - bP^2 \qquad (2.33)$$

Linear compression will also give the bulk modulus for single crystal materials of cubic symmetry but crystals of lower symmetry will show different compressibilities along each axis.

Table 2.11. Compressibilities of some solids at 30 °C[a]

	$10^2 \beta_T (\text{kbar}^{-1})$	$10^6 b (\text{kbar}^{-2})$
Li	0.886	101
Be	0.0868	4.03
B	0.0558	2.28
C	0.018	
Al	0.0137	5.19
Ti	0.0809	−0.124
Fe	0.0595	2.18
Zn $\{\parallel$	0.132	5.53
Cryst $\{\perp$	0.0195	1.15
Ag	0.100	4.57
Te $\{\parallel$	−0.0418	−9.98
Cryst $\{\perp$	0.28	5.48
Ba	1.039	134
LiCl	0.34	33.6
NaF	0.211	18.4
NaCl	0.4261	52.4
NaBr	0.507	64.6
KBr	1.044	106
KI	0.853	157
RbBr	0.793	138
RbI	0.956	205
Silica		
Pyrex glass		

[a] The data adapted from reference 7 fits the regression equation:

$$-\frac{\Delta V}{V_0} = \beta_T P - bP^2 \quad (P \text{ in kbar})$$

From the table it will be seen that compressibilities of solids are 1–2 orders of magnitude less than liquids. Among the elements, the most striking feature is the strong periodicity displayed, fig. 2.10, with alkali metals some 100 times more

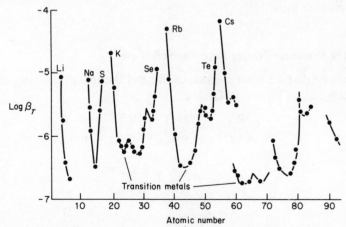

Fig. 2.10. Periodicity in compressibilities of solid elements[7]

compressible than transition metals and main group elements in between. This suggests that a reduction of the effective diameters of the atoms occurs which accounts for the much smaller compressibilities compared to liquids in which initially a reduction in free space between molecules occurs.

The alkali metals in particular are relatively highly compressible since their outer s^1 electronic configuration indicates a low electron density at the van der Waals radius and consequently a smaller repulsive interaction than in those elements with external filled p- and d-shells. The compressibility is mainly determined by compression of the delocalized electrons in the metallic structure.

Borelius expressed the compressional volume change of a solid ΔV, in the empirical relationship eq. (2.34) (Borelius' Law)[81-83]

$$(A + P)(B + \Delta V) = AB \tag{2.34}$$

where α, β are constants characteristic of the solid having dimensions of pressure and volume respectively; α has the connotation of 'internal pressure' while β represents the maximum possible compression and is known as the free volume (Table 2.12).

Table 2.12. Borelius constants for some metals

	A (kbar)	B/V_0		A (kbar)	B/V_0
Cu	29	0.23	Ag	206	0.21
Al	176	0.25	Pb	81	0.21
Li	56	0.50	Na	25	0.38
K	14	0.43	Rb	9	0.44
Cs	7	0.41	Si	150	0.17

Information on the behaviour of solids under extreme conditions of temperature and pressure has been obtained by means of shock waves during the passage of which, 50 kbar, 1000° may be attained.[84] The most precise equations of state are discussed by Ruoff.[258]

2.1.10   The Pressure–Temperature Coefficients, γ_v

This may be derived from compressibility and thermal expansivity data and calculated by eq. (2.35) due to Gibson[9]:

$$\gamma_v = \left(\frac{\partial P}{\partial T}\right)_v = -\left(\frac{\partial V}{\partial T}\right)_p \bigg/ \left(\frac{\partial V}{\partial P}\right)_T = \frac{\alpha}{\beta_T} \tag{2.35}$$

$$= \frac{1}{\bar{v}}\left(\frac{\partial V}{\partial T}\right)_p \frac{B + P}{0.4343C}$$

$$= \frac{\alpha(B + P)}{0.4343C} \tag{2.36}$$

where

$$\bar{v} = \text{specific volume (cm}^3\text{ g}^{-1}\text{)} \, (1/\rho)$$

B, C constants in the Tait equation (2.11)

$$\alpha = \text{thermal expansivity } (K^{-1}) = \frac{1}{\bar{v}}\left(\frac{\partial V}{\partial T}\right)_p$$

In principle this equation permits the calculation of pressures developed upon heating a liquid in a rigid confining vessel (section 1.1.8) although it is likely to overestimate the practical value on account of expansion of the vessel itself and the change of γ_v with pressure. Some values of the quantity γ_v for some organic liquids are calculated below.

Benzene $$\gamma_v = \frac{0.001237 \times 970}{1.1451 \times 0.4343 \times 0.2159} = 11.2 \, \text{bar K}^{-1}$$

Acetone $$\gamma_v = \frac{0.001487 \times 815}{1.2754 \times 0.4343 \times 0.2356} = 9.3$$

Aniline $$\gamma_v = \frac{0.000855 \times 2006}{0.9829 \times 0.4343 \times 0.2159} = 18.6$$

This leads to a useful expression for the internal energy E_i, of a liquid as a function of volume; since

$$P(\text{N m}^{-2}) = \left(\frac{\partial E_i}{\partial V}\right)_s \, (\text{J m}^{-3}) \tag{2.37}$$

it may be shown that

$$\left(\frac{\partial E_i}{\partial V}\right)_T = T\gamma_v - P \tag{2.38}$$

The thermal pressure coefficient may, of course, be directly measured by constraining a material in a fixed volume and directly measuring pressure as the temperature is varied. This type of experiment has been performed at high temperatures with molten salts.[220] In general, γ_v is a quantity likely to be needed in converting thermodynamic parameters at constant pressure to the equivalent quantities at constant volume. Thus, for specific heats:

$$C_v = C_p - TV\alpha\gamma_v \tag{2.39}$$

and excess energies:

$$(U)_v^E = (H^E)_p - TV^E\gamma_v \tag{2.40}$$

Thermal expansivity of liquids normally falls with pressure; water below $40°$ is an exception, fig. 2.26.

2.1.11 Internal Pressure of a Liquid

On a microscopic scale, a molecule within a liquid is under constant random bombardment from neighbouring molecules, the collisions which may be inelastic, and are considered, in effect, to give rise to a pressure quite apart from the hydrostatic pressure due to gravity. This net internal pressure, P_i, will be the resultant of attractive and repulsive forces and thermal components. The energy-volume coefficient $(\partial E/\partial V)_T$ has the connotation of internal pressure according to Hildebrand, where

$$P_i = \left(\frac{\partial E}{\partial V}\right)_T = T\left(\frac{\partial P}{\partial T}\right)_v - P = (T\gamma_v - P)$$

$$\approx T\gamma_v \qquad (2.41)$$

The pressure–temperature coefficient, γ_v, is obtained from thermal expansivity and compressibility data and hence P_i may be evaluated. Values of P_i are large, of the order of several kbar and diminish with temperature. Representative data are shown in Table 2.13. It would seem that the repulsive term is increasing more rapidly than the attractive component as the volume of the liquid is diminished isothermally. The decrease of $\partial E/\partial V$ with temperature has also been associated with an increase in disorder which might result in a decrease of the attractive interactions if these are directional (donor–acceptor or hydrogen bonds or dipole–dipole types).[9,79]

Hildebrand[86] found that the value of γ is a function of the molar volume that is, the greater the concentration of mass in unit volume, the greater the internal pressure exerted.

Internal pressure of a different kind, mechanical pressure P_M exerted by the solvent internally, has been defined by Trotter[87] as the change in energy of the solvent with respect to volume analogous to eq. (2.41):

$$P_M = \left(\frac{\partial W}{\partial V}\right)_T \qquad (2.42)$$

Using a statistical hard sphere theory of fluids, P_M may be evaluated according to eq. (2.43)

$$P_M = \frac{kT}{\pi b^2 a}[6y/(1-y)][4r/a-1] + [18y^2/(1-y)^2][2(r/a)-1] + 1 \quad (2.43)$$

where $y = \pi a^3 \rho/6$; a = solvent molecular diameter, b = solute molecular diameter (which may be identical with a), ρ = density. According to this, the internal pressure on a solute molecule decreases as its size increases, being 8 kbar for a helium atom and $ca.$ 3 kbar for a cyclohexane molecule, each in benzene. P_M seems not to be related to P_i (defined as T_y) (Table 2.13). For mixtures of liquids, Biron proposed an additivity relationship for γ_v, eq. (2.44)[88] (Table 2.14)

$$\gamma_v(AB) = \frac{\gamma_v(A) \cdot \gamma_v(B)}{\gamma_v(A)N_A + \gamma_v(B)N_B} \qquad (2.44)$$

where N are mole fractions. The experimental values, however, are usually smaller than predicted by 2–5%.[86] The physical significance of P_i and P_M, however, remains uncertain and their identification with the extra-thermodynamic concept of an internal pressure is contentious.

Table 2.13. Internal pressures of some liquids[87]

	P_i(bar)	P_M(bar)
Heptane	2510	2310
Octane	2900	
Benzene	4050	1550
Toluene	3780	
m-Xylene	3640	
p-Xylene		
Chloroform	3830	
Bromoform	5600	
Carbon tetrachloride	3640	3320
Dichloromethane		2170
1,2-Dibromoethane	4700	
Iodobenzene	4600	
Diethyl ether	3100	
Carbon disulphide	5400	1470
Mercury	13,050	

Table 2.14. Values of the pressure–temperature coefficient, $\gamma = (\partial P/\partial T)_v$

	T(K)	γ_v(bar K^{-1})	$P_i = T\gamma_v$(bar)[a]
Heptane	298	8.41	2506
	308	7.93	2442
Carbon tetrachloride	298	11.15	3322
	308	10.52	3240
Carbon disulphide	298	12.31	3668
	308	11.63	3582
Benzene	298	12.23	3644
	308	11.49	3540
Dichloroethane	298	13.88	4136
	308	13.11	4038
Dibromoethane	298	14.80	4410
	308	14.05	4327
Bromoform	298	14.92	4446
	308	14.15	4358
Acetone	298	11.00	3278
	308	10.33	3182

[a] Internal pressure of the liquid.

2.1.12 Adiabatic Compressibilities, β_s

Since on adiabatic compression, the temperature rises and the volume change for unit pressure rises, the compressibility is less than under isothermal conditions, thus

$$\beta_T - \beta_s = \frac{v(\mathrm{d}v/\mathrm{d}T)^2 \cdot \rho \cdot T}{C_p} \tag{2.45}$$

and

$$\beta_T/\beta_s = C_p/C_v$$

Adiabatic compressibilities are usually obtained from measurements of sound velocity v at ultrasonic frequencies for which $\beta_s = 1/v^2 \rho$ or

$$v = (1/\rho\beta_s)^{1/2} = (\gamma/\rho\beta_T)^{1/2} \tag{2.46}$$

Some values of v and β_s are given in Table 2.15.

Table 2.15. Ultrasonic sound velocities[a] in pure liquids, 30°.[114]

	$v(\text{m s}^{-1})$		$v(\text{m s}^{-1})$
Pentane	981	Methanol	1120
Hexane	1082	Ethanol	1160
Isopentane	958	n-Propanol	1188
Heptane	1115	2-Propanol	1150
Octane	1150	1-Butanol	1225
Octene-1	1150	Isobutanol	1250
Cyclohexane	1234	Tert. butanol	1233
Decahydronaphthalene	1365		
Benzene	1270	Acetic acid	1120
Toluene	1280	Acetone	1150
Xylenes	1320	Acetonitrile	1265
		Ethyl acetate	1160
Carbon disulphide	1127	Dioxan	1360
Carbon tetrachloride	905	Diethyl ether	970
Chloroform	969		
Bromoethane	865	Aniline	1615
1,2-Dibromoethane	985	o-Toluidine	1635
1,2-Dichloroethane	1175	Anisole	1460
Bromomethane	905	Diethylphthalate	1430
Dichloromethane	1048	Nitrobenzene	1440
Dibromomethane	945	Pyridine	1400
Diiodomethane	954	Formamide	1610
Chlorobenzene	1250	Nitromethane	1310
Bromobenzene	1140	Acetophenone	1515
Iodobenzene	1088		

Uncertainty in many of these values of at least 5%

[a] $$1\,\text{bar} = 10^6\,\frac{\text{dyn}}{\text{cm}^2} = 10^6\,\frac{\text{cm g}}{\text{cm}^2\,\text{s}^2} = 10^6\,\frac{\text{g}}{\text{cm s}^2} \quad \text{and} \quad 1\,\text{kbar}^{-1} = \frac{10^{-9}\,\text{s}^2\,\text{cm}}{\text{g}}$$

If v is expressed in cm s^{-1}, ρ in g cm^{-3}, β_s will be in $10^9\,\text{kbar}^{-1}$. Thus, for ethanol,

$$\beta_s = 1/v^2 \rho = 1/(116000^2 \times 0.78097) = 95.1 \times 10^{-3}\,\text{kbar}^{-1}$$

(cf. $\beta_T = 111 \times 10^{-3}\,\text{kbar}^{-1}$).

2.1.13 Compressibility, Surface Tension, and the Parachor

Empirical observations have shown that for many organic liquids an inverse relationship exists between surface tension, σ (at 1 bar) and isothermal compressibility, thus Richards and Mathews[89] reported

$$\beta_T \sigma^{4/3} = \text{constant} \tag{2.47}$$

(for β_T in $1/\text{kg}\,\text{cm}^{-2}$ and σ in $\text{dyn}\,\text{cm}^{-1}$ the constant is approximately 5.4×10^{-3}).
McGowan favoured

$$\beta_T \sigma^{2/3} = \text{constant} \qquad (2.48)$$

while Tyrer[91] incorporated temperature dependence, eq. (2.49),

$$\frac{\beta_T \sigma^{4/3}}{T^{1/3}} = \text{constant} \qquad (2.49)$$

Since the parachor, $[P]$, a property defined by Sugden[92] as

$$[P] = \frac{M\sigma^{1/4}}{(\rho_1 - \rho_g)} \qquad (2.50)$$

(M = molecular weight, ρ_1 and ρ_g the densities of liquid and vapour, respectively at the same temperature), and which is an additive property of the number and type of covalent bonds present, contains the surface tension, the relation may be written as

$$[P] = \frac{M}{21(\rho_1 - \rho_g)\beta_T^{1/6}} \qquad (2.51)$$

and temperature variation may also be incorporated,

$$\beta_T = \frac{1.33 \times 10^{-8} [P]^{6/5}}{k^{9/5}(T_c - T)^{9/5}} \qquad (2.52)$$

where T_c is the critical temperature and k the Eötvös constant.[93] While these empirical equations are of doubtful significance in understanding liquid behaviour, they may have some use in predicting approximate compressibility or other data which is not available experimentally. Care is needed in their use in particular to maintain the correct units which are exemplified below. These equations point to a relationship between the cube root of the bulk modulus and the square root of the surface tension and suggests that similar forces are operating in three dimensions for compressibility phenomena and in two for surface properties.

Calculation of Compressibility Using Eq. (2.51)

(i) benzene

$$M = 78.108\,\text{g mol}^{-1}$$

$$\rho_1(25°) = 0.8734\,\text{g cm}^{-3}$$

$$[P]^* = (6 \times 46.35) + (6 \times 24.7) - (12 \times 18.6) = 203.1$$

$$\beta_T_{(25°)} = \left[\frac{78.108}{20.7 \times 0.8734 \times 203.1}\right]^6 = 92.6 \times 10^{-12}\,\text{cm}^2\,\text{dyn}^{-1}$$

$$\beta_T \text{ (experimental)} = 96.5^{94}$$

* For the purposes of this equation, parachors are calculated by the numbers given by McGowan[95]; $[P] = \sum^i nP_i - (18.6m)$ where P_i are the atomic contributions as below and m the number of covalent bonds of any type in the molecule; values of P_i are: C, 46.35; H, 24.7; N, 40.8; O, 35.25; Cl, 59.4. Values for all elements are given in ref. 95.

(ii) cyclohexane

$$\beta_T = \left[\frac{84.216}{20.7 \times 0.7743 \times 239.7} \right]^6 = 111 \times 10^{12}$$
$$(25)$$

experimental $= 114$[94]

The disadvantage of this formula is the extreme sensitivity of the value to all the numbers involved, on account of the sixth power relationship and the arbitrary constant averaged at 20.7 is somewhat variable with structure. The reliance is generally no better than 10%.

The effect of pressure on the surface tension itself has been investigated. Such measurements are necessarily performed under a gas atmosphere so that the interfacial surface tension is affected by solution of the gas. Values of σ fall with pressure (for aqueous solutions) the effect depending on the nature of the gas in the order ethane $> CO_2 >$ methane $>$ argon.[137]

2.1.14 Effect of Pressure on Refractive Index and Dieletric Constant

These two experimental quantities are related to the electronic polarization of molecules by an alternating electric field, the high frequency component of a light wave (refractive index, n) or a low frequency a.c. field (dielectric constant, ε) such that $\varepsilon \equiv n^2$ and consequently are convenient to consider together. The effect is proportional to the quantity of material in the light path or capacitor and hence is a volume property.

The measurement of ε is straightforward, involving the determination of capacitance[148] or permittivity[147] with the test liquid as dielectric. The capacitance cell may be placed in a high pressure cylinder with provision for separating the test substance from the pressure transmitting medium if necessary such as by mercury or a bellows.[122,128] External electrical leads then permit readings to be taken as a function of pressure.[96] The measurements may be extended to the microwave frequency range[97,98] and to very low temperatures.[99] Simultaneous measurement of ε and density under pressure was made by Eckert.[100]

Several techniques for the determination of refractive index under pressure have been reported. Usually, an optical cell is required. A prism mounted internally for observing the condition of minimum deviation may be used[101] or the optical cell may be incorporated into an interferometer,[102,103,255] a method which has the advantage of rapid measurements in order to obtain adiabatic changes in refractive index.[104] A simple technique used by Gibson[105] consisted of placing several glass specimens of differing refractive index inside the cell then elevating the pressure until each in turn disappeared from view as matching of refractive index increase with pressure (Tables 2.16–2.18), and this is related to the refractive index increase with pressure (Tables 2.12, 2.13), and this is related to the decrease in volume which occurs. Thus, Newton observed the relationship $(n^2 - 1)/\rho = $ constant,[106] and the term $\varepsilon/\rho = $ constant applies to liquids (excepting water).

The empirical equation of Eykman[107] which relates ε or n to the change in volume due to thermal expansion also applies to compression (eq. (2.53) and eq. (2.54))

$$\bar{v}\frac{(n^2-1)}{(n^2+0.4)} = \text{constant} \tag{2.53}$$

$$\bar{v}\frac{\varepsilon-1}{\varepsilon+0.4} = \text{constant} \tag{2.54}$$

where $\bar{v}$ is the specific volume $(1/\rho)$. By this means it is possible to estimate compressions from refractive index data to quite high precision.[108] The Eykman constant for refractive index of benzene at 589 nm is 0.7506 and at 436 nm, 0.7814.[105] Precise data for water and some organic liquids have been shown to correlate well with densities calculated from sophisticated equations of state such as the Keane equation (section 2.1.5).[255] Rosen has related dielectric constants of many organic liquids to their volume using eq. (2.55)[109] and it has been shown that $\ln\varepsilon$ is linearly related to $\ln v$ over a range of many kbar:[110]

$$\left(\frac{\varepsilon_p+2}{\varepsilon_p-1}\right) = mv_p+b \tag{2.55}$$

where m, b are constants for the compound.

A detailed investigation shows that the index of refraction for a non-polar liquid such as benzene is an almost linear function of specific volume and that the same line may define points at different temperatures (fig. 2.11) whereas with more structured liquids such as water slightly different lines connect plots at different temperatures showing that there is a structural difference created in water by increasing the temperature and pressure at constant volume (fig. 2.12).[102]

In addition to temperature and pressure, frequency (wavelength) is a further variable. The dielectric constant is independent of frequency above a few kHz but the refractive index is frequency dependent (dispersion). This may be seen from the coefficients in the regression equation fitting refractive index data for ethanol and for water:

$$n = a+10^{-5}bP+10^{-9}cP^2$$

		a	b	c
Ethanol	579 nm	1.3606	3.837	-11.8
	406	1.3703	4.150	-15.60
Water	579	1.3331	1.410	-1.40
	406	1.3424	1.454	-1.49

In both cases, the refractive index at short wavelength is more sensitive to pressure and the effect is more marked for ethanol than for water.[111,112] A similar correlation has been established for ethanol–water mixtures.[113]

If ε and n are volume properties their variation with pressure should fit a relationship similar to the Tait equation (2.14) or other of the empirical P–V

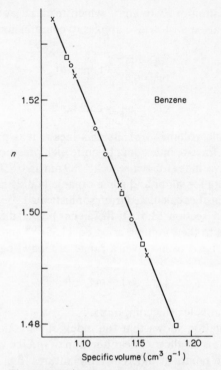

Fig. 2.11. Plot showing that the effect of pressure on refractive index is due to increased density: ○ 24.8, × 34.5, □ 54.3

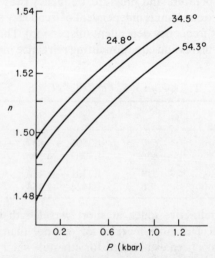

Fig. 2.12. Effect of pressure on the refractive index of water

equations of state. Thus it is found that data fit equations of the type, eq. (2.57), (2.58):[101, 114]*

$$1 - \frac{\varepsilon_1}{\varepsilon_P} = A\varepsilon_1 \log\left(\frac{B+P}{B+1}\right) \tag{2.57}$$

$$1 - \frac{n_1^2}{n_P^2} = A'n_1^2 \log\left(\frac{B+P}{B+1}\right) \tag{2.58}$$

where $\varepsilon_1, \varepsilon_P, n_1, n_P$ are dielectric constants and refractive indices at 1 bar and P respectively; A, A' are constants not identical with C of eq. (2.14), while B is the same B as in the Tait equation. The theory of these pressure effects has been discussed by Whalley.[115] From the chemical viewpoint, the effect of pressure on dielectric constant is extremely important since it determines the magnitude of the volume change known as electrostriction of the solvent in the vicinity of an ion. This effect is frequently the major cause of rate changes of chemical reactions under pressure. Tables 2.16 and 2.17 set out values of refractive index and

Table 2.16. Effect of pressure on refractive indices of some liquids at 5875 nm[102]

(a) Water

P(bar)	T(°C)	1.56	7.64	24.8	34.5	54.34
1		1.33439	1.33423	1.33293	1.33177	1.32866
200		3745	3722	3568	3435	3140
400		4052	4015	3852	3704	3194
600		4345	4305	4115	3972	3648
800		4635	4590	4370	4225	3890
1000		4922	4830	4530	4465	4130

(b) Carbon tetrachloride

P(bar)	T(°C)	24.80	34.50	54.34
1		1.45791	1.45197	1.44067
200		4672	4620	4524
400		4757	4712	4623
600		4830	4789	4706
800		4894	4852	4782
1000		4952	4912	4848

(c) Benzene

1		1.49859	1.49221	1.47910
200		5080	5108	4903
400		5162	5108	4995
600		5236	5188	5077
800		—	5255	5157
1000		—	5317	5244

*A high precision equation of state for the dielectric constant of water to 550 and 5 kbar has been given by Franck.[254]

Table 2.17. Effect of pressure on dielectric constant of organic liquids

	$T(K)$	ε_i: 1 bar	0.5 kbar	1 kbar	2 kbar	Reference
Hexane	293	1.8887	1.9557	2.0002	2.065	123
	303	1.8658	1.9390	1.9700		
Benzene	293	2.2832	2.3479			
	303	2.2533	2.3218	2.3720		
CCl_4	293	2.2375	2.3070	2.3573		
	303	2.2080	2.282	2.3361		
Chlorobenzene	293	5.6895	5.881	6.034		
	303	5.437	5.636	5.790		
Methanol	293	33.60	35.27	36.55		
	303	30.83	32.52	33.76		
2-Octanol	289	8.75	9.62	10.26	11.21	119
	332	5.45	5.90	6.32	7.07	
3-Octanol	273	7.55	9.45	10.20	12.60	
	296	5.54	6.54	7.31	8.65	
2-Methyl-3-	288	3.24	3.42	3.61	3.95	
heptanol	320	3.58	3.79	3.98	4.31	
Benzyl alcohol	293	13.740	14.17	14.54		123
	303	12.31	12.74	13.10		
Acetone	293	21.21	22.53	23.46		
	303	19.75	21.10	22.07		
Nitrobenzene	293	35.72	36.98			
	303	33.05	34.30	35.35		
Perfluoro-	289	1.847	(0.11 kbar) 1.869			124
heptane	311.6	1.812	(0.115 kbar) 1.837			
Water	298.8	78.31	(0.144 kbar) 78.75			

dielectric constant at different pressures for some common solvents. Table 2.18 gives data for organic liquids in the form of constants ε_0, A' and B' in eq. (2.59):[116, 117, 123]

$$\frac{1}{\varepsilon_0} - \frac{1}{\varepsilon_p} = A' \log \frac{(B' + P)}{(B' + P_0)} \tag{2.59}$$

or, for $P_0 = 1$ bar,

$$1 - \frac{\varepsilon_0}{\varepsilon_p} = A' \varepsilon_0 \log \frac{(B' + P)}{B'} \tag{2.60}$$

Differentiation of (2.59) gives

$$\frac{\partial(1/\varepsilon)}{\partial P} = \frac{1}{\varepsilon_0^2}\left(\frac{\partial \varepsilon}{\partial P}\right) = \frac{A'}{B' + P} = \text{(at 1 bar)}\ \frac{A'}{B'} \tag{2.61}$$

This quantity $\Phi = 1/\varepsilon_0^2(\partial \varepsilon/\partial P)$ is important as it appears in the Drude–Nernst equation which defines solvent electrostriction. Some values appear in Table 2.18. The pressure coefficient of dielectric constant, $\partial \varepsilon/\partial P$, is also readily calculated from eq. (2.62):

$$(\partial \varepsilon/\partial P)_{p \to 0} = \varepsilon_0^2 \cdot A'/B' \tag{2.62}$$

although some measure of disagreement is apparent when compared to values quoted by Hamann.[125]

2.1.15 Solvent Electrostriction

When an ion is created in a liquid medium of dipolar molecules, strong electrostatic forces tend to pull solvent molecules close around the charge so that a reduction in volume results, the effect being known as electrostriction. This is the effect responsible for the demonstrable volume contraction which occurs in, for example, a Menshutkin reaction:[127]

$$\Delta \bar{V} = -46 \text{ cm}^3 \text{ mol}^{-1}$$

Table 2.18. Pressure effects on dielectric constants; constants in eq. (2.67):[104, 105]

$$1 - \frac{\varepsilon_0}{\varepsilon_p} = A' \varepsilon_0 \log\left(\frac{B' + P}{B' + 1}\right) \text{ [P in bar]}$$

Liquid	$T(°C)$	ε_0	$10^3 A'$	B'	$10^6 \Phi$
n-Hexane	20.0	1.8887	64.52	540	119
	35.0	1.8658	64.45	461	140
	50.0	1.8410	64.95	402	161
Benzene	20.0	2.2832	59.48	829	72
	35.0	2.2533	60.52	761	79
	50.0	2.2229	66.23	754	88
Carbon tetrachloride	20.0	2.2375	63.78	791	81
	35.0	2.2080	69.48	789	88
	50.0	2.1773	66.03	646	102
Chlorobenzene	20.0	5.6895	38.97	1232	32
	35.0	5.4370	36.86	986	39
	50.0	5.1990	38.27	882	43
Methanol	20.0	33.60	7.245	875	8.3
	35.0	30.83	6.983	659	10.6
	50.0	28.24	7.749	592	13.0
Benzyl alcohol	20.0	13.740	22.35	1951	11.4
	35.0	12.309	20.25	1347	15.0
	50.0	11.036	17.17	824	20.8
Acetone	20.0	21.24	11.00	651	16.9
	35.0	19.75	12.35	594	20.8
	50.0	18.34	12.35	443	27.8
Nitrobenzene	20.0	35.72	9.882	1996	4.95
	35.0	33.05	7.813	1278	6.11
	50.0	30.60	7.665	1032	7.42
Chloroform	30.0	4.6362	46.47	739	63
	50.0	4.3183	43.43	483	90

Table 2.18 (*cont.*)

Liquid	$T(^{\circ}C)$	ε_0	$10^3 A'$	B'	$10^6 \Phi$
Fluorbenzene	30.0	5.2693	30.50	559	54.5
	50.0	4.9279	39.28	580	67.7
Bromobenzene	30.0	5.2876	29.48	976	30.2
	50.0	5.0196	26.33	682	38.6
Iodobenzene	30.0	4.5494	26.96	1205	24.8
	50.0	4.3818	25.91	904	28.7
o-Chlorotoluene	30.0	4.6380	36.94	873	42.3
	50.0	4.3978	35.44	666	53.2
m-Chlorotoluene	30.0	5.6061	25.52	786	32.5
	50.0	5.3101	24.58	574	42.8
p-Chlorotoluene	30.0	6.1413	31.69	1115	28.4
	50.0	5.7969	24.89	635	39.2
o-Dichlorobenzene	30.0	9.8425	24.82	1102	22.5
	50.0	9.0534	23.52	806	29.2
m-Dichlorobenzene	30.0	4.9165	32.20	895	36.0
	50.0	4.6681	29.20	639	45.7
Methylchloroform	30.0	6.9886	30.50	601	50.7
	50.0	6.4101	31.37	455	69.0
1,1-Dichloroethane	30.0	9.9008	21.49	568	37.8
	50.0	8.9859	22.13	403	54.9
1,2-Dichloroethane	30.0	10.152	40.32	978	41.2
	50.0	9.2244	35.64	628	56.7
1-Chlorobutane	30.0	6.6940	30.31	594	51.0
	50.0	6.4228	30.67	443	69.0
Chlorocyclohexane	30.0	7.9505	30.53	1086	28.1
	50.0	7.3095	29.66	836	35.5
Bromocyclohexane	30.0	8.0026	22.33	856	26.1
	50.0	7.4988	24.56	803	30.6
Cyclohexanone	30.0	15.421	9.267	718	12.9
	50.0	14.262	10.63	640	16.6
2-Butanone	30.0	17.675	14.97	750	20.0
	50.0	16.038	12.30	402	30.5
Propionitrile	30.0	27.901	5.995	636	9.4
	50.0	25.716	6.188	469	13.2
Benzonitrile	30.0	24.899	6.077	1047	5.8
	50.0	23.122	5.814	715	8.1
Nitromethane	30.0	35.780	8.168	1420	5.7
	50.0	32.671	8.348	1143	7.3
Methylenechloride	30.0	8.6381	24.88	628	39.6
Toluene	30.0	2.3614	51.92	727	71.4
Ethanol	0	27.7	10.94	1884	5.8
	20.0	25.71	10.35	1255	8.2
	30.0	23.3	9.44	1290	7.3
Diethyl ether	20.0	4.328	85.65	588	145
Carbon disulphide	20.0	2.647	95.05	1275	74.5
	30.0	2.63	83.8	1013	82.7
Pentane	30.0	1.82	95.0	709	134
Isobutanol	30.0	17.3	14.57	1034	14.0
Water	20.0	80.79	5.00	2963	1.68

The volume of electrostriction, ΔV^e, for an ion, charge q, radius r, is the change in partial molar volume when transferrred to a medium (continuous) of dielectric constant ε from one for which $\varepsilon = 1$ and is given by the Drude–Nernst equation (2.63)[128]

$$\Delta V^e = -\frac{N_0 q^2}{2r} \cdot \frac{1}{\varepsilon^2} \frac{\partial \varepsilon}{\partial P} = -\frac{N_0 q^2 \, \Phi}{2r} \tag{2.63}$$

where N_0 is the Avogadro number. If q is in e.s.u. r in cm and P in dyn cm^{-2} ($= 10^{-6}$ bar) then ΔV^e will be in $\text{cm}^3 \, \text{mol}^{-1}$. Values of $1/\varepsilon^2 \cdot (\partial \varepsilon / \partial P)$ are taken from Table 2.18.* We can examine how well this equation accords with the example above. The separate contributions for the methylpyridinium ion ($r = 2.5 \, \text{Å}$) and the iodide ion ($r = 2.2 \, \text{Å}$) will be

$$\Delta V^e(\text{Mepy}^+) = -\frac{(4.803 \times 10^{-10})^2 \times 8.28 \times 10^{-6} \times 10^{-6}}{2 \times 2.5 \times 10^{-8}}$$

$$= -23 \, \text{cm}^3 \, \text{mol}^{-1}$$

$$\Delta V^e(I^-) = -\frac{(4.803 \times 10^{-10})^2 \times 8.28 \times 10^{-6} \times 10^{-6}}{2 \times 2.2 \times 10^{-8}}$$

$$= -26 \, \text{cm}^3 \, \text{mol}^{-1}$$

$$\Delta V^e_{\text{total}} = -49 \, \text{cm}^3 \, \text{mol}^{-1}$$

which is remarkably close to the experimental value considering the approximations and assumptions made.

Charge neutralization is accompanied by a volume expansion of the solvation shell, i.e. relaxation of electrostriction. Values of Φ on which the amount of electrostriction largely depends increase the less 'polar' the solvent. This is because the electric field in the vicinity of an ion is transmitted to greater distances through a medium of low dielectric constant than high and consequently affects a greater volume of the nearby solvent:[129] Φ also increases with temperature, properties which must be borne in mind when interpreting volumes of activation for reactions, Chapter 4, or the partial molar volume of ions.[130-132] From another viewpoint, electrostriction is a pressure directed isotropically towards the ion such that

$$P \approx \frac{\chi^2 \varepsilon}{8\pi} \tag{2.64}$$

where χ is the electric field strength due to the ion.

*For a given solvent, temperature we can write $\Delta V^e = -Aq^2/r$ where A is a constant $= 8.0$ for water at $25°$.

2.1.16 The Effect of Pressure on Optical Rotation

Few studies have been reported on this although rotation and rotatory dispersion have been examined up to 4 kbar for several crystals, thus[133]

	$d\alpha/dP^0$ $(\text{mm}^{-1}\text{kbar}^{-1})$ at 589 nm
α-Quartz	−0.17
Sodium chlorate	−0.05
Benzil	+0.99
Ethylenediamine sulphate	+0.14

The value of $d\alpha/dP$ increases towards the blue end of the spectrum for crystals with a negative coefficient but decreases towards the blue for those with a positive one.

The rotations of chiral molecules in solution depend upon their concentrations and it would be expected that this would be the major effect upon pressurization. However, changes in the solvation of the species could also affect $[\alpha]$.

Birefringence, which is induced in an amorphous material such as a plastic by non-isotropic stress, is found to disappear or relax much more slowly if hydrostatic pressure is simultaneously applied.[216] Relaxation requires molecular motion so that this phenomenon is related to the increase in viscosity which accompanies pressurization.

2.2 KINETIC PROPERTIES OF LIQUIDS

The mobility of the molecules in a liquid affects the most diverse properties including viscosity, diffusion, electrochemical transport, and thermal conductivity. A model for molecular motion which is often used view the migration of a molecule or group of molecules as an activation process. First, the migrating entity must possess sufficient energy to move neighbours from its path when translation or rotation may occur. This energy is described as an activation energy and is obtained from the temperature coefficient of the process considered. However, according to this model, the displacement of neighbouring molecules will cause a local change in volume of the system which is describable as an activation volume, $\Delta V^{\neq}$ analogous to the quantity derived in Section 4.1.[218] This is obtained from the pressure coefficient of the appropriate transport property. On the whole, molecular movement is made more difficult by an increase in pressure and the effects on the following properties can be understood on this basis.

2.2.1 The Effect of Pressure on Viscosity, η

Many types of viscometer may be used under pressure, all depending upon measurement of the rate of flow of liquid through a constriction or the rate of movement of an object such as a steel sphere through the fluid,[59] which gives relative viscosity directly. When used within a steel high pressure vessel, some arrangement for signalling the arrival of the dropping or rolling ball (short-circuiting of electrical contacts) and for resetting and starting the ball (inversion of the whole apparatus or an electromagnet) needs to be provided.[7, 134-137]

Viscosities of liquids increase with pressure (Table 2.19) with an upward curvature of the lines since the interlocking of molecules which increases the viscosity becomes more prevalent the closer they approach.

The coefficient ϕ, where

$$\phi = \frac{1}{\eta} \cdot \left(\frac{\partial \eta}{\partial P} \right)_T \tag{2.66}$$

also increases with pressure so that if the liquid does not freeze (i.e. undergo a phase change) it eventually reaches a vitreous state (fig. 2.13). Water below 20° is

Table 2.19. The effect of pressure on the viscosities of liquids $\eta_P/\eta_{1\,bar}$ at 30°

	Relative viscosity	
	$P = 1\,kbar$	$4\,kbar$
Pentane	2.06	7.03
Isopentane	2.01	7.84
Hexane	2.15	8.2
Octane	2.12	12.25
Methylcyclohexane	2.44	18.8
Benzene	2.22	—
Toluene	1.95	7.89
o-Xylene	2.05	
m-Xylene	1.95	9.27
Dichlorodifluoromethane	—	4.77
Carbon tetrachloride	2.24	
Carbon disulphide	1.44	3.23
Chloroethane	1.75	4.46
Bromoethane	1.67	4.28
Iodoethane	1.65	4.53
Chlorobenzene	1.79	7.36
Bromobenzene	1.83	7.89
Methanol	1.47	2.96
Ethanol	1.58	4.14
1-Propanol	1.92	6.86
Isopropanol	2.20	9.60
1-Butanol	2.09	8.60
Isobutanol	2.44	16.0
1-Pentanol	2.19	11.5
Glycerol	1.82	8.63
Eugenol	3.48	187.0
Acetone	1.68	4.03
Ethyl acetate	1.81	6.58
Diethyl ether	2.11	6.20
Water	3.27	

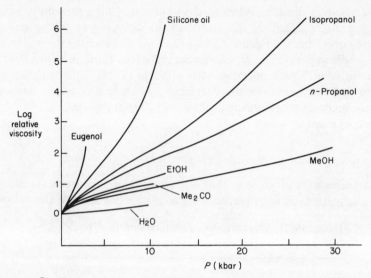

Fig. 2.13. Viscosities of some liquids as a function of pressure. (After Bridgman[7])

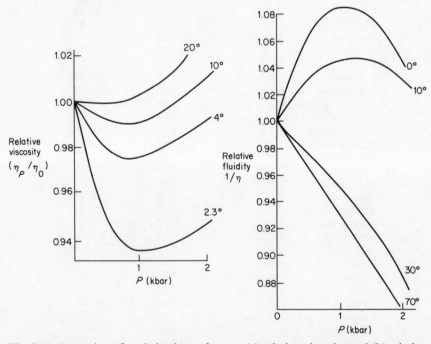

Fig. 2.14. Anomalous flow behaviour of water: (a) relative viscosity and (b) relative fluidity ($= 1/\eta$) at various pressures. (After Brummer and Gancy[197])

exceptional in that its viscosity decreases with pressure up to 1 kbar above which a normal increase ensures, fig. 2.14. This is associated with the breaking of hydrogen bonds with increasing pressure. As the temperature is raised, both η and α fall. The anomalous behaviour is even more apparent if fluidity, $1/\eta$, is plotted against P, fig. 2.14.

Viscosity is an important property of lubricants and, in addition to studies carried out on pure liquids by Bridgman[7] and other workers, [137-142] measurements have been made on petroleum products.[144-147] Many empirical relationships relating pressure and viscosity have been developed[148-153] but the theory of viscosity is complex. Eyring[154] developed a model in which the resistance to flow is related to the difficulty experienced by a molecule in undergoing rotation or other motion. This is associated with the need to create a local increase in volume, ΔV, an activation process, and consequent increase in energy; thus

$$\phi = \frac{\Delta V}{RT} \tag{2.68}$$

This concept explains the increase in viscosity with pressure in terms of the increasing difficulty of creating a local volume expansion.[154-156] The same decrease in mobility is also responsible for decreasing rates of bimolecular reactions[157] as viscosity increases.

The following generalizations may be made; in a homologous series, an increase in chain length is associated with an increase in absolute viscosity but ϕ remains fairly constant. The viscosity and especially the coefficient ϕ increase considerably with chain branching or the introduction of rigid substituents. Hydrogen-bonding groups such as —OH, —NH$_2$ tend to be associated with high values of ϕ while the introduction of ether groups tends to lower ϕ.[138] It is usually taken that fluidity ($= 1/\eta$), and is a fairly linear function of molar volume, is a measure of the free volume in the liquid.

A series of careful studies of the isomeric octanols has been made[119-122] (Table 2.20) showing that the pressure coefficient of viscosity increases markedly

Table 2.20. Viscosity properties of some isomeric alcohols at 10 °C[119-122]

	$\dfrac{\eta_{1\,kbar}}{\eta_0}$	$\dfrac{\eta_{4\,kbar}}{\eta_0}$	$\dfrac{\Delta H_{4\,kbar}}{\Delta H_0}$	$\dfrac{\Delta V_{1\,kbar}}{\Delta V_0}$	$\dfrac{\tau_{1\,kbar}}{\tau_0}$[a]
2-Octanol	4.0	62	1.59	0.688	—
3-Octanol	4.9	135	1.97	0.753	3.7
5-Methyl-3-heptanol	5.0	475	2.72	0.862	—
6-Methyl-3-heptanol	6.8	416	2.16	0.780	7.5
2-Methyl-3-heptanol	6.75	545	2.23	0.840	3.3
1-Phenyl-1-propanol	6.7	350 (3 kbar)			

[a] Relative relaxation time.

with branching. The enthalpies of activation* for viscous flow have been calculated from the temperature dependence of η at different pressures and are shown to increase with pressure, while the volumes of activation (obtained from the pressure coefficient) decrease with pressure (Table 2.21) except in the case of water. Ring structures are particularly prone to an increase in viscosity. These

Table 2.21. Volumes of activation for viscous flow, $\Delta V_v^{\neq}$ $(P \to 0)^{218}$ and for diffusion, $\Delta V_D^{\neq}$ $(P \to 0)$

$$\Delta V_v^{\neq} = RT(\partial \ln \eta / \partial P)_T \quad (2.69); \quad \Delta V_D^{\neq} = -RT(\partial \ln D / \partial P)_T \quad (2.72)$$

Liquid	$\Delta V_v^{\neq}$	$\Delta V_D^{\neq}$ $(cm^3\,mol^{-1})$	Reference
Mercury	0.5	0.0	170
Water	0.16		
(at 10 kbar)	1.4		
Glycerol	15		
Methanol	8.3		
(at 10 kbar)	3.2		
Ethanol	9.5		
Propan-1-ol	15.3		
Propan-2-ol	19.6		
Butan-1-ol	18.4		
2-Methylpropan-1-ol	21.7		
Diethyl ether	20		
Acetone	14.7, 15.9	13.0	
Bromoethane	11.4		
Iodoethane	11.0		
1-Bromobutane	14.3		
Chlorobenzene	13.8		
Bromobenzene	14.4		
Carbon tetrachloride	23.1		
Nitromethane		13.6	
Carbon disulphide	8.3		
Benzene	20.4	20.0	
Toluene	22.0		
Cyclohexane	36.3, 30.8	28	
Methylcyclohexane	28.6		
Pentane	14.8		
2-Methylbutane	19.2, 23.8	34	
Hexane	16.0		
Octane	19.9		
$NaNO_3$ melt, 623 K		5.3	219
KNO_3 melt, 657 K		10.0	
$RbNO_3$ melt, 628 K		9.6	
$CsNO_3$ melt, 709 K		13.2	

* The enthalpy $\Delta H^{\neq}$ and volume $\Delta V^{\neq}$ of activation for viscous flow are given by the equations

$$\left(\frac{\partial \ln \eta}{\partial P}\right)_T = \frac{\Delta V^{\neq}}{RT}, \quad \left(\frac{\partial \ln \eta}{\partial 1/T}\right)_P = \frac{\Delta H^{\neq}}{R} \qquad (2.69)$$

where η is the viscosity (actually the kinematic viscosity).

results can be interpreted as showing that association equilibria shift in favour of associated species under pressure, and that molecular rotation is curtailed. The latter is confirmed by increases in the relaxation times at high pressure (section 2.6). The viscosity behaviour of mixtures of non-interacting liquids falls close to the expected average value and that of some mixtures of complex hydrocarbons is very similar to the properties of pure compounds embodying the structural features of both components.[158] Mixtures of water with a series of alcohols may show unexpected properties, for example the pressure coefficient of viscosity falls with temperature for aqueous mixtures of ethanol and 2-propanol whereas those of the individual components rises. Presumably this is related to changes in the liquid structure.[159]

2.2.2 Diffusion in Liquids

The Stokes–Einstein equation, eq. (2.70), relates the coefficient of diffusion, D, to the viscosity of a liquid:

$$D = \frac{kT}{3\pi\sigma\eta} \tag{2.70}$$

where σ is the diameter of the molecule diffusing. Thus it would be predicted that

$$\frac{D_P}{D_0} = \frac{\eta_0}{\eta_P} \tag{2.71}$$

and this relationship has been found to be obeyed for several liquids (self-diffusion).[160, 161] This implies that diffusion is reduced by increasing viscosity at high pressures and the the two phenomena have a common physical basis. The rates of bimolecular processes may eventaully be subject to diffusion control at sufficiently high pressures, i.e. the slow step in the reaction becomes the diffusion together of the reagents. This can introduce serious errors into activation volume measurements (section 2.12). By the same token, reactions in a solvent cavity (cage effects) may be promoted by working at high viscosities. The diffusion of solute or solvent molecules in a liquid may be measured using radioactive tracers. Thus, for mercuric chloride in butanol the diffusion coefficients falls smoothly with pressure according to eq. (2.70).[162 – 163] Water and aqueous solutions on the other hand behave erratically. A typical result for the self-diffusion of water in an electrolyte solution is given in fig. 2.15. However, measurements of self-diffusion in water by a nuclear magnetic resonance spin–echo technique seem to indicate a smooth decrease with pressure up to 10 kbar.[164] The question is discussed further in Chapter 6.

Diffusion, like viscosity, may be treated as an activation process. The passage of a molecule through the surrounding medium in a random path will also depend upon it possessing the necessary kinetic energy and so the activation energy for diffusion E_D may be defined,

$$E_D = \left[R \cdot \frac{\partial (\ln D)}{\partial (1/T)} \right]_P \tag{2.73}$$

and the activation volume, $\Delta V_D^{\neq}$, as

$$\Delta V_D^{\neq} = -RT(\partial \ln D/\partial P)_T \qquad (2.74)$$

Volumes of activation for self-diffusion may be obtained from NMR spin–echo measurements [170] and may be very large for organic solvents, Table 2.21. The value for water is an order of magnitude less suggesting a different mechanism of

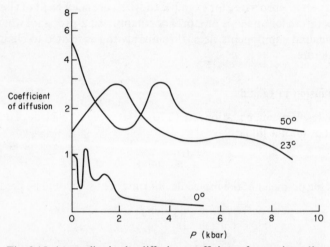

Fig. 2.15. Anomalies in the diffusion coefficient of water in sodium sulphate solution. (After Drickamer et al.[162])

transport is operating. Diffusion of a solute species may be studied but the activation energies obtained give information primarily concerning the solvent. Diffusion of Ag^+ through molten alkali metal nitrates (observed using chrono-potentiometry) gives activation volumes which increase with the mass of the metal ion.[219]

2.2.3 Effect of Pressure on Thermal Conductivity, k

In the absence of convection, heat conduction at a molecular level depends upon molecular collisions and the transfer of translational, rotational, and vibrational momentum from one to another. The closer the molecules are packed the more frequent will be the collisions and the more effective the transfer of energy. It would be expected that thermal conductivity, k, would be a function of density, ρ, which is indeed the case.[166–168] To a reasonable approximation,

$$\frac{d(k_{rel})}{d(\rho_{rel})} \approx q \qquad (2.75)$$

where q is a constant. Some values of q are given in Table 2.22 so that it may be used to estimate effects on analogous compounds.

An approximate relationship which holds for hydrocarbons between normal melting and boiling points is[169] (eq. (2.76))

$$k/k_0 = 1 + 0.2P(\text{kbar}) \tag{2.76}$$

More exact relationships have been published by Weber[7]

$$\frac{k}{\rho C_p} = \text{constant} \quad \text{or better} \quad \frac{k}{\rho C_p} \cdot \left(\frac{m}{\rho}\right)^{1/2} \tag{2.77}$$

though, as Bridgman remarks, no particular significance can be attached to the form of this equation.

Table 2.22. Effect of pressure on thermal conductivity

	k_p/k_0 (30°, $P = 12$ kbar)	q
Pentane	2.481	3.403
Carbon disulphide	1.962	3.11
Diethyl ether	2.451	3.76
Acetone	1.864 (9 kbar)	2.95
Methanol	2.097	3.40
Ethanol	2.122	3.40
Isopropanol	2.150	3.87
Butan-1-ol	2.008	3.57
Isopentanol	2.069	3.75
Bromoethane	1.928	2.82
Iodoethane	1.724	2.28

Usually, the thermal conductivities of solids are greater than those of the corresponding liquids. A further relationship is with sound velocity, v, and molecular spacing, δ, $(m/\rho)^{1/2}$ m = molecular mass, g per molecule)[7]

$$k = \frac{2Rv}{\delta^2} \tag{2.78}$$

(R = gas constant). This relationship would seem to indicate a similar mechanism for transference of heat energy and of a longitudinal wave front through a liquid.

2.3. THE EFFECT OF PRESSURE ON PHASE CHANGE

According to the general principle that an equilibrium tends towards the state with the lesser volume on application of pressure, phase changes, solid–liquid, or solid–solid will tend to take place at a lower temperature if a volume contraction is involved and conversely. Since for most substances the liquid density is less than that of the solid at atmospheric pressure (some exceptions to this are water, bismuth, gallium, and germanium for which $\Delta S_{L \to S}$ is positive), melting points are depressed under pressure and the melting point line on the P–T graph has a positive slope though usually decreasing towards higher pressures, fig. 1.54.

110

Phase diagrams may be simple as in the case of nitrobenzene or complex when two or more solid phases are capable of existing under appropriate conditions and differing in their crystalline structure. Carbon tetrachloride and acetic acid are examples. Water shows this type of behaviour to an unusual degree. Many high pressure forms of ice are known and the initial slope of the P–V curve near atmospheric pressure is negative though it becomes more normal above ~ 2 kbar.

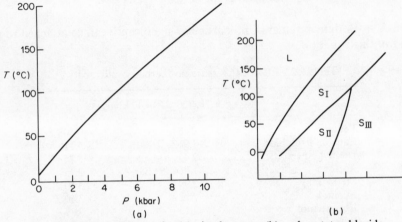

Fig. 2.16. Phase diagrams for (a) nitrobenzene, (b) carbon tetrachloride

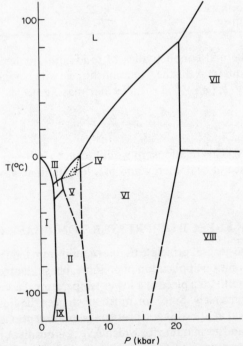

Fig. 2.17. Phase diagram of water showing regions of stability of nine forms of ice. (After Fletcher[185])

The onset of freezing is apparent from the sudden decrease in compressibility showing as a discontinuity in the P–V graph. Crystallization may be observed if the sample is contained in an optical cell. It is necessary in an isothermal melting point determination to approach the transition pressure with small increments allowing thermal equilibration to remove the heat of compression and, at the transition point, latent heat. The latent heat emission may serve to detect freezing by the use of differential scanning calorimetry (DSC) for which equipment may be constructed to operate at high pressure.[177, 178] In a DSC experiment, the pressure

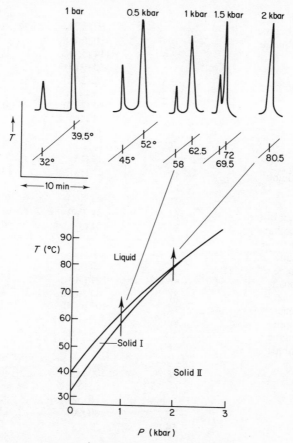

Fig. 2.18. Construction of the phase diagram of n-$C_{21}H_{44}$ by DSC. (After Kamphausen[178])

is maintained constant and temperature increased by a constant rate of heating. At the transition temperature appropriate to the pressure used there is a sharp temperature excursion which is recorded as also is the temperature of the calorimeter. The results for a higher paraffin are shown in fig. 2.18 which has two solid phases converging at around 2 kbar.

Table 2.23. Freezing temperatures, T_f and pressures, P_f or some liquids

	$T_f,^{\circ}C$ (1 kbar)	$T_f\,^{\circ}C$ (P kbar)	P_f (25 °C) kbar
Pentane		0 (12)	
Cyclohexane	32.5	96.6 (4)	0.355
Methylcyclohexane			13.8
Hexane		10.2 (30)	
Heptane			11.1
Octane			5.3
Decane			3.0
2,2-Dimethylbutane		50 (5.3)	
2,3-Dimethylbutane		0 (6.1)	
Benzene	33.4	114 (5.1)	0.72
Toluene		30 (9.6)	
o-Xylene	−3.5	30 (2.51)	2.29
m-Xylene	−25.2	0.4 (2.28)	3.54
p-Xylene	46.0	80 (2.13)	0.36
Tetralin		18 (3.2)	
Cumene		25 (40)	
Mesitylene			3.5
Bromoethane	−108	5 (15)	
Dichloromethane	−83	42 (15)	13.0
Chloroform	−45.2	32.4 (6)	5.5
Bromoform	31.5	94.7 (4)	0.715
Carbon tetrachloride	14.2	102.7 (4)	1.462
1,2-Dibromoethane	34.0	74 (3)	0.63
Pentachloroethane	−6.3		
1-Bromobutane		0 (12)	
1-Chlorobutane		0 (12)	
1-Bromopropane	−98	197 (40)	14
Chlorobenzene	−28	84.5 (10)	4.55
Bromobenzene	−12	107.6 (10)	3.26
1-Bromopentane		0 (12)	
1-Chloropentane		0 (12)	
Bromonaphthalene	6.1		
Methanol			30
Ethanol	−108	109 (35)	20
1-Propanol		25 (50)	
2-Propanol		25 (50)	
1-Butanol	−77.2	80 (20)	11.7
Tert-butanol	58.1		0.001
1-Pentanol			50
Tert. pentanol	13.4		
Cyclohexanol	62.3		
Ethanediol		0 (3.2)	
Benzyl alcohol	0.2		
o-Cresol	47.7		
m-Cresol	25.6		
Menthol	60		
Formic acid	20.6		
Acetic acid	37	69.4 (3)	0.39

I–II

Table 2.23. (cont.)

	T_f °C (1 kbar)	T_f °C (P kbar)	P_f (25 °C) kbar
Propanoic acid	−1.2		
Butanoic acid	13.8		
Nitrobenzene	13.5 (27.9)	48.1 (2)	0.84
m-Nitrotoluene	40.6		
1,2-Dimethoxybenzene	41.5		0.139
Cyanobenzene	7.6		
Aniline	13.5	140 (10)	1.62
N,N-Dimethylaniline	26.3	—	0.93
Acetone		20 (8)	
Diethyl ether		35 (12)	
Dioxan	23	68 (6)	1.2
Dipentyl ether			7.5
Furan	−73	−1.6 (9.14)	—
Formamide	10.8		
Methyl benzoate	31.8		0.65
Ethyl acetate			12.1
Carbon disulphide	−98		
Hexamethyl disiloxane	−37.5	40.2 (4)	3.3
Water	−9.0		
Sodium	106	167 (10)	
Potassium	78	170 (10)	
Mercury		21.9 (12)	

2.3.1 Melting Point Relationships

From a practical point of view it is often of importance to know whether a liquid will solidify in a high pressure experiment and, furthermore, a phase transition may conveniently be used to calibrate a pressure gauge (Chapter 1.3.2). Phase transition data are given in Tables 2.23 and 2.24. The melting curve for many substances may be defined at least over a modest pressure range by a quadratic function, eq. (2.79):

$$T_p = T_0 + aP + bP^2 \qquad (2.79)$$

where T_p, T_0 are transition temperatures at P, 1 bar respectively, and a, b are constants. Another equation which has been used for representing the data is eq. (2.80):[130]

$$\log(P+m) = n \log T_p - \log p \qquad (2.80)$$

where m, n, and p are constants (Table 2.24).

The melting point curve is defined by the Clausius–Clapeyron equation, eq. (2.81), which applies to any equilibrium:

$$\frac{dP}{dT_p} = \frac{\Delta H_L}{T_0 \Delta V} \qquad (2.81)$$

where ΔH_L is the latent heat of fusion (J g^{-1}), T_m, T_0 (K), ΔV the change in volume on melting (m^3 g^{-1}) then P is in Pa. Thus the sign of the slope of the melting point curve is determined by the sign of ΔV and its magnitude by the combination of latent heat and volume terms. Since the compressibilities of solids are usually less than those of liquids, $\Delta V (= V_{liquid} - V_{solid})$ falls with pressure and thus dP/dT also falls.

Table 2.24. Freezing curves of some liquids defined by eq. (2.79) Data from references 184, 186

$$T_f = T_0 + aP + bP^{2\,a} \quad (2.79)$$

	T_0	$10^4\,a$	$-10^8\,b$
Cyclohexane	6.63	532	(0)
p-Xylene	13.2	343.8	171
Benzene	5.43	283	198
p-Chlorotoluene	6.85	265.9	122
p-Bromotoluene	26.5	291.1	131
Aniline	−6.1	203	112
Acetophenone	19.2	235 (243)	152 (162)
p-Cresol	33.3	228	63
Formic acid	7.75	127.6	80
Tert. butanol	24.9	355	397
Tert. pentanol	−8.45	235	152
Carbon tetrachloride	−23	350	147
1,2-Dibromoethane	10	260	133
1,4-Dichlorobenzene	52.9	275	134
4-Bromotoluene	28	301	140
4-Chlorotoluene	6.9	275	130
p-Cresol	33.3	236	67
1,2-Dimethoxybenzene	23.3	224	108
Naphthalene	80	376	192

^{a} T(°C) and P (bar).

Table 2.25. Freezing curves of some liquids defined by eq. (2.80)[187]

	$\log(P+m) = n\log T - \log p$		
	m	n	p
Carbon tetrachloride	3150	2.08	1.498
Chloroform	6050	1.83	0.466
Bromobenzene	5000	2.42	2.070
Nitrobenzene	6000	1.97	1.037
Methyl oxalate	6010	2.35	2.13

Other melting laws have been formulated from theoretical considerations such as Lindemann[59,179] (eq. (2.82)) and Clapeyron (eq. (2.83)):[180]

$$T_p = \frac{C.h}{k} . Mv^{2/3} \quad (2.82)$$

where C is a constant, h, k Planck's and Boltzmann's constants, v is a frequency associated with the lattice, M the molecular weight, and V the volume change per mole; V, S are volumes and entropies of liquid and solid phases at the transition temperature in eq. (2.83)

$$\frac{dT_p}{dP} = \frac{V_{liq} - V_s}{S_{liq} - S_s} \tag{2.83}$$

or the equation of Kraut and Kennedy,[181]

$$T_p = T_0\left(1 + C'\left(\frac{\Delta V}{V_0}\right)_T\right) \tag{2.84}$$

with C' a constant ($C' = 1.325$ (Li), 6.260 (Na), 8.668 (K), 13.124 (Rb), 3.3209 (Fe)).

The value of such equations has been examined by Babb who shows that experimental data are reproduced accurately only at moderate compressions.[182]

On the other hand, the Simon equation, eq. (2.85), has been shown to fit the melting point data of a number of organic compounds (Table 2.26) with acceptable precision:[183]

$$\frac{P}{a} = \left(\frac{T_p}{T_0}\right)^c - 1 \tag{2.85}$$

where a and c are constants.

Table 2.26. Freezing curves of some liquids defined by eq. (2.85)[183]

	T_0 (K)	$\dfrac{P}{a} = \left(\dfrac{T_p}{T_0}\right)^c - 1$	
		a (bar)	c
Propane	85.3	7180	1.283
Butane	134.5	3634	2.210
Pentane	143.5	6600	1.649
Isobutane	128.2	4246	2.478
Isopentane	112.5	5916	1.563
Ethylene	103.4	3275	1.811
Propylene	86.0	3196	2.821
Dichlorodifluoro-methane	117.9	3288	2.231

The onset of freezing as pressure is raised or temperature lowered is detected by the evolution of latent heat or by a discontinuity in the pressure–volume curve (as shown in piston movement). For metals a discontinuity in resistance measurements may be used.

The latent heat of a substance usually increases with pressure although, as shown in fig. 2.17 and fig. 2.18, there is no simple monotonic trend and quite widely different behaviour may be observed.

2.3.2 Effects of Pressure upon Specific Heats

It has been found for several organic liquids that the specific heat diminishes with pressure for about 2–3 kbar after which it rises in a rather irregular fashion.

116

The pressure coefficients of principal specific heats are obtainable from P–V–T data by the relationships

$$\left(\frac{\partial C_p}{\partial P}\right)_T = -T\left(\frac{\partial^2 V}{\partial T^2}\right)_P \tag{2.86}$$

$$(C_v - C_p) = T\left(\frac{\partial V}{\partial T}\right)_P^2 \bigg/ \frac{\partial V}{\partial P} \tag{2.87}$$

The complex variation of $\partial C_p/\partial P$ is not easily interpreted in terms of physical changes in the liquid. Specific interactions, dependent upon the nature of liquid, are undoubtedly of importance.

Multi-component systems also show pressure-dependent behaviour, the phase change now being that accompanying miscibility from separate phases. The benzene–water[188] system is illustrative of the vast amount of data available, fig. 2.19. In general, solubilities of liquids or solids in a solvent will be increased by pressure whenever a reduction of volume occurs during solution, the usual

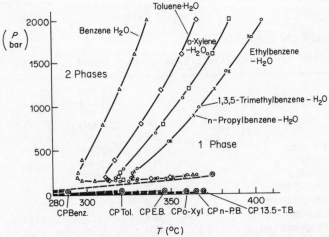

Fig. 2.19. Phase diagrams for mixtures of water with aromatic hydrocarbons[188] (CP = critical point). (Reproduced by permission of John Wiley and Sons, Inc.)

situation. Few direct measurements of solubility of a solid under pressure have been made. Sodium chloride is initially more soluble in water under pressure, fig. 2.20, but a maximum is reached at 4 kbar.[246] In general, the solubility at a given pressure, S_p, may be obtained from eq. (2.98):

$$\ln(S_p/S_1) = -\frac{P\bar{V}'}{RT} \tag{2.88}$$

where S_1 is the solubility at 1 bar and $\bar{V}'$ the averaged partial molar volume of the solute over the pressure range $1-P$.

Water is known to behave anomalously as a solvent for non-polar as well as ionic solutes. The solubilities of hydrocarbons and halocarbons in general show a maximum around 1–2 kbar, i.e. ΔV changes sign from negative to positive and this is particularly marked at low temperature. The discontinuity of solubility, viscosity, spin-lattice relaxation, self-diffusion, chemical shift, and such physical properties seems to indicate a profound structural change in the solvent at moderate pressures, not observed at temperatures above about 30°.[257]

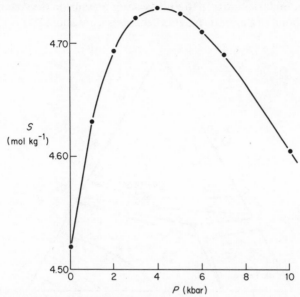

Fig. 2.20. Solubility of sodium chloride in water as a function of pressure[246]

2.3.3 Geochemical Phase Equilibria

The earth may be aptly described as a high temperature–high pressure laboratory since extreme conditions are generated within its interior. A rough guide to the temperatures and pressure from the surface to the core is

	Crust	Upper mantle	Lower mantle	Core
Depth (km)	0–50	50–400	400–2500	2500–3000
P (kbar)	0.001–20	20–130	130–900	900–>1000
T(°C)	0–1350	1350–3800	~3800	>4000?
Density (g cm^{-3})	2.6–3.0	3.0–4.5	4.5–8.0	8.0–>10

It is a familiar fact that many minerals, subsequently brought near the surface, were necessarily formed under high temperatures and pressures. Diamond is a prime example but many siliceous minerals, particularly those containing volatiles—H_2O, CO_2, H_2S for example, can only have been formed deep in the crust. The subject is of extreme complexity and space permits only a few examples. Laboratory simulation of petrogenic processes may be carried out at

6 kbar/800° or up to 1 kbar/1300° using the kind of vessels shown in fig. 1.24 which can only approach natural conditions to relatively small depths within the crust.[189]

The granite system is basically an equilibrium between Na_2O–K_2O–Al_2O_3–SiO_2 and may be depicted as the three-phase system, fig. 2.21. Phase boundaries are seen to shift as water pressure in the system is changed. The melting point of a granitic magma decreases from about 800 at 1 bar to 625 at

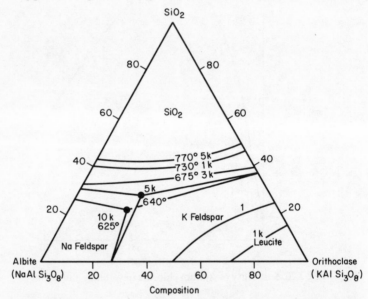

Fig. 2.21. Phase diagram of minerals present in granite showing some effects of pressure and temperature on stabilities of the various phases. (After Edgar and Platt[189].) (Numbers refer to pressure: 10 k = 10 kbar)

10 kbar water pressure corresponding to conditions at 10 km depth. The stabilities of individual phases change; leucite ($KAlSi_2O_6$) is rapidly destabilized by pressure while jadeite ($NaAlSi_2O_6$) becomes stable around 20 kbar. The presence or absence of these minerals in a granite can point to conditions during its formation. Between the crust and upper mantle is a sharp boundary (the Mohorovicic discontinuity) which apparently marks a change in rock composition to higher density types. From studies of inclusions in rocks originating in the mantle, such as the diamond-bearing Kimberlite pipes, it is believed that these are the garnet-containing minerals, eclogite, or peridotite. An equation for garnet formation from spinel or crustal basalt indicates a product favoured by pressure:

$$\underset{\text{spinel}}{MgAlO_4} + \underset{\text{enstatite}}{3.2MgSiO_3} + \underset{\text{diopside}}{0.4CaMgSi_2O_6} \rightleftharpoons$$
$$\underset{\text{garnet}}{Mg_{2.6}Ca_{0.4}Al_2Si_3O_{12}} + Mg_2SiO_4$$

Such transformations have been simulated in the laboratory.[190] The entrainment of volatile molecules by silicate minerals requires pressure. In the following examples, the vapour pressure of water is 2 kbar at the temperature indicated when equilibrium is established:[191]

$$Mg(OH)_2 \rightleftharpoons MgO + H_2O \ (660\,°C)$$

$$Al(OH)_3 \rightleftharpoons AlOOH + H_2O \ (135\,°C)$$

$$2AlOOH \rightleftharpoons Al_2O_3 + H_2O \ (375\,°C)$$

$$\underset{\text{muscovite}}{KAlSi_3O_{10}(OH)_2} \rightleftharpoons NaAlSi_3O_8 + Al_2O_3 + H_2O \ (715\,°C)$$

$$\underset{\text{annite}}{3KFe_3AlSi_2O_{10}(OH)_2} + 2O_2 \rightleftharpoons 3KAlSi_3O_8 + \underset{\text{magnetite}}{Fe_3O_4} + \underset{\text{hematite}}{3Fe_2O_3} + 3H_2O$$

The control of concentrations of volatiles is crucial in setting up such equilibria and is normally achieved by adding solid components which will decompose to the vapour in question. Under geological conditions, water will produce a fixed fugacity of oxygen and hydrogen by decomposition,

$$2H_2O \rightleftharpoons H_2 + O_2$$

and other systems which control oxygen (known as buffers) include

$$2Fe + SiO_2 + O_2 \longrightarrow Fe_2SiO_4$$

$$2Ni + O_2 \longrightarrow 2NiO$$

The water equilibrium can be used to supply a hydrogen buffer; the sample is placed in a palladium vessel immersed in water and hydrogen is capable of diffusing in. Unusual forms of quartz, coesite, and stishovite (density 4.26, cf. quartz, 2.66 g cm^{-3}) are formed at 90 kbar, 700° (anvil apparatus)[192] and have been shown to be associated with meteor impact craters.[193]

2.3.4 High Pressure Polytypes of ABX$_3$ Compounds

There exist a great many crystalline compounds of this formula in which A is a relatively large metal ion, radius 0.9–1.9 Å, B is a smaller transition metal ion, $r = 0.5$–0.9 Å, and X is oxygen or fluorine. The most symmetric lattice is the cubic perovskite structure, fig. 2.22, but occurs only at atmospheric pressure for a fairly narrow range of cation sizes given by

$$\frac{(r_A + r_X)}{\sqrt{2}(r_B + r_X)} = 0.75\text{--}1.0 \tag{2.89}$$

Many ABX$_3$ compounds crystallize in other polymorphs at atmospheric pressure which differ in the stacking arrangements of the octahedral units, fig. 2.22. The least dense are hexagonal (denoted H) then rhombic (R) and finally the cubic (C) structures. High pressure and temperature may permit the formation of denser, more symmetric structures which may be recovered on cooling under pressure.

120

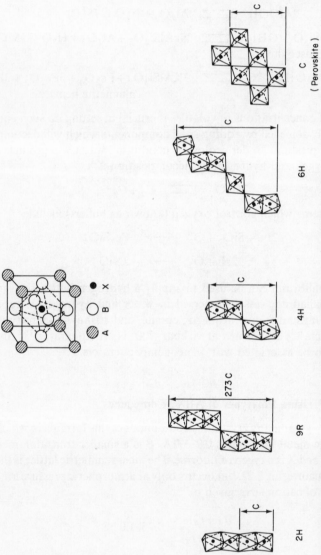

Fig. 2.22. Coordination geometry of ABX_3 compounds and stacking arrangements in the various crystalline polytypes[194]. (Reproduced by permission Academic Press, Inc.)

For example, barium ruthenate undergoes the following transformations at 1000°:

BaRuO$_3$	9R	→	4H	→	6H
pressure	atmos- pheric		15		30 kbar
density, g cm^{-3}	4.16		4.23		4.33

With other compounds, such as LuMnO$_3$, the hexagonal form at low pressure is transformed to cubic at 45 kbar.[194] Yet others are only stable at all at high pressure though they may be recovered on cooling, for example;

$$Bi_2O_3 + Cr_2O_3 \xrightarrow[700]{40\,kbar} 2BiCrO_3$$

2.4 PRESSURE EFFECTS ON ELECTRICAL PHENOMENA

The conduction of current through an electrolyte solution occurs by ionic transport and this has many features in common with diffusion. Conductivity may be treated as an activation process whereby the migrating ion must possess sufficient 'activation energy' to enable it to create a space in which to move. An activation volume for conductivity may also be defined which should be comparable with those for diffusion or viscosity and, since they are positive, should lead to a reduction in mobility of the ions under pressure. Other transport and electrical processes which depend on mobility will be similarly affected. Equilibrium processes, however, such as cell e.m.f. and interfacial properties may depend upon volume changes for the reactions in question (Chapter 3). While most information comes from aqueous solutions a great deal of research has been concerned with molten salts which behave in a far simpler manner, and a little is known about non-aqueous solvents.

2.4.1 The Effect of Pressure on Electrical Conductivity, Λ

While studies on the phenomena have a history going back over a century[7, 195] it is relatively recently that pressure effects on conduction of electrolyte solutions have received an interpretation.

The measurement of conductivity is straightforward. A cell of design, for example as shown in fig. 1.38, is immersed in an oil-filled pressure vessel with electrical leads projecting and the resistance of the system measured by a bridge. Some problems can arise with cracking of the platinum–glass seals of the electrodes due to differential contraction above about 2 kbar. The use of a PTFE vessel removes this difficulty but substitutes another in that the cell is less dimensionally constant under pressure. For kinetic studies, one is interested in a change of conductance with time and this is not a problem. For studies of ion transport and of ion-solvent interactions, one needs to know the specific conductance extrapolated to infinite dilution and must take into account changes in the electrical conductivity of the solvent under pressure, which for the most part will be water. Aqueous solutions have been most thoroughly investigated although they turn out to be highly anomalous.[197] The conductivity of many strong electrolytes initially increases with pressure, reaching a maximum and eventually decreasing fig. 2.23. The extent to which an increase occurs depends on

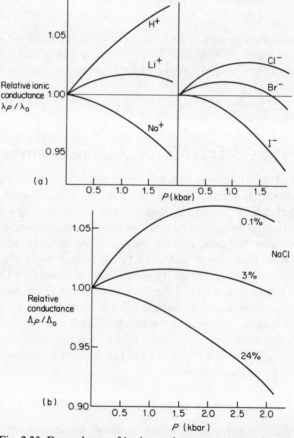

Fig. 2.23. Dependence of ionic conductance on pressure: (a) effects for different ions, (b) effect of concentration. (After Brummer and Gancy[197])

the conducting ion, its concentration, and on the temperature. An increase in conductivity is believed to be due to changes in the solvent structure, in the vicinity of the ion. The situation is additionally complicated since the single ion conductivities show different degrees of anomalous behaviour reflecting different extents of involvement in solvent structure. The small ions H^+ and Li^+ show a greater pressure dependence of conductivity than do, say, Cs^+ or I^-. The effect of pressure is also dependent upon the concentration of the electrolyte, this again depending upon the specific case, fig. 2.24, and presumably is due to ion–ion

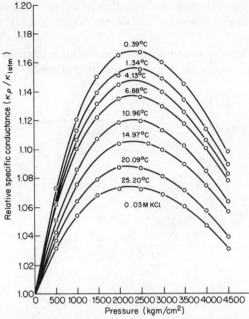

Fig. 2.24. Relative specific conductance of 0.003-M
potassium chloride as a function of pressure[201]

interactions and aggregation. The conductivities of weak electrolytes are
additionally affected by pressure-induced changes in the species present, pressure
tending to increase ionization (Chapter 3) on account of the volume diminution
with which it is accompanied. Conductivities of carboxylate ions[227] and
ammonium ions[228,231] fall with pressure. Proton conductivity is a special case
involving a charge-relay mechanism,

which appears to be facilitated by pressure.[226] It seems that the rate-determining
step is the rotation of a water molecule into the correct orientation to receive the
proton.[165] Both H_3O^+ and OH^- show an 'excess conductance' above that
expected for ions of their size due to this effect. The passage of an ion through a
medium should be governed by the Stokes–Einstein equation (eq. (2.70)) and
hence by the viscosity. In fact the product of conductivity and viscosity is
approximately constant (Walden's Rule, eq. (2.90))

$$\Lambda\eta = \text{constant} \qquad (2.90)$$

Although there is often a similarity between the pressure dependencies of Λ°
and η, fig. 2.25,[113,202] the Walden product, $\Lambda^\circ\eta$, often changes appreciably ith

124

pressure, increasing or decreasing according to the species in an unpredictable manner attributable to specific solvation which causes the ion size to vary.[231-233]

Electrical conductivity is usually described as an activation process in which the ability of charge to migrate depends upon the availability of thermal energy. It

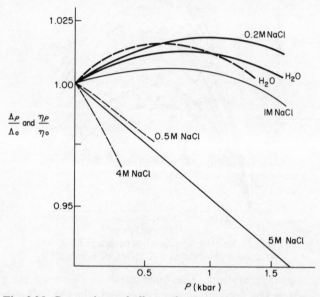

Fig. 2.25. Comparison of effects of pressure on relative molal conductivity (solid lines) and on viscosity (dashed lines) for sodium chloride solutions, 25 °C. (After Molenat[202])

is then possible to define energies, entropies, and volumes of activation as differences between these quantities in initial and transition states (Chapter 4). The volume of activation, $\Delta V^{\neq}$, for conduction is given by eq. (2.91):

$$\Delta V^{\neq} = -RT\left(\frac{\partial \ln \Lambda^{\circ}}{\partial P}\right)_T - \tfrac{2}{3}\beta_T \qquad (2.91)$$

in which the molar conductivity Λ° may be replaced by a single ion conductivity, λ°. The model assumes this volume change to comprise components for 'hole-formation' in the medium, $\Delta V_{h}^{\neq}$ and for the ionic 'jump', $\Delta V_{j}^{\neq}$ although, as Hills points out,[237] this distinction is largely artificial and is a device to account for the evident work which has to be done by the migrating ion against the volume of the solvent and the 'friction' of intermolecular motion. Activation volumes have been measured in various media (Table 2.27) and, where the data are available, appear to be of a similar magnitude to those determined from viscosity or diffusion. In water it appears that an electrolyte may enhance the structured component of water (as do NaCl or KF) or may diminish it (KI, KCl, NH₄Cl).[164]

Conductivity in non-aqueous media should be free of this complication although equilibria between free ions and ion-pairs might vitiate a simple

Table 2.27. Volumes of activation for ionic conductance, $\Delta V_{\Lambda}^{\neq}$ (eq. (2.93))

Electrolyte	Solvent	$T(°C)$	$\Delta V^{\neq}$ (cm^3 mol^{-1})	Reference
KCl	Water	20	-1.5	205, 237
		30	-0.5	
		50	0	
		75	$+0.04$	
HCl	Water	25	-2.3	
		45	-1.0	
		75	-0.42	
NaNO$_3$	MeOH	40	4	
Bu$_4$N$^+$ picrate$^-$	MeOH	40	7	
	PhNO$_2$	25	14.5	
		40	13.7	
		60	12.7	
nPr$_4$N$^+$ picrate$^-$	PhNO$_2$	25	14.2	
		40	13.4	
		60	12.7	
Et$_4$N$^+$ picrate	PhNO$_2$	40	16	206
Me$_4$N$^+$ picrate		40	16	
	Acetone			
LiCl		643	0	208
LiBr		596	0	
LiI		525	0	
NaCl		825	3.0	
NaBr		774	3.1	
NaI		706	4.8	
KCl		792	6.2	
KBr		742	7.2	
KI		716	9.7	
RbCl		797	8.0	
RbBr		802	9.1	
RbI		798	11.4	
CsCl		671	9.3	
CsBr		683	10.6	
CsI		679	13.5	
AgCl		472	1.7	
AgBr		454	1.3	
LiNO$_3$		270	0	210
NaNO$_3$		350	3.2	
KNO$_3$		350	6.0	
RbNO$_3$		350	7	
CsNO$_3$[a]		450	7	
HgCl$_2$[a]		300	-50[a]	214
HgBr$_2$		300	-65[a]	
HgI$_2$[a]		300	-105[a]	212
AlI$_3$		223	-81[a]	213
GaI$_3$		270	-117[a]	
InI$_3$		252	-11[a]	
BiI$_3$		447	-6.0[a]	
CdI$_2$		388	-0.4[a]	
I$_2$		139	-24[a]	

[a] Although measured by conductivity these values are activation volumes for ionization in the melt.

126

interpretation. Relatively little data are available in solvents other than water although qualitative early studies showed that the cell resistance of solutions of tetraethylammonium chloride in a variety of organic solvents increased with pressure, although to widely varying degrees.[234] In methanol, specific conductances of sodium bromide and piperidinium bromide become very similar as pressure is applied suggesting that increasing solvation has a levelling effect on the effective sizes of the two cations.[235,236] In dimethylformamide also the distinction between formally large (tetrabutylammonium picrate) and small (potassium bromide) ions diminishes with pressure,[207] and activation volumes, initially 11.5 and 10.5 cm^3 mol^{-1} respectively, approach 8.3 cm^3 mol^{-1} at several kbar. Isobaric and isochoric energies of activation, E_p and E_v, were determined and, from eq. (2.92),

$$(E_p^{\neq} - E_v^{\neq} \sim 5.4\,\text{kJ mol}^{-1} = \Delta V_j(P + P_i) + 3.67\,RT^2\,\alpha \qquad (2.92)$$

the jump volume, ΔV_j, was inferred as ca. 4.5 cm^3 mol^{-1}. Nitrobenzene gives conducting solutions of quaternary ammonium picrates whose conductivities fall with pressure[200,204,205] and for which $E_p^{\neq} - E_v^{\neq} \sim 8.5\,\text{kJ mol}^{-1}$. Activation volumes are considerably larger in this solvent than in methanol or acetone.[238]

Despite the experimental difficulties, ionic conductances of many fused salts have been measured. Fused alkali metal halides for example behave as simple fluids whose properties are easier to interpret than are those of solutions. Conductances of all salts fall smoothly with pressure but the activation volumes for conductance diminish through the series Cs–Li and fall with halide ion size also, a reflection of the volume displacement of neighbouring ions necessary for an ion to migrate through the fluid (Table 2.27).[208,209] Similar conclusions are found to apply to alkali metal nitrates[210-211] but the pressure effect on molten silver nitrate conduction is very much smaller than that of KNO$_3$ whose radius it resembles, $\Delta V_\Lambda^{\neq} = 1.5$ compared with 6.2 cm^3 mol^{-1}. This may reflect partial covalency in the silver salt, which diminishes with pressure. The same phenomenon is undoubtedly responsible for conduction in halides of Hg, Cd, Ga, In,

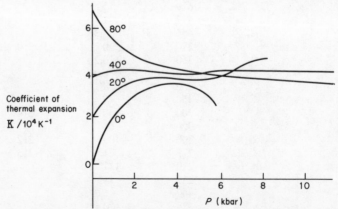

Fig. 2.26. Thermal expansivity of water showing anomalous behaviour at low temperatures. (After Lawson and Hughes[203])

and other B-group metals, which initially increases rapidly with pressure, fig. 2.28 above. This is due to ionization of the almost entirely covalent material which is greatly assisted by pressure due to electrostriction of the surrounding medium (q.v.).[212] At a sufficiently high pressure, ionization is complete and conductivity falls since transport becomes limiting. The maximum for conductivity is observed at 9.5 kbar for HgI_2 and 7 kbar for CdI_2. A similar situation is experienced with water whose self-ionization increases enormously with pressure. It becomes wholly dissociated at around $1000°$, 100 kbar expressed as

$$\log K_\omega \approx -3108/T - 3.55$$

Activation volumes measured[215] from conductivity commence highly negative (GaI_3 holds the record, $\Delta V_\Lambda^{\neq} = -136\,cm^3\,mol^{-1}$) but eventually at very high pressures become small and positive like KCl,[213, 214].

(a) $\quad 2HgX_2 \;\rightleftharpoons\; HgX^+ + HgX_3^-$

(b) $\quad 3HgX_2 \;\rightleftharpoons\; 2HgX^+ + HgX_4^=$

$$\Delta V_\Lambda^{\neq\,cm^3\,mol^{-1}} = -50 \;(X = Cl)$$
$$-65 \;(X = Br)$$
$$-105 \;(X = I)$$

The wide variation in values of $\Delta V_\Lambda^{\neq}$ for these compounds reflects the ease with which they undergo ionization, their covalent bond-strengths, and the stabilities

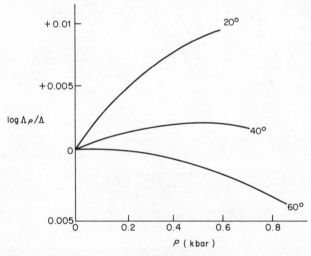

Fig. 2.27. Relative conductance of 0.002 M KCl at different temperatures

of the ions formed. The amount of electrostriction afforded by the surrounding medium also contributes and the nature of the equilibrium needs to be taken into account, i.e. the relative importance of eqs (a) and (b), since multiply-charged ions contribute to the electrostriction by a z^2 term.

Transport numbers have also been measured under pressure by a suitably adapted moving boundary method;[192, 193] values for potassium chloride are given below:

P (bar)	1	1000	2000
$t(K^+)$	0.490	0.481	0.477
$t(Cl^-)$	0.510	0.519	0.523

The trends for the effect on mobility of pressure is thus in opposite directions for anion and cation in this case but this is not necessarily general, and is not found in KI or KBr.[247]

The effect of concentration on conductivity is expressed by the Debye–Hückel–Onsager equation, (2.93)[189]

$$\Lambda = \Lambda^\infty - (B_1 \Lambda^\infty + B_2) I^{1/2} \tag{2.93}$$

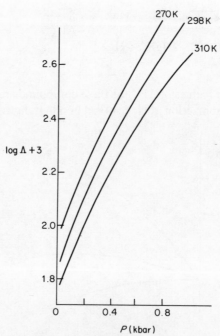

Fig. 2.28. Conductance of molten GaI_3. (After Cleaver et al.[213])

Table 2.28

P (bar)	$B_1{}^a$	$B_2{}^a$
1	0.2297	60.38
1000	0.2157	59.37
2000	0.2038	56.05

^{a} Appropriate to 1:1 electrolytes.

where B_1, B_2 are constants derived from solvent properties and temperature and I is the ionic strength. Values of B_1 and B_2 for water at 25° have been measured as a function of pressure, Table 2.28.

2.4.2 The Effect of Pressure on Cell Potentials

An electrolytic cell e.m.f. is set up by a difference in chemical potential between reagents and products in a reversible reaction involing electron transfer. It is likely to be pressure dependent since the chemical potentials are, and the pressure coefficient will be, given by

$$\left(\frac{\partial E}{\partial P}\right)_T = \frac{1}{nF} \cdot \left(\frac{\partial \Delta G}{\partial P}\right)_T = -\frac{1}{nF}(\textstyle\sum^{\text{prods}} \bar{V} - \sum^{\text{rgts}} \bar{V}) \tag{2.94}$$

This is simply a case of equilibrium, the theory of which is dealt with in Chapter 3. The volume change for the cell reaction, then, determines the variation of e.m.f. with pressure. It is relatively easy to construct a suitable apparatus for investigating e.m.f. under pressure[237] but if liquid junction potentials are present the significance of the results may be clouded by uncertainty as to the effect of pressure on this. For the cell,[239]

$$\text{Tl, Hg} \mid \text{TlCNS, KCNS} \parallel \text{KCl, TlCl} \mid \text{Tl, Hg}$$
$$(\text{Tl}^+ + \text{Hg} \; \rightleftarrows \; \text{Hg}_2^{2+} + \text{Tl})$$

$\partial E/\partial P = 3 \times 10^{-6}$ V bar^{-1}, so the effect is quite small. The coefficient of pressure for

$$\text{Pt, H}_2 \mid \text{H}_3\text{O}^+ \mid \text{Hg}_2\text{Cl}_2 \text{ Hg}$$
$$(\text{H}_2 \; \rightleftarrows \; 2\text{H}^+ + 2\text{e}^-)$$

is much larger, ca. 100×10^{-6} V bar^{-1} but this is largely attributable to an increase in hydrogen solubility and hence activity.[240] If a_{H_2} is maintained at a fixed value it becomes 8×10^{-6} V bar^{-1}.[241] The glass electrode may be used under pressure[242] and adapted to measurement of dissociation volumes of weak acids and bases from concentration cells (Chapter 3).

2.4.3 Electrode Processes

If an electrochemical reaction is allowed to occur at a charged surface, a rate, k_F, may be defined

$$k_F = k_F^0 \exp\left(-\frac{a\omega nF}{RT}\right) \tag{2.95}$$

where a is an asymmetry factor and ω the overpotential used. Application of transition state formalism (chapter 4) to this gives,

$$\frac{d \ln k_F}{dP} = -\frac{\Delta V_F^{\neq}}{RT} \tag{2.96}$$

where $\Delta V_F^{\neq}$ is the volume difference between reagents and transition state. However, it is difficult to assign a precise value to $\Delta V_F^{\neq}$ because of its dependence on the overpotential. The sign may be taken as the result of change in charge during the electrode process. The rate of discharge of H^+ is decreased by pressure ($\Delta V_F^{\neq} \sim +5\,cm^3\,mol^{-1}$) which is a result of relaxation of electrostriction.

A certain amount of polarography has been carried out under pressure. The height of the current step in the discharge wave for Cu^{++} is reduced by pressure indicating increasing difficulty in reducing the ion. This may in part be due to an increase in the dielectric constant but also a decrease in the ionic charge as seen by the equation:[244]

$$Cu^{++} + 2Cl^- + Hg \longrightarrow CuCl_2^- + Hg^+$$

Changes in the diffusion rates of the ions may affect the nature of a polarogram, and also the effects of pressure on adsorption and other complex interactions associated with the metal-solution interface—the electric double layer.[245] On the other hand, other workers have found little or no dependence of limiting current on pressure for reductions of Tl^+ Zn^{++}, Cd^{++} H^+, Cu^{++} Ni^{++} and In^{++} under conditions up to 3 kbar and at elevated temperatures.[250-252]

REFERENCES

1. A. D. Buckinham, E. Lippert, and S. Bratos (Eds.), *Organic Liquids*, John Wiley, New York (1975).
2. E. H. Amagat, *Comptes. rend.*, **108**, 228, (1889).
3. P. W. Bridgman, *Proc. Am. Acad. Arts. Sc.*, **74**, 21 (1942).
4. P. W. Bridgman, *Proc. Am. Acad. Arts. Sc.*, **74**, 425 (1942).
5. P. W. Bridgman, *Proc. Am. Acad. Arts. Sc.*, **77**, 115 (1949).
6. P. W. Bridgman, *Phys. Rev.*, **60**, 351 (1941).
7. P. W. Bridgman, *The Physics of High Pressures*, Bell, London (1949).
8. R. E. Gibson and J. F. Kincaid, *J. Amer. Chem. Soc.*, **60**, 511 (1938).
9. R. E. Gibson and O. H. Loeffler, *J. Amer. Chem. Soc.*, **61**, 2515 (1939).
10. R. E. Gibson, *J. Phys. Chem.*, **43**, 207 (1939).
11. R. E. Gibson and O. H. Loeffler, *J. Amer. Chem. Soc.*, **63**, 898 (1941).
12. K. E. Weale, *J. Chem. Soc.*, 794 (1938).
13. H. E. Edjulee, D. M. Newitt, and K. E. Weale, *J. Chem. Soc.*, 3086 (1951).
14. A. T. J. Haywood, *J. Phys. (D)*, **4**, 938 (1971).
15. E. H. Amagat, *Ann. chim. phys.*, **29**, 68, 505 (1893).
16. D. Tyrer, *J. Chem. Soc.*, 1675 (1913).
17. D. Tyrer, *J. Chem. Soc.*, 2534 (1914).
18. B. C. Sakladis and J. Coates, *Bulletin No. 46*, Louisiana State Univ. Engineering Exptl. Station, Baton Rouge, La. (1954).
19. G. Tammann and W. Jellinghaus, *Z. anorg. chem.*, **174**, 225 (1928).
20. M. D. Pena and M. L. McGlashan, *Trans. Farad. Soc.*, **55**, 2018 (1959).
21. P. W. Bridgeman, *Proc. Am. Acad. Arts. Sc.*, **48**, 309 (1912).
22. G. S. Kell and E. Whalley, *Phil. Trans.*, **258**, 565 (1965).
23. L. H. Adams, *J. Amer. Chem. Soc.*, **53**, 3769 (1931).
24. J. H. Hyde, *Proc. Roy. Soc. (A)*, **97**, 2408 (1920).
25. L. A. K. Staveley, W. I. Tupman, and K. R. Hart, *Trans. Farad. Soc.*, **51**, 2415 (1959).
26. P. W. Bridgman, *J. Chem. Phys.*, **3**, 597 (1935).

27. W. G. Cutler, R. H. Miekle, W. Webb, and R. W. Schiessler, *J. Chem. Phys.*, **29**, 727 (1958).
28. M. Millet and G. Jenner, *J. chim. phys.*, **67**, 1667, 1766 (1970).
29. W. M. Madigosky, *Rev. Sci. Instr.*, **37**, 227 (1966).
30. A. T. J. Hayward, *J. Phys. (D), Appl. Phys.*, **4**, 951 (1965).
31. P. W. Bridgeman, *Proc. Amer. Acad. Arts. Sc.*, **74**, 399 (1942).
32. H. W. Schamp, J. R. Hastings, and S. Wussman, *Phys. Fluids*, **8**, 8 (1965).
33. G. Aime, *Ann. chim. phys.*, **8**, 257 (1843).
34. D. M. Newitt and K. E. Weale, *J. Chem. Soc.*, 3092 (1951).
35. L. H. Adams, *J. Amer. Chem. Soc.*, **53**, 3769 (1931).
36. H. E. Edjulee, D. M. Newitt, and K. E. Weale, *J. Chem. Soc.*, 3086 (1951).
37. E. B. Freyer, J. C. Hubbard, and D. H. Andrews, *J. Amer. Chem. Soc.*, **51**, 7598 (1929).
38. J. C. Hubbard and A. L. Loomis, *Phil. Mag.*, **5**, 117 (1928).
39. L. A. Davis and R. B. Gordon, *J. Chem. Phys.*, **46**, 2650 (1967).
40. P. W. Bridgman, *Rev. Mod. Phys.*, **7**, 1 (1935).
41. P. I. Gold, *Chem. Eng. (N.Y.)* June 30, (1969), p. 129.
42. Y. Wada, *J. Phys. Soc. Jap.*, **4**, 280 (1949).
43. R. V. G. Rao, *Current Sci. (India)*, **9**, 584 (1940).
44. R. V. G. Rao and J. C. M. Li, *Z. physik (Leipzig)*, **213**, 166 (1960).
45. M. R. Rao, *J. Chem. Phys.*, **14**, 699 (1946).
46. H. Reiss, H. L. Frisch, and G. Lebowitz, *J. Chem. Phys.*, **31**, 369 (1959).
47. N. S. Sneider and T. M. Herrington, *J. Chem. Phys.*, **47**, 2248 (1967).
48. R. M. Gibbons, *Mol. Phys.*, **18**, 809 (1970).
49. M. B. Ewing and K. N Marsh, *J. Chem. Thermodyn.*, **9**, 357, 371, 651, 863 (1977).
50. P. G. Tait, Voyage of *HMS Challenger, Reports of Physics and Chemistry*, **2**, iv, 76 (1888).
51. P. G. Tait, *Ann. Phys. Beibl.*, **13**, 442 (1889).
52. P. G. Tait, *Scientific papers*, Cambridge University, **2**, 1 (1900).
53. A. Z. Wahl, *Z. phys. Chem.*, **99**, 234 (1921).
54. G. A. Holder and E. Whalley, *Trans. Farad. Soc.*, **58**, 2095 (1962).
55. R. Ginell, *J. Chem. Phys.*, **34**, 1249 (1961).
56. L. J. Hudleston, *Trans. Farad Soc.*, **33**, 97 (1937).
57. K. E. Bett, K. E. Weale, and D. M. Newitt, *Brit. J. Appl. Phys.*, **5**, 243 (1954).
58. M. Millet and G. Jenner, *J. Chim. Phys. Physico-Chim. Biol.*, **67**, 1667, 1766 (1970). J. R. McDonald, *Rev. Mod. Phys.*, **38**, 669 (1966).
59. J. R. Partington, *Advanced Treatise on Physical Chemistry*, Vol. II, (1951).
60. D. Harrison and E. A. Moelwyn-Hughes, *Proc. Roy. Soc.*, **239A**, 230 (1957).
61. O. L. Anderson, *J. Phys. Chem.*, **27**, 547 (1966).
62. J. R. Macdonald, *Rev. Mod. Phys.*, **38**, 669 (1966).
63. E. Whalley, *Experimental Thermodynamics Vol. II*, I.U.P.A.C.; Butterworth, London (1968).
64. O. Tumlirz, *Schr. Balkankomm. Akad. Wiss, Wien*, **118**, 293 (1909).
65. G. Tammann and E. Schwartzkopf, *Z. anorg. algem. Chem.*, **174**, 215 (1928).
66. T. Peczalski, *Compt. rend.*, **157**, 584, 770 (1913).
67. E. Wilhelm, R. Schano, G. Becker, G. H. Findenegg, and F. Kohler, *Trans. Farad. Soc.*, **65**, 1443 (1969).
68. E. Wilhelm, *Monatsh. Chem.*, **105**, 291 (1974).
69. E. Wilhelm, *J. Chem. Phys.*, **63**, 3379 (1975).
70. M. D. Pena and G. Tardajos, *J. Chem. Thermod.*, **10**, 19 (1975).
71. M. B. Ewing and K. N. Marl, *J. Chem. Thermod.*, **10**, 267 (1978).
72. G. J. Hills, *Soc. Exp. Biol. Symposium XXVI* Cambridge University Press (1972).
73. J. D. Bernal and R. Fowler, *J. Chem. Phys.*, **1**, 515 (1933).
74. P. A. Egelstaff, *Ann. Rev. Phys. Chem.*, 159 (1973).
75. J. S. Dohler and J. O. Hurschfelder, *J. Chem. Phys.*, **32**, 330 (1960).

132

76. V. S. Venkatasubramanian, *J. Chem. Phys.*, **26**, 965 (1957).
77. A. Wohl, *Z. physik. Chem.*, **99**, 234 (1921).
78. R. Ginell, *J. Chem. Phys.*, **34**, 1249 (1961).
79. R. E. Gibson and O. H. Loeffler, *J. Amer. Chem. Soc.*, **61**, 2877 (1939).
80. G. S. Kell, *J. Chem. Eng. Data*, **12**, 67 (1967).
81. G. Borelius and A. Saudin, *Arkiv. Fysik.*, **10**, 187 (1955).
82. G. Borelius, *Arkiv Fysik*, **16**, 413 (1960); **28**, 499 (1965); **30A**, 267 (1969).
83. G. Borelius, *Solid State Physics*, **6**, 65 (1958).
84. M. H. Rice, R. G. McQueen, and J. M. Walsh, *Solid State Phys.*, **6**, 1 (1958).
85. J. H. Hildebrand, *Phys. Rev.*, **34**, 984 (1929).
86. W. Westwater, H. W. Frantz, and J. H. Hildebrand, *Phys. Rev.*, **31**, 135 (1928).
87. P. J. Trotter, *J. Amer. Chem. Soc.*, **88**, 5721 (1966).
88. T. Biron, *J. Russ. Chem. Soc.*, **44**, 1264 (1912).
89. T. W. Richards and J. H. Mathews, *Z. physik. Chem.*, **61**, 449 (1908).
90. J. C. McGowan, *Rec. Trav. Chim.*, **76**, 155 (1957).
91. J. R. Partington, *Advanced Treatise on Physical Chemistry*, Vol. 2, p. 98 (1951).
92. S. Sugden, *J. Chem. Soc.*, **125**, 32 (1924).
93. F. Partington, *Advanced Treatise on Physical Chemistry*, Vol. 2, 161, Longmans, London (1951).
94. G. A. Holder and E. Whalley, *Can. J. Chem.*, 2095 (1962).
95. J. C. McGowan, *Rec. Trav. Chim.*, **75**, 193 (1956).
96. C. P. Smyth, *Dielectric Behaviour and Structure*, McGraw-Hill, New York (1955).
97. W. E. Danforth, *Phys. Rev.*, **38**, 1224 (1931).
98. B. Bowen, R. C. Miller, C. E. Milner, and H. L. Cogan, *J. Phys. Chem.*, **65**, 2065 (1961).
99. J. W. Stewart, *J. Phys. Chem.*, **40**, 3297 (1964).
100. L. G. S. Scharnack and C. A. Eckert, *J. Phys. Chem.*, **74**, 3014 (1970).
101. J. S. Rosen, *J. Opt. Soc. Amer.*, **37**, 932 (1947).
102. R. M. Waxler and C. E. Weir, *J. Res. Nat. Bur. Stand.*, **67A**, 163 (1963).
103. W. C. Rontgen and L. Zehnder, *Ann. Physik.*, **44**, 24 (1891); F. Himstedt and I. Wertheimer, *Ann. Physik*, **67**, 395, (1922); I. Eisele, *Ann. Physik*, **76**, 396 (1925).
104. V. Raman and K. S. Venkataraman, *Proc. Roy. Soc.*, **171A**, 137 (1939).
105. R. E. Gibson and J. F. Kincaid, *J. Amer. Chem. Soc.*, **60**, 511 (1938).
106. S. S. Kurtz and A. L. Ward, *J. Franklin Inst.*, **222,** 563 (1936); **224**, 583 (1937).
107. J. F. Eykman, *Rec. trav. chim.*, **14**, 201 (1895).
108. T. C. Poulter, C. Richey, and C. A. Benz, *Phys. Rev.*, **41**, 366 (1932).
109. J. S. Rosen, *J. Chem. Phys.*, **17**, 1192 (1949).
110. I. S. Jacobs and A. W. Lawson, *J. Chem. Phys.*, **20**, 1162 (1952).
111. F. E. Poindexter and J. S. Rosen, *Phys. Rev.*, **45**, 760 (1934).
112. F. E. Poindexter and J. S. Rosen, *Phys. Rev.*, **47**, 202 (1935).
113. F. E. Poindexter and J. S. Rosen, *Phys. Rev.*, **43**, 760 (1933).
114. B. B. Owen and S. R. Brinkley, *Phys. Rev.*, **64**, 32 (1943).
115. E. Whalley, *Adv. High Pres. Res.*, **1**, 143 (1966).
116. H. Hartmenn, A. Newmann, and A. P. Schmidt, *Ber. Bunsenges*, **72**, 877 (1968).
117. B. B. Owen and S. R. Brinkley, *J. Chem. Ed.*, **21**, 59 (1944).
118. E. Whalley, *Adv. Phys. Org. Chem.*, **2**, 93 (1964).
119. G. P. Johari and W. Dannhauser, *J. Chem. Phys.*, **48**, 5114 (1968).
120. G. P. Johari and W. Dannhauser, *J. Chem. Phys.*, **50**, 1862 (1969).
121. G. P. Johari and W. Dannhauser, *J. Chem. Phys.*, **51**, 1626 (1969).
122. G. P. Johari and W. Dannhauser, *Phys. Chem. Liquids*, **3**, 1 (1972).
123. H. Hartmann and A. Newmann, *Zeit. phys. Chem. N.F.*, **44**, 204 (1965).
124. B. J. Alder, E. W. Hatcock, J. H. Hildebrand, and H. Watts, *J. Chem. Phys.*, **22**, 1060 (1954).
125. S. D. Hamann, *Mod. Aspects of Electrochem.*, **9**, 47 (1974).

126. B. C. Sakiadis and J. Coates, *Bulletin No. 46*, Lousiana State Univ. Engineering Exptl. Station, Baton Rouge, La. (1954).
127. W. J. LeNoble and T. Asano, *J. Amer. Chem. Soc.*, **97**, 1778 (1975).
128. P. Drude and W. Nernst, *Z. physik. Chem.*, **15**, 79 (1894).
129. W. J. LeNoble, *Prog. Phys. Org. Chem.*, **5**, 302 (1973).
130. P. Mukerjee, *J. Phys. Chem.*, **65**, 740, 744 (1961).
131. E. Gluekamf, *Trans. Farad. Soc.*, **61**, 914 (1965).
132. S. W. Benson and C. S. Copeland, *J. Phys. Chem.*, **67**, 1194 (1963).
133. M. B. Myers and K. Vedam, *J. Opt. Soc. Amer.*, **55**, 1180 (1965).
134. E. M. Greist, W. Webb, and R. W. Schersler, *J. Chem. Phys.*, **29**, 711 (1958).
135. M. D. Hersey and H. S. Shore, *Mech. Eng.*, **50**, 221 (1928).
136. B. H. Sage, *Ind. Eng. Chem. Anal. Ed.*, **5**, 261 (1933).
137. S. E. Babb and G. V. Scott, *J. Chem. Phys.*, **40**, 3666 (1966).
138. E. Kuss, *Chem-ingr. Tech.*, **37**, 465 (1965).
139. D. W. McCall, D. C. Douglass, and E. W. Anderson, *J. Chem. Phys.*, **31**, 1555 (1959).
140. W. Gambill, *Chem. Eng.*, **66**(3), 123 (1959).
141. H. H. Reamer, G. Cokelet, and B. H. Sage, *Anal. Chem.*, **31**, 1422 (1959).
142. W. Weber, *Rheol. Acta.*, **14**, 1012 (1975).
143. R. A. Horne and R. A. Courant, *J. Phys. Chem., Ithaca*, **69**, 2224 (1965).
144. R. W. Schiessler, *Amer. Chem. Soc. Div. Petrol. Chem. preprints*, **1**, No. 4 (1956).
145. B. Kauzel, *Hydrocarbon Process. Petrol. Refiner.*, **44**, 120 (1965).
146. L. T. Carmichael, V. Berry, and B. H. Sage, *J. Chem. Eng. Data*, **9**, 411 (1964).
147. G. P. Fresco, E. E. Klaus, and E. J. Tewkesbury, *J. Lubric. Technol.*, **91**, 451 (1969).
148. G. M. Panchenkov, *J. Phys. Chem. USSR*, **21**, 187 (1947).
149. H. G. de Carvalho, *Anais. accoc. quim. Brasil*, **8**, 49 (1949).
150. E. Kuss, *Z. angew. Phys.*, **7**, 372 (1955).
151. R. T. Sanderson, *Mech. Eng.*, **71**, 349 (1949).
152. G. M. Panchenkov, *Doklady Akad. Nauk, SSSR*, **63**, 701 (1948).
153. A. K. Mukherjee, *Trans. Ind. Inst. Chem. Engrs.*, **2**, 36 (1948).
154. A. Bondi, *J. Chem. Phys.*, **15**, 527 (1947).
155. A. Bondi, *J. Chem. Phys.*, **14**, 592 (1946).
156. E. Kuss, *Angew. Chem., Int. Ed.*, **4**, 944 (1965).
157. S. D. Hamann, *Trans. Farad. Soc.*, **54**, 507 (1958).
158. R. W. Schiessler, *Proc. Amer. Petr. Inst.*, **26**(3), 254 (1946).
159. W. Weber, *Rheol. Acta*, **14**, 1012 (1975).
160. G. B. Benedek and E. M. Purcell, *J. Chem. Phys.*, **22**, 2003 (1954).
161. N. H. Naelitreib and J. Pekt, *J. Chem. Phys.*, **24**, 746 (1956).
162. R. B. Cuddeback, R. C. Koeller, and H. G. Drickamer, *J. Chem. Phys.*, **21**, 589 (1953).
163. R. B. Cuddeback, R. C. Koeller, and H. G. Drickamer, *J. Chem. Phys.*, **21**, 597 (1953).
164. G. J. Hills, *Soc. Exptl. Biol. Symp. XXVI*, Chapter 1, Cambridge University Press (1972).
165. K. M. Horne, B. R. Myers, and G. R. Frysinger, *J. Chem. Phys.*, **39**, 2666 (1963).
166. P. W. Bridgman, *Proc. Amer. Acad. Arts Sc.*, **59**, 141 (1923).
167. L. Riedel, *Chem-ingr., Tech.*, **23**, 321 (1951).
168. J. M. Lenoir, *Petr. Refiner.*, **36**, 162 (1957).
169. Inst. Chem. and Mech. Engrs., *Engineering Sciences Data Iten 750009*, Engineering Sciences Data Unit, London, June (1975).
170. D. W. McCall, D. C. Douglas, and E. W. Anderson, *J. Chem. Phys.*, **31**, 1555 (1959).
171. F. J. Millero, G. Ward, F. K. Lepple, and E. V. Hoff, *J. Phys. Chem.*, **78**, 1636 (1974).
172. F. J. Millero, R. W. Curry, and W. Drost-Hansen, *J. Chem. Eng. Data*, **14**, 422 (1969).
173. G. S. Kell and E. Whalley, *Phil. Trans. Roy. Soc.*, **258**, 565 (1965).
174. G. S. Kell, *J. Chem. Eng. Data*, **15**, 119 (1970).
175. G. Gotze, R. Jockers, and G. M. Schnieder, *Proc. 4th Int. Conf. on Chem. Thermod.*, 57, Montpellier (1975).

134

176. P. Engels and G. M. Schneider, *Ber. Bunsenges.*, **76**, 1239 (1972).
177. M. Kamphausen, *Rev. Sci. Instr.*, **46**, 837 (1975).
178. M. Kamphausen, G. M. Schneider, W. Spratte, and A. Wurflinger, *Proc. 4th Int. Conf. Chem. Thermod.*, **6**, Montpellier (1975).
179. F. A. Lindemann, *Physik. Z.*, **11**, 609 (1910).
180. K. Mukherjee, *Phys. Rev. Letters*, **17**, 1252 (1966).
181. E. A. Krout and G. C. Kennedy, *Phys. Rev.*, **151**, 668 (1966).
182. S. E. Babb, *Phys. Rev. Letters*, **17**, 1250 (1966).
183. L. E. Reeves, G. J. Scott, and S. E. Babb, *J. Chem. Phys.*, **40**, 3662 (1964).
184. Landolt-Bornstein, *Tables of Physical Constants*, 6th ed., Vol. II, part 2, Springer Verlag, Berlin.
185. N. Fletcher, *Chemical Physics of Ice*, Cambridge University Press, (1970).
186. *International Critical Tables*, Vol. II; McGraw-Hill, (1929).
187. F. Simon and Glatzel, *Z. anorg. Chem.*, **178**, 309 (1929).
188. G. M. Schnieder, *Adv. Chem. Phys. XVII*, 1 (1970).
189. A. D. Edgar and R. G. Platt, *High Temp. High Pres.*, **3**, 1 (1971).
190. I. D. McGregor and A. E. Ringwood, *Carnegie Inst. Washington Year Book*, **63**, 161 (1964).
191. S. P. Clark, *Geol. Soc. Amer. Memoir*, **97**, sect. 15 (1966).
192. B. C. Tofield, *Nature*, **272**, 714 (1978).
193. J. E. J. Martini, *Nature*, **272**, 715 (1978).
194. J. B. Goodenough, J. A. Kafalas, and J. M. Lange, *Preparative Methods in Solid State Chemistry*, Academic Press, New York (1972).
195. E. Cohen, *Physico-Chemical Metamorphosis and Problems in Piezo-Chemistry*, McGraw-Hill, New York (1926).
196. S. D. Hamann, *Physico-Chemical Effects of Pressure*, Butterworth, London (1957).
197. S. D. Brummer and A. D. Gancy, *Water and Aqueous Solutions*, R. A. Horne (Ed.), John Wiley, New York (1972).
198. S. B. Brummer and A. B. Gancy, *Electrochem. Symp.* (1972), J. B. Berkowitz (Ed.), *Electrochemical Soc.*, Princeton, N.J. p. 76 (1973).
199. A. G. Stearne and H. Eyring, *J. Chem. Phys.*, **5**, 113, (1937).
200. S. D. Brummer and G. J. Hills, *Trans. Farad. Soc.*, **57**, 1816, 1823, (1961).
201. R. A. Horne and R. P. Young, *J. Phys. Chem.*, **71**, 3824 (1967).
202. J. Molenat, *J. chim. phys.*, **67**, 368 (1970).
203. A. W. Lawson and A. J. Hughes, *High Pressure Physics and Chemistry*, R. S. Bradley (Ed.), **I**, 207, Academic Press, New York (1963).
204. F. Barreira and G. J. Hills, *Trans. Farad. Soc.*, **64**, 1359 (1968).
205. S. B. Brummer and G. J. Hills, *Trans. Farad. Soc.*, **57**, 1826 (1961).
206. P. M. Chandhuri, G. P. Mather, R. A. Stayer and G. Long, *Amer. Inst. Chem. Eng. J.* **17**, 1003 (1971).
207. S. B. Brummer, *J. Chem. Phys.*, **42**, 1636 (1965).
208. B. Cleaver, S. I. Smedley, and P. N. Spencer, *J. Chem. Soc., Faraday I*, **68**, 1720 (1972).
209. J. E. Bannard and G. J. Hills, *High Temp. High Pres.*, **1**, 571 (1969).
210. A. F. M. Barton, B. Cleaver, and G. J. Hills, *Trans. Farad Soc.*, **64**, 208 (1968).
211. J. E. Bannard, A. F. M. Barton, and G. J. Hills, *High Temp. High Pres.*, **3**, 65 (1971).
212. J. E. Bannard and G. Treiber, *High Temp. High Pres.*, **5**, 177 (1973).
213. B. Cleaver, P. N. Spencer, and M. A. Quddus, *J. Chem. Soc. Faraday I*, **74**, 686 (1978).
214. B. Cleaver and S. I. Smedley, *Trans. Farad. Soc.*, **67**, 1115 (1971).
215. W. Holtzapfel, *J. Chem. Phys.*, **50**, 4424 (1969).
216. R. M. Waxler and L. H. Adams, *J. Res. Nat. Bur. Stand*, **65A**, 283 (1961).
217. T. S. Brun, H. Høiland, and E. Vikingstad, *J. Coll. Interface. Soc.*, **63**, 89 (1978).
218. A. E. Stearne and H. Eyring, *Chem. Rev.*, **41**, 509 (1941).
219. B. Cleaver and C. Herdlicka, *J.C.S. Faraday I*, **72**, 1861 (1976).
220. B. Cleaver, B. C. J. Neil, and P. N. Spencer, *Rev. Sci. Instr.*, **42**, 578 (1971).

221. H. Høiland, J. A. Ringseth, and E. Vikingstad, *J. Sol. Chem.*, **7**, 515 (1978).
222. O. Kiyohara, J-P. E. Grolier, and G. C. Benson, *Can. J. Chem.*, **52**, 2287 (1974).
223. R. J. Fort and W. R. Moore, *Trans. Farad. Soc.*, **61**, 2102 (1965).
224. S. Cabani, G. Conti, and E. Matteoli, *J. Sol. Chem.*, **8**, 11 (1979).
225. S. Harada and T. Nakagawa, *J. Sol. Chem.*, **8**, 267 (1979).
226. M. Nakahara and J. Osugi, *Rev. Phys. Chem. Jap.*, **47**, 1 (1977).
227. M. Nakahara and J. Osugi, *Rev. Phys. Chem. Jap.*, **45**, 1 (1975).
228. M. Ueno, M. Nakahara, and J. Osugi, *Rev. Phys. Chem. Jap.*, **45**, 9, 17 (1975).
229. J. Osugi, M. Nakahara, Y. Matubara, and K. Shimizu, *Rev. Phys. Chem. Jap.*, **46**, 7, 18 (1976).
230. F. T. Wall and J. Berkowitz, *J. Phys. Chem.*, **62**, 87 (1958).
231. M. Ueno, *Rev. Phys. Chem. Jap.*, **45**, 61 (1975).
232. E. Inada, *Rev. Phys. Chem. Jap.*, **46**, 19 (1976).
233. E. Inada, *Rev. Phys. Chem. Jap.*, **48**, 72 (1978).
234. E. W. Schmidt, *Z. phys. Chem.*, **75**, 305 (1911).
235. S. D. Hamann and W. Strauss, *Trans. Farad. Soc.*, **51**, 1684 (1955); S. D. Strauss, *Austr. J. Chem.*, **10**, 359 (1957).
236. S. D. Hamann and W. Strauss, *Disc. Farad. Soc.*, **22**, 70 (1956).
237. G. J. Hills and P. J. Ovenden, *Adv. Electrochem. Electrochem. Engring*, **4**, 185 (1966).
238. I. Ishihara, *Rev. Phys. Chem. Jap.*, **47**, 102 (1977).
239. E. Cohen and K. Piebenbrock, *Z. physik. Chem.*, **170A**, 145 (1934).
240. W. R. Hainsworth, H. J. Rowley, and D. A. McInnes, *J. Amer. Chem. Soc.*, **46**, 1437, (1924).
241. G. J. Hills and D. R. Kinnibrugh (quoted in ref. 237).
242. A. Distèche, *Soc. Exptl. Biol. Symp. XXVI*, Cambridge University Press (1972).
243. G. J. Hills, *Adv. High Pres. Res.*, Vol. 2, R. S. Bradley (Ed.), 225, Academic Press, New York (1969).
244. A. H. Ewald and S. C. Lim, *J. Phys. Chem.*, **61**, 1443 (1957).
245. G. J. Hills, *J. Phys. Chem.*, **73**, 3591 (1969).
246. L. H. Adams and R. E. Hall, *J. Phys. Chem.*, Ithaca, **35**, 2145 (1931).
247. J. Osugi, M. Nakahara, Y. Matsubara, and K. Shimizu, *Rev. Phys. Chem. Jap.*, **45**, 23 (1975).
248. E. Whalley, 'The compression of liquids', *Experimental Thermodynamics*, Vol. II, I.U.P.A.C., Butterworth, London (1975).
249. E. Whalley, 'The compression of liquids', in *Proc. Nobel Symposium: 'Chemistry and Geochemistry of Solutions at High Temperatures and Pressures'*, Butterworth, London (1979).
250. S. G. Mairanovskii, M. G. Gonikberg, and A. A. Opekunov, *Dokl. Akad. Nauk. SSSR* **123**, 312 (1958).
251. G. J. Hills and R. Payne, *Trans. Farad. Soc.*, **61**, 316, 326 (1965).
252. K. Hayashi and I. Kono, *Jap. J. Physiol.*, **8**, 246 (1958).
253. P. Pruzan and L. Ter Minassian, *Proc. 6th AIRAPT Conf.* (Boulder), 368, Plenum, London (1979).
254. M. Uematsu and E. U. Franck, *Proc. 6th AIRAPT Conf.* (Boulder), 415, Plenum, London (1979).
255. K. Vedam and P. Limsuwan, *Proc. 6th AIRAPT Conf.* (Boulder), 421, Plenum, London, (1979).
256. A. Keane, *Austr. J. Phys.*, **7**, 322 (1954).
257. K. Susuki, Y. Taniguchi, and M. Tsuchiya, *Proc. 6th AIRAPT Conf.* (Boulder), 548, Plenum, London (1979).
258. A. L. Ruoff and L. C. Chhabildas, *Proc. 6th AIRAPT Conf.* (Boulder), 19, Plenum, London (1979).

Chapter 3

Pressure Effects on Equilibrium Processes

3.1 THE THERMODYNAMICS OF CHEMICAL EQUILIBRIA[1,2]

If reagents A, B ... are in equilibrium with X, Y ... as expressed by the equation

$$v_A A + v_B B ... \underset{\longleftarrow}{\overset{\longrightarrow}{}} v_X X + v_Y Y ... \tag{3.1}$$

where v_i are the stoichiometric numbers of molecules so related, the thermodynamic expression of equilibrium requires equality of free energy on either side of the equation. Thus

$$v_A \mu_A + v_B \mu_B ... = v_X \mu_X + v_Y \mu_Y ... \tag{3.2}$$

or

$$\sum v_i \mu_i = \sum v_j \mu_j \tag{3.3}$$

where μ_i, μ_j are chemical potentials (molar free energies), the summation being made on each separate side of the equation. We may replace μ_i by $RT \ln \bar{a}_{i(j)}$ where $\bar{a}$ is the absolute activity of the species, whence

$$\frac{\prod \bar{a}_j^{v_j}}{\prod \bar{a}_i^{v_i}} = \frac{\bar{a}_x^{v_x} \cdot \bar{a}_y^{v_y} \cdots}{\bar{a}_A^{v_A} \cdot \bar{a}_B^{v_B} \cdots} = 1 \tag{3.4}$$

More adaptable to chemical problems is the relative activity of a species, a_i, which is related to the chemical potential, thus

$$\mu_i = \mu_i^0 + RT \ln a_i \tag{3.5}$$

μ_i^0 being the chemical potential of pure i in its standard state and is a constant, a property of i.

For dilute solution in which Raoult's law is obeyed, we may replace a_i by the mole fraction x_i where

$$x_i = \frac{n_i}{n_{\text{total}}}$$

and n_i is the number of moles of the ith species and n_{total} the sum total of the number of moles of all species present in solution including solvent. Thus

$$\mu_i = \mu_i^0 + RT \ln x_i \tag{3.6}$$

or, in more concentrated solution where significant deviations to Raoult's Law are observed,

$$\mu_i = \mu_i^0 + RT \ln f_i x_i \tag{3.7}$$

where f_i are activity coefficients appropriate to ideal mixtures, therefore

$$\frac{a_X^X a_Y^Y \cdots}{a_A^A a_B^B \cdots} \exp \frac{(v_A \mu_A^0 + v_B \mu_B^0 \cdots - v_X \mu_X^0 + v_Y \mu_Y^0 \cdots)}{RT} = K_x \tag{3.8}$$

where K_x is the equilibrium constant in mole fraction units, or

$$\frac{\prod (x_j \cdot f_j)^{v_j}}{\prod (x_i \cdot f_i)^{v_i}} = K_x \tag{3.9}$$

and this is the usual expression of the chemical equilibrium constant, which is not, strictly speaking, constant but depends on temperature and pressure. The question now arises as to how K may be affected by pressure.[3,4] The temperature dependence is shown in eq. (3.10), derived from the Gibbs–Helmholtz equation,

$$\left(\frac{d \ln K}{dT} \right)_P = -\frac{\Delta H}{RT^2} \tag{3.10}$$

The pressure dependence may be derived as follows: since

$$\left(\frac{\partial \mu_i^0}{\partial P}\right)_T = V_i^0 \tag{3.11}$$

(section 2.1.12) where V_i^0 is the molar volume of i in its pure, standard state, and

$$-RT\frac{\partial \ln K_x}{\partial P} = v_X\frac{\partial \mu_X^0}{\partial P} + v_Y\frac{\partial \mu_Y^0}{\partial P} \ldots - v_A\frac{\partial \mu_A^0}{\partial P} - v_B\frac{\partial \mu_B^0}{\partial P} \tag{3.12}$$

$$= v_X\,\partial V_X^0 + v_Y\,\partial V_Y^0 \ldots - v_A\,\partial V_A^0 - v_B\,\partial V_B^0 \ldots \tag{3.13}$$

$$-RT\frac{\partial \ln K_x}{\partial P} = \Delta V^0 \tag{3.14}$$

where ΔV^0 is the difference in molar volumes of products over reagents, i.e. the volume change if pure reagents reacted to completion at constant pressure (1 bar) and temperature.

In dilute solution the analogous expression, (3.14), holds: it is not convenient to use mole fractions since values will be very small and not easily obtained with high accuracy. It is best, however, to choose a scale which is independent of pressure (and volume) such as molality, m_i, defined as the number of moles of the ith components per kg of solvent whence

$$m_i = x_i \cdot \frac{1000}{M_s \cdot x_s} \tag{3.15}$$

where M_s is the molecular weight of the solvent and x_s its mole fraction (~ 1 in dilute solution). Thus the molality scale is proportional to mole fraction but the values are increased (in a given solvent) by a constant, the number of moles in 1 kg of solvent. We may now rewrite eq. (3.6) as follows:

$$\mu_i = \mu_i^\infty + RT\ln \gamma m_i \tag{3.16}$$

where μ_i^∞ is the chemical potential of the species at infinite dilution and γ, the activity coefficient related to ideal dilute solution. Similarly,

$$\frac{\partial \mu_i^\infty}{\partial P} = \bar{V}_i^\infty \tag{3.17}$$

where $\bar{V}_i^\infty$ is the partial molar volume of the solute i at infinite dilution* in pure solvent defined as [5]

$$\bar{V}_i^\infty = \left(\frac{\partial V}{\partial m_i}\right)_{T,P,m_j} = \left(\frac{\partial \mu^\infty}{\partial P}\right)_{T,m_i,m_j} \tag{3.18}$$

A relationship analogous to (3.13) follows,

$$-RT\frac{\partial \ln K_m}{\partial P} = v_X\frac{\partial \mu_X^\infty}{\partial P} + v_Y\frac{\partial \mu_Y^\infty}{\partial P} \ldots - v_A\frac{\partial \mu_A^\infty}{\partial P} - v_B\frac{\partial \mu_B^\infty}{\partial P} \tag{3.19}$$

where K_m is the equilibrium constant derived from molal concentrations and will be identical to K_x only when there are the same number of concentration terms in numerator and denominator of 3.9:

$$K_m = \frac{m_X^{\nu X} m_Y^{\nu Y} \cdots}{m_A^{\nu A} m_B^{\nu B} \cdots} \frac{\gamma_X \gamma_Y \cdots}{\gamma_A \gamma_B} \tag{3.20}$$

and

$$-\left(\frac{\partial RT \ln K_m}{\partial P}\right) = \Delta \bar{V}^0 \tag{3.21}$$

where $\Delta \bar{V}^{0*}$ is the difference between total partial molar volumes of products and those of reagents at $P \to 0$ (i.e. 1 bar). This is an important relationship and enables one to determine the effect of a pressure change on the equilibrium constant in dilute solution from a knowledge of the partial molar volumes of the components, quantities which may be separately evaluated and tabulated. This is precisely analogous to the treatment of an ideal gas equilibrium.

In practice, while activity data of many inorganic salts in aqueous solution are known, systematic information is generally not available for organic compounds or other solvents and consequently the activity coefficient term may be omitted when K_m is replaced by the somewhat concentration-dependent 'equilibrium constant', K'_m. However, analogously to eq. (3.21) is the expression eq. (3.22) derived by Planck:[6]

$$-\frac{\partial (RT \ln K'_m)}{\partial P} = \Delta \bar{V} \tag{3.22}$$

where $\Delta \bar{V}$ is the difference in partial molar volumes measured under the same conditions (concentration) as K' rather than at infinite dilution and is thus more practically useful.

In the discussion above, the use of molal or mole fraction concentration units avoided problems encountered due to compression of the solution. By contrast, the molar scale of concentrations, c_i, is pressure dependent but is very commonly used. A correction for compression must be added if the molar scale is used. Thus,

$$-\frac{\partial (RT \ln K_c)}{\partial P} = \Delta \bar{V} + (\nu_A + \nu_B \cdots - \nu_X - \nu_Y \cdots) RT \beta_T \tag{3.23}\dagger$$

where K_c, the equilibrium constant on the molar scale, is given by

$$K_c = \frac{c_A^{\nu A} \cdot c_B^{\nu B} \cdots}{c_X^{\nu X} \cdot c_Y^{\nu Y} \cdots} \cdot \frac{\gamma_A \gamma_B \cdots}{\gamma_X \gamma_Y \cdots} \tag{3.24}$$

and the correction is RT times the compressibility of the solvent multiplied by the change in number of molecules in the stoichiometric equation.

---

* The superscript ^{0} will be omitted for clarity but, unless otherwise stated, partial molar volumes, $\bar{V}$, will be understood to be limiting quantities.

† R, the gas constant in $cm^3 \cdot bar \, K^{-1} = 80.98$, P in bar, β in bar^{-1} then $\Delta \bar{V}$ will be in $cm^3 \, mol^{-1}$.

3.1.1 The Effect of Pressure on the Equilibrium Constant

The measurement of concentrations of species involved in an equilibrium as a function of pressure will lead to a measure of K and $\Delta \bar{V}$ according to eq. (3.22). The change of equilibrium constant with pressure will be positive or negative according to whether the volume change for the reaction is negative or positive, respectively, fig. 3.1.

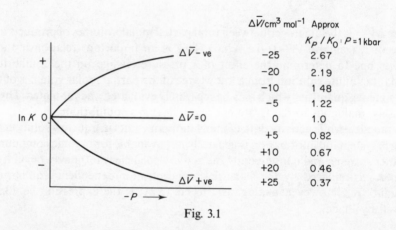

	$\Delta \bar{V}/\text{cm}^3\,\text{mol}^{-1}$	Approx $K_P/K_0 : P=1\,\text{kbar}$
$\Delta \bar{V}$ - ve	−25	2.67
	−20	2.19
	−10	1 48
	−5	1.22
$\Delta \bar{V}=0$	0	1.0
	+5	0.82
	+10	0.67
	+20	0.46
$\Delta \bar{V}$ + ve	+25	0.37

Fig. 3.1

In principle, any means of determining concentration may be used such as spectrophotometry using an optical high pressure cell, conductivity or other. If the equilibrium is established extremely slowly, or if it may be frozen by rapid cooling, the high pressure reactor may be sampled and the analysis performed externally before concentrations have appreciably changed.

3.1.2 Dissociation Constants of Weak Acids and Bases

Conductivity is frequently used to determine the extent of dissociation,[7]

$$HA+S \underset{K_A}{\rightleftarrows} A^- + HS^+$$

where S is a protic solvent, usually water. Ionization is increased by pressure, the effect on some neutral acids and bases is shown in fig. 3.2.

The glass electrode, with pressure equalized across the glass membrane, may be used at high pressure and appears to behave normally to several kbar.[8-10] Thus, the pH (concentration of solvated hydrogen ions) of buffer solutions containing a weak acid and its salt may be measured and

$$\text{pH} = \text{pK}_A + \log \frac{c_{A^-}}{c_{HA}} \tag{3.25}$$

The change in pK_A with pressure may be taken to follow the change in pH since at the fairly high concentrations of acid and salt used, the ratio c_{A^-}/c_{HA} will not

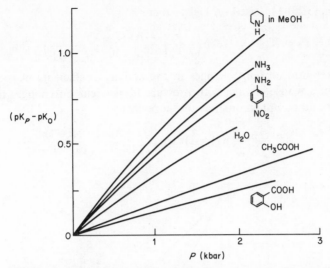

Fig. 3.2. Effect of pressure on ionization constants of neutral acids and bases (water, 25° unless otherwise indicated). (After Hamann[4])

change appreciably with pressure and will be that initially determined by weighing the constituents. The other standard method of measuring dissociation constants, the indicator method, is also applicable to high pressures. Into a solution containing acid and conjugate base is placed an indicator at a concentration sufficient for its spectrum to be observed and of range such that it is only partly dissociated. An indicator is a weak acid with considerably different spectral properties (i.e. colour) between its acidic (InH) and basic (In$^-$) forms. The ratio of these two, c_{In-}/c_{InH}, may be determined by spectrophotometry,[11-15] thus,

$$pH = pK_{InH} + \log \frac{c_{In-}}{c_{InH}} \qquad (3.26)$$

and the pH of the solution is obtained (ignoring activity coefficients) from eq. (3.26) knowing the dissociation constant of the indicator. In this case, the concentration of the indicator is so low that it exerts no buffering action. For a solution containing both the weak acid HA under examination and the indicator of known pK_A,

$$\frac{d(pH)}{dP} = \frac{d(pK_A(HA))}{dP} = \frac{d(pK_A(InH)}{dP} + \frac{d(c_{In}^-/c_{InH})}{dP} \qquad (3.27)$$

The variation of pK_A of the indicator with pressure ($= \Delta \bar{V}$ for ionization) must be separately determined. Values of $\Delta \bar{V}$ for some indicators as well as for other weak acids are given in Table 3.1. and a typical spectral plot shown in fig. 3.3.

The dissociation constant (expressed in molalities) of weak acids in water shows a pressure dependence which may be expressed by the semi-empirical equation, (3.28) 11, based on Debye theory:

$$K_m(P = P) = K_m(P = 1)\exp\left(\frac{-\Delta\bar{V}(P-1)}{RT[1+9.2\times 10^{-5}(P-1)]}\right) \quad P \text{ in bar} \quad (3.28)$$

This takes into account changes in the activity coefficients of species in the equilibrium whose change with pressure (dependent on changes in dielectric constant and compression), has the same factor:

$$\gamma(P = P) = \gamma(P = 1)\exp\frac{-2.801^{1/2}(P-1)}{RT[1+9.2\times 10^{-5}(P-1)]}$$

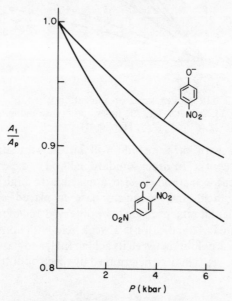

Fig. 3.3. Increase in absorbance with pressure of some phenolate ions in buffered solution under pressure. (After Neuman et al.[14])

3.1.3 The Use of Partial Molar Volumes

Equation (3.22) shows that the pressure dependence of an equilibrium constant and the volume change occurring when molar quantities of reagents are converted to products, give essentially the same information. We may therefore investigate the pressure dependence of K by determining $\Delta\bar{V}$, an experiment requiring no high pressure apparatus. The volume change for the reaction, if an equilibrium is being considered, is best determined by summing algebraically the separate partial molar volumes of the compounds involved.

The partial molar volume, $\bar{V}^0$, of a compound is the volume change which occurs when 1 mole is dissolved in a very large volume of solvent so that the

Table 3.1. Partial molar volumes,[17] $\bar{V}^0$(cm^3 mol^{-1}) in dilute solution: ions in water at 25°, relative to H$^+$(aq.) = 0[b]

	$\bar{V}^0$		$\bar{V}^0$		$\bar{V}^0$
H$^+$	0.0	OH$^-$	−4.04	HCO$_2^-$	32.4
Li$^+$	−0.88	F$^-$	−1.16	ClO$_4^-$	44.12
Na$^+$	−1.21	Cl$^-$	17.83	BF$_4^-$	44.18
K$^+$	9.02	Br$^-$	24.71	SO$_4^=$	13.98
Ca^{++}	−17.85	I$^-$	36.22	S$^=$	−8.2
Ba^{++}	−12.47	OPh$^-$	68.7	AsO$_4^\equiv$	−15.6
Al^{+++}	−42.2	OAc$^-$	40.46	Ph$_4$B$^-$	277.6
NH$_4^+$	17.86	OCOEt$^-$	54.0		
MeNH$_3^+$	36.11	OCOPr$^-$	70.40		
EtNH$_3^+$	52.94				
PrNH$_3^+$	69.44				
Me$_4$N^{+} [a]	89.57				
Et$_4$N$^+$6	149.1				
nPr$_4$N$^+$	214.4				
nBu$_4$N$^+$	275.7				
Ph$_4$As$^+$	300.6				
(CH$_2$)$_n$N$^+$CH$_2$)$_m$					
$n = 4$	152				
5	180				
6	207				

[a] Ammonium ions, reference 32.
[b] $\bar{V}^0$(H$^+$) may be taken as −5.3 cm^3.

solution is effectively at infinite dilution.[5] In practice, measurements are made at a series of finite concentrations and extrapolated to zero concentration, the observed values being 'apparent partial molar volumes; usually symbolized Φ_v. Values are dependent upon the solvent and the temperature. For many purposes, such as for obtaining d ln $K/$dP for finite concentrations, it is sufficient to use values of Φ_v at similar concentrations. Tables of partial molar volumes are available[15-18] though most of the collected information refers to electrolytes in water (Tables 3.1–3.5).[19]

Values of $\bar{V}^0$ may be positive or negative. In the latter case, the volume contraction in the vicinity of the solute is greater than the intrinsic volume of the solute itself. Much can be learnt about solute–solvent interactions by measurements of partial molar volumes.

3.1.4 Measurement of Partial Molar Volumes

In principle it is necessary to determine the change in volume which occurs on mixing (to give a solution) a known amount of solute (and of known volume) with a large amount of solvent, since

$$\bar{V}_i^0 = \left(\frac{\partial V}{\partial n_i}\right)_{\substack{T,P \\ c^i \to 0}} \tag{3.29}$$

Table 3.2. Partial molar volumes of ionic solutes in various solvents[17,19] (at 25°); cm³ mol⁻¹

Solute	H_2O	MeOH	$HCONH_2$	$HCONMe_2$	MeCONHMe	$(CH_2OH)_2$	HMPT	[cyclic carbonate]	Me_2SO	MeCN	HCOOH
LiCl	17.1	-3.8	19.6	19.5					4.2		
LiI	35.3	1.0					20.7		28.4	4.9	
NaCl	16.62	-3.3	21.1		26.0				11.6		
NaBr	23.50	5.1	28.0	6.6	32.6	27.3			19.7		15.5
NaI	35.01	12	39.8	21.5	42.0	38.4	29.8	27.7	32.8	7.3	31.5
KCl	26.85	7.3	32.0						20.0		18.5
KBr	33.73	15.5	38.9		40.2						23.4
KI	45.24	21.5	50.7	32.8	45.4	46.2	41.6	40.3	40.5	15.7	34.5
CsCl	39.2		42.3	40.4							
CsI	57.6		61.1						45.8		
NaOMe		2.1									
NaOPh		50.5									
$NaOC_6H_4\text{-}pBr$		69.0									
KSCN	44.72	28.2									
NH_4Br	42.57	20.8									
Me_4NCl	107.4	83.0									
Et_4NCl	166.9	140.7									
Et_4NBr	173.8	148.0									
nPr_4NBr	239.1	220.0									
nBu_4NI	311.9		322		322.9						
$nPent_4NI$			394		391.1						
$nHex_4NI$					461.2						
Piperidine HBr		91.2									103.1 (in AcOH)
Pyridine HBr		71.3									
$NaBPh_4$	277	241	289	280			283	281	290	255	
$KBPh_4$			326	300				295		271	
Ph_4AsCl	319									285	
Ph_4AsBr	329			318			281	308	304	292	
Ph_4AsBPh_4	579	527	594	570			564	575	588	557	

Table 3.3. Temperature effect on partial molar volumes of ions in water[17] (relative to $\bar{V}^0(H_{aq}^+) = 0$ at each temperature: $cm^3\,mol^{-1}$)

[a]	$T(°C)$:	0	25	50	75	100	200
H^+		0	0	0	0	0	0
Na^+		−3.51	−1.21	−0.30	0.8	0.8	−0.1
K^+		7.17	9.02	9.57	10.3	9.5	7.3
NH_4^+		17.47	17.86	19.2	19.5	19.5	20.0
Me_4N^+		88.59	89.57	91.2	95.8		
nBu_4N^+		271.1	275.7	285.0	304.1		
Mg^{++}		−21.81	−21.17	−20.90	−21.70	−23.4	−37.0
Ca^{++}		−19.84	−17.85	−18.22	−19.1	−20.0	−30.7
OH^-		−6.8	−4.04	−4.35	−5.2		
Cl^-		16.45	17.83	18.00	17.4	16.0	0.5
Br^-		23.06	24.71	25.49	25.1	24.9	13.0
$SO_4^=$		11.1	13.98	16.03	13.8	11.5	−22.9
$\bar{V}^0(H_{aq}^+)^a$		−5.1	−5.3	−5.9	−6.6	−7.5	−13

[a] Estimated from $\bar{V} = -5.1 - 0.008T(°C) - 1.7 \times 10^{-4}\,T^2$.

This may be carried out in a dilatometer such as fig. 3.4.[18] The solute is held in the cap contained by the magnetically located plate. When this is removed by an external magnet, the volume change is observed by the level in the capillary, stirring to ensure mixing. Thermostatting to better than 0.002° is required for high accuracy. Alternatively, a series of solute additions of known volume may be made using a micrometer syringe, measuring the volume change in the solution by the capillary level. In this way, several values of ΔV at a series of concentrations may be obtained for one experiment.[20]

Measurement of the density of a solution and comparison with the density of the solvent can yield the partial molar volume and this method is probably the most convenient and rapid at present, eq. (3.30):

$$\bar{V}^0_{c \to 0} = \frac{1000(\rho_0 - \rho)}{\rho_0 c} + \frac{M}{\rho_0} \qquad (3.30)$$

(ρ, ρ_0 are the densities of solution and solvent, respectively; c is the molar concentration of the solute, M its molecular weight).

Very high precision is necessary in the density measurements, but one part in 10^7 may be obtained by the use of a differential float method,[21] fig. 3.5, in which the difference in weight between two glass bulbs submerged in solvent is compared with the difference when one is submerged in solution. The difference in buoyancy may be finally measured as a current supplying a solenoid which maintains the balance.[22-24] Very accurate direct reading densitometers are available commerically which employ the principle of a submerged tuning fork whose frequency is determined by the density of the medium in which it lies.[25-27] This innovation, which is relatively recent, should make the determination of partial molar volumes and reaction volumes so easy that these measurements will

Table 3.4. Partial molar volumes of non-electrolytes

	Solvent[a]	Temperature (°C)	$\bar{V}^0(cm^3\,mol^{-1})$	Reference
Hexane	B	25	135	
	C	25	135	
Cyclohexane	B	25	113	
	C	25	113	
Benzene	B	25	89.3	
	C	25	89	
Methanol	A	20[b]	38.05	34
Ethanol	A	20	54.97	
n-Propanol	A	20	70.20	
Isopropanol	A	20	71.73	
n-Butanol	A	20	85.77	
Isobutanol	A	20	86	
Sec. Butanol	A	20	86.63	
n-Pentanol	A	20	101.92	
Benzyl alcohol	A	20	100.82	
Acetic acid	A		51.9	35
Propionic acid	A		67.9	
Butanoic acid	A		84.6	
Pentanoic acid	A		100.5	
Hexanoic acid	A		116.0	
Methyl iodide	B	25	64.2	18
	C	25	64.95	
	D	25	63.8	
	E	25	62.8	
Pyridine	B	25	78.25	
	C	25	79.6	
	D		80.08	
	E	25	79.45	
Pyridinium chloride	B	25	63.69	13
2,6-Dimethyl pyridinium chloride	B	25	97.72	
2,6-Diethyl pyridinium chloride	B	25	128.4	
2,6-Diisopropyl pyridinium chloride	B	25	164.3	
2,6-Dit.butyl pyridinium chloride	B	25	196.0	

[a] A, water; B, methanol; C, acetone; D, benzene; E, carbon tetrachloride.
[b] Data also at 1°, 40°, and 50°.

become routinely used. In addition, the technique of using ionic vibration potential measurements allows the estimation of partial molar volumes of individual ions without making dubious assumptions.[22,23,28,29]

The additivity principle applies to partial molar volumes which may be estimated by the summation of volumes of component molecular fragments. This meets with greatest success if the increments are assessed for use within a given homologous series, Table 3.5, and is a principle known as Traube's Rule.[30, 31]

Table 3.5. Structural additivity data for partial molar volumes of organic compounds

1. Alkanes, 20°

$$\bar{V}^0 = 12.08N_1 + 9.65N_2 + 6.9N_3 + 2.0N_4 + 2.1N_5 + Y + 2.7$$

where N_1 number of carbon atoms in open chains
N_2 number of carbon atoms in rings
N_3 number of carbon atoms at ring junctions
N_4 number of carbon atoms in double bonds
N_5 number of hydrogen atoms
Y = ring correction; 4-membered, $+1.4$;
5-, $+0.7$; 6-, 0.0; 7-, -0.7

2. Carboxylic acids, water, 25°

each COOH	25.9 cm^3
CH$_3$	26.1
CH$_2$	16.0
CH$_2$COO$^-$	33.7

3. Missenard's equation:[36]

$$\bar{V}^0 = A + 16.49m + B/m$$

A, B are constants characteristic of homologous series, m is the number of carbons:

	A	B
Alkanes	29.96	29
RCl	38.15	1.9
RBr	42.5	-2.3
RI	48.95	-3.3
ROH	26.2	-2.3
RCOOH	43	-2.7
HCOOR	49.3	-4.2

4. Ammonium ions in solution;[37,38]

N$^+$	11.7
CH$_2$	13.8
CH$_3$	18.1

Alcohols in aqueous solution:[9]

$\Delta\Delta\bar{V}^0$ per CH$_2$:	1	20	40	50	
	15.6	16.1	16.5	16.8	°C

Amines in aqeous solution[38]

	$\Delta\Delta\bar{V}^0$ per CH$_2$
Primary n-aliphatic	16.2
Secondary n-aliphatic	16.0
Secondary cyclic	14.9
N-Methyl cyclic	13.3

The effect of concentration on values of Φ_v may be seen in fig. 3.6. Neutral non-electrolytes are usually simple to deal with and it is sufficient to extrapolate the linear graph of Φ_v against c. For electrolytes, the raw data may need to be corrected for changes in the species present, i.e. ion-pairing ionic dissociation or salt hydrolysis.[32] The corrected data may also show a linear relationship. For

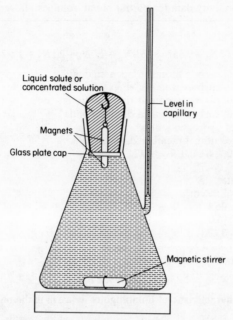

Fig. 3.4. Dilatometer to measure volume change on mixing of two liquids. (After Eckert *et al.*[18])

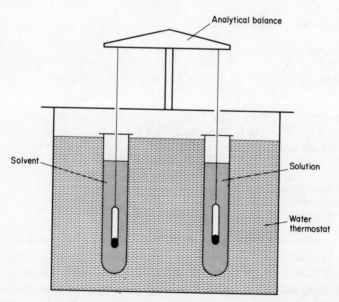

Fig. 3.5. Differential buoyancy balance for precise density measurements

electrolytes, such as highly dissociated salts, the Redlich–Meyer equation, eq. (3.31),[119] is used:

$$\Phi_v = \bar{V}^0 + S_v c^{1/2} + hc \qquad (3.31)$$

where c is the concentration, and S_v and h are constants. Of these, S_v depends on charge type and has the values 1.868 for 1:1 and 9.706 cm^3 1$^{1/2}$ mol$^{3/2}$ for 1:2 electrolytes in aqueous solution at 25°. A plot of c against $\Phi_v - S_v c^{1/2}$ is linear to about 1 M and gives $\bar{V}^\circ$ as the intercept.

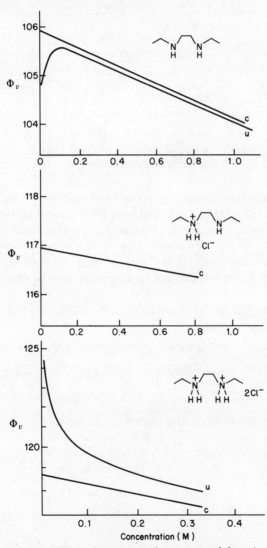

Fig. 3.6. Effect of concentration on partial molar volumes of diethylenetriamine in three charge states; c = corrected for dissociation; u = uncorrected. (After Cabani et al.[32])

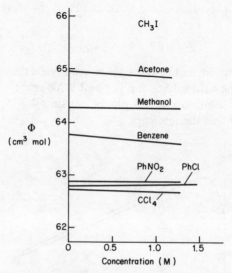

Fig. 3.7. Effect of solvent on partial molar
volumes of methyl iodide

The two methods of measuring volumes of reaction for equilibria permit a check to be made that they do, indeed, give the same values. Hamann has made measurements using both approaches and reports very similar values although there may be a systematic difference. Almost all volumes of ionization obtained from density measurements are slightly more negative than those from pressure dependence[29,33] though the extent to which this may be experimental error is uncertain.

The use of partial molar volumes may be illustrated with reference to the dissociation of acetic acid:

$$CH_3COOH \rightleftharpoons CH_3COO^- \equiv +(H^+)_{aq}$$

$$\bar{V}^0 (cm^3\,mol^{-1})\,50.0 \qquad 42.3 \qquad -42$$

$$\Delta \bar{V} = (42.3 - 4.2) - 50.0 = -11.9\,cm^3\,mol^{-1}$$

(experimental value from conductivity at high pressure[39,40] = −11.5), thus

$$\frac{\partial \ln K_A}{\partial P} = \frac{-\Delta \bar{V}}{RT} = -\frac{-11.9\,(cm^3\,mol^{-1})}{83.12\,(cm^3\,bar\,K^{-1}\,mol^{-1}) \times 298\,(K)}$$

$$= +0.464 \times 10^{-3}\,bar^{-1}$$

The effect of pressure on the dissociation constant of acetic acid may be calculated:

$$CH_3COOH; \quad K_A(25°) = 1.75 \times 10^{-5}\,M$$

$$1\,bar \quad \ln K_A = -10.95$$

$$1 \,\text{kbar} \quad \ln K_{\text{A}} = -10.95 + (1000 \times 0.464 \times 10^{-3})$$

$$= -10.486$$

$$K_{\text{A}} = 2.79 \times 10^{-5} \text{ at } 1 \,\text{kbar}$$

This corresponds to 0.53% dissocation compared with 0.42% at 1 bar, but overestimates the effect since $\partial \ln K/\partial P$ is not constant but falls with pressure. The fact may be accommodated by fitting experimental data to a quadratic or higher order polynominal ($\ln K(P) = \ln K \,(1\,\text{bar}) + a'P + b'P^2 \dots$) or to eq. (3.32) obtained by the integration of eq. (3.8),

$$RT\ln\frac{K_p}{K_0} = -\Delta\bar{V}^0 P - \frac{P^2}{2}\frac{\partial\Delta\bar{V}^0}{\partial P} + \frac{P^3}{6}\frac{\partial^2\Delta\bar{V}^0}{\partial P^2} - \dots \tag{3.32}$$

where K_p refers to pressure P and $K_0, \Delta\bar{V}^0$ refers to 1 bar. It is necessary to carry out accurate high pressure experiments in order to evaluate the pressure derivatives of ΔV and so it is more convenient to use one of the semi-empirical approaches which have been proposed.[41] Owen and Brinkley used the extended Tait equation[43] to derive eq. (3.33) which fits ionic equilibria in aqueous solution to about 1 kbar:[42]

$$RT\ln\frac{K_p}{K_0} = -\Delta V_0 P + \Delta\bar{\beta}. B(P - B\ln^{(B+P)/B}) \tag{3.33}$$

where B is the constant of the Tait equation for the solvent ($= 2996$ bar for water at 25°) and $\Delta\bar{\beta}$ is the change in partial molar compressibilities of solutes between reagents and products. These are taken as compressibilities of pure compounds and assumed to be pressure independent. Values of $\bar{\beta}$ for ionic compounds can be obtained from the additive values in Table 3.6.

Table 3.6. Partial molar compressibilities of ions, $\bar{\beta}(10^4 \times \text{cm}^3 \,\text{mol}^{-1}\,\text{bar}^{-1})$

H^+	0 (ref)	Mg^{++}	-83	Br^-	$+2$
Li^+	-34	Ca^{++}	-71	I^-	$+18$
Na^+	-42	Cu^{++}	-62	OAc^-	-10
K^+	-37	OH^-	-44	NO_3^-	$+7$
NH_4^+	-11	Cl^-	-8	HCO_3^-	$+2$

The application of eq. (3.33) to the dissociation of acetic acid at 25° and 1000 bar is given below:

$$(83.12 \times 298)\ln\frac{K_p}{(1.75 \times 10^{-5})} = -(-11.5 \times 1000)$$

$$+(-17 \times 10^{-4})(2996 \times 1000) - \left(2996^2 \ln\frac{3996}{2996}\right)$$

whence $K_p = 2.46 \times 10^{-5}$ at 1 kbar (0.49% dissociation). El'yanov and Gonikberg[44] found the simple linear free energy relationship, eq. (3.34), to hold fairly well up to 12 kbar:

$$\log \frac{K_p}{K_0} = -\frac{\psi \Delta V_0}{T} \tag{3.34}$$

where ψ is a constant dependent upon P but essentially independent of the acid under consideration (Table 3.7).

It has been shown that ψ may be given by eq. (3.35):[122]

$$\psi = P/(1 + 9.2 \times 10^{-5} P) \, RT \ln 10 \tag{3.35}$$

hence

$$RT \ln \frac{K_p}{K_0} = -\Delta V_0 \, P/(1 + (9.2 \times 10^{-5} P)) \tag{3.36}$$

This empirical equation has been justified theoretically by developing the Born theory of solvation, taking into account the pressure dependence of dielectric constant of the medium.[41] In the revised version due to Nakahara, the constant,

Table 3.7. Values of ψ from eq. (3.34)[44]

P (kbar)	ψ (mol cm^{-3} K^{-1})	P (kbar)	ψ (mol cm^{-3} K^{-1})
0	0	4	15.3
1	4.78	5	17.9
2	8.82	10	27.2
3	12.3	12	29.8

ψ', in eq. (3.37) is temperature dependent and values are available for other solvents, Table 3.8:

$$RT \ln \frac{K_p}{K_0} = \frac{-\Delta \bar{V}^0 P}{(1 + \psi' P)} \tag{3.37}$$

This predicts, in effect, that for a given temperature and solvent, the effect of pressure on all ionic dissociation equilibria will be proportional to the volume of reaction at 1 bar.

A further relationship, eq. (3.38), derived by the same authors, enables the pressure effects for the dissociation of weak acids to be calculated knowing only ΔV_0 and the Tait B value for the solvent:

$$RT \ln \frac{K_p}{K_0} = -\Delta \bar{V}_0^0 \, B \ln \frac{B+P}{B} \tag{3.38}$$

The equation has been tested for water, methanol, and formamide and is said to be satisfactory to 12 kbar.

A somewhat similar treatment due to North[45] gives the relationship, eq. (3.39),

$$\frac{RT}{(P-1)} \ln \frac{K_p}{K_0} = -\Delta V_0 + n \bar{V}_w A \left(1 - \frac{B+P}{P-1} \ln \frac{B+P}{B+1} \right) \tag{3.39}$$

Table 3.8. Values of ψ' from eq. (3.37)

Solvent	$T=$	10	25	45	65
Water		8.98	9.46	11.67	14.24
Formamide			4.63	5.64	

where $\bar{V}_w$ is the partial molar volume of water, A and B constants of the Tait equation and n the hydration number, the number of water molecules involved as a result of the ionization process. According to this concept, there is a relationship between volume of reaction and hydration number, the former being the more negative and the greater number of water molecules involved.[46] For carboxylic acids at 25° n is around 3.6, somewhat smaller for the lower members, and larger at higher temperatures.

3.1.5 The Pressure Derivatives of $\Delta\bar{V}$

By differentiating eq. (3.35) once and twice we obtain respectively the volume change at pressure P, $\Delta\bar{V}_p$, and the pressure dependence of $\Delta\bar{V}_p$ at P, κ_p:

$$\Delta V_p = \frac{\Delta V_0}{(1+bP)^2} \tag{3.40}$$

$$\kappa_p = \frac{2b\,\Delta V_0}{(1+bP)^3} \tag{3.41}$$

where $b = 9.2 \times 10^{-5}\,\mathrm{bar}^{-1}$ (for water as solvent). Again, these functions are independent of the solute. Thus, for acetic acid, $\Delta\bar{V}^0 = -11.7\,\mathrm{cm}^3\,\mathrm{mol}^{-1}$, at 1000 bar:

$$\Delta\bar{V}_p = -11.9/(1+(9.2\times10^{-5}\times1000))^2$$
$$= -9.97\,\mathrm{cm}^3\,\mathrm{mol}^{-1}$$

that is, at high pressure the volume change on ionization is less negative. Values of κ are of the order $-3 \times 10^{-3}\,\mathrm{cm}^3\,\mathrm{mol}^{-1}\,\mathrm{bar}^{-1}$ for carboxylic acids at 25°.[46]

3.1.6 The Relationship of Volume and Entropy Change

The equilibrium constant is related to the Gibbs energy change for the reaction, whose pressure coefficient is the volume change:

$$-\Delta G = RT\ln K \quad \text{and} \quad -\frac{\partial\Delta G}{\partial P} = \Delta\bar{V} \tag{3.42}$$

and the Gibbs energy in turn is related to heat (enthalpy) and entropy changes,

$$\Delta G = \Delta H - T\Delta S \tag{3.43}$$

so it may be asked whether a change in K as a result of raising the pressure is a result of a change in the enthalpy or entropy of the process. The thermodynamic

154

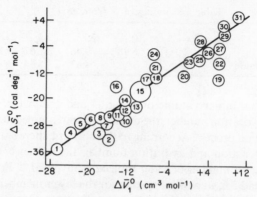

Fig. 3.8. Graph of $\Delta \bar{S}_i^0$ vs. $\Delta \bar{V}_i^0$ for aqueous acids at 25°. Numbered circles correspond to data for ionization of various acids as follows (note that points for several acids fall within some circles): (1) HCO_3^-; (2) third citric; (3) second malonic; (4) $H_2PO_4^-$; (5) H_2O (1 m); (6) HSO_4^-; (7) phenol; (8) trimethylacetic; (9) second oxalic; (10) second succinic; (11) butyric and isobutyric; (12) propionic; (13) acetic; (14) second ε-aminocraproic; (15) formic, benzoic, m-methoxybenzoic, p-methoxybenzoic, p-nitrophenol, first succinic, chloroacetic, bromoacetic, lactic, glycolic, and second citric; (16) first phosphoric; (17) m-nitrobenzoic and p-nitrobenzoic; (18) cyanoacetic; (19) $(CH_3)_3NH^+$; (20) second hydroxyproline and second proline; (21) first salicylic and first citric; (22) $(CH_3)_2NH_2^+$; (23) second glycine and second alanine; (24) first glycine, first alanine, first hydroxyproline, and first proline; (25) diethylammonium ion and piperidinium ion; (26) $H_2N(CH_2)_2NH_3^+$; (27) $CH_3NH_3^+$ and pyridinium ion; (28) ethylammonium ion, propylammonium ion, and first ε-aminocaproic acid; (29) NH_4^+; (30) anilinium ion; (31) $^+H_3N(CH_2)_2NH_3^+$.[47] (Reprinted with permission from *J. Phys. Chem*, **69**. Copyright 1965, the American Chemical Society)

quantities for a number of acid equilibria are shown in Table 3.9 below and it may be seen that there is a rough parallelism between $\Delta \bar{V}$ and ΔS while ΔH is not related. This correlation was pointed out by Hepler[47] who summarized the data for some thirty carboxylic and other acids, fig. 3.8. The best straight line gives

$$\Delta S(J\ K^{-1}) = -29.2 + 4.4\ \Delta \bar{V}(cm^3\ mol^{-1})$$

Other authors have noticed this relationship[32,48-51] though the slope may differ slightly for different series of acids.

For such a series of similar reactions, the relationship points to a common cause for the variation of $\Delta \bar{V}$ and ΔS, almost certainly related to the solvation changes brought about by ionization. The more tightly is water held in the solvation sphere of the ions compared with neutral species the more negative are both the volume and the entropy changes.

3.1.7 Characteristics of Partial Molar Volumes

The partial molar volumes of individual ions increase with atomic weight within a group of the periodic table but values do not resemble crystallographic ones. Negative volumes are common, especially for cations, indicating a strong net contraction of solvent in the vicinity of the ion and this is particularly marked for multi-valent cations in accordance with the z^2 dependence of the electrostriction volume. In water at least, anions appear to produce much less electrostriction than do cations even though the Drude–Nernst theory makes no distinction between the sign of the charge and the volume of electrostriction. Clearly in this solvent a different mechanism of interaction between ions and solvent is occurring and is believed to result in the destruction of ice-like organization in the vicinity of the ion. The presence of ions and other solutes has a substantial effect on $\bar{V}$ [52].

The effect of temperature on $\bar{V}$ in water is erratic, but there seems to be a general increase to a maximum volume somewhere between 25° and 50°. It is likely that this is related to the maximum in compressibility of solvent which occurs at about 30°. Hamann has shown the partial molar volumes of electrolytes to be related to the compressibility of the solvent[29] and it has been shown that the volumes of Ph_4As^+ and Ph_4B^-, in a series of dipolar aprotic solvents, correlate well with compressibilities.[25] As expected, partial molar volumes of electrolytes are highly solvent dependent, the values reflecting a complex interplay of dielectric constant specific interactions and compressibility dependence.[53] Much less data are available concerning volumes of non-electrolytes, though some representative values were given in Table 3.4. Structural additivity rules have been deduced for some homologous series, Table 3.5. There is a small solvent isotope effect between values of $\bar{V}^0$ in H_2O and D_2O which presumably reflects differences in the degree of structural change induced in the solvent by the solute molecule and differences in hydrogen bonding ability of the two solvent species.[54]

3.2 PRESSURE EFFECTS ON CHEMICAL EQUILIBRIA

3.2.1 Acid–base Dissociation

One of the most thoroughly studied of all equilibria are the dissociations of weak acids and bases in water:

$$HA + H_2O \rightleftharpoons H_3O^+ + A^-, \quad K_A = \frac{[H^+][A^-]}{[HA]}$$

$$B + H_2O \rightleftharpoons BH^+ + OH^-, \quad K_B = \frac{[BH^+][OH^-]}{[B]}$$

Table 3.9. Volumes of ionization for weak acids and bases in water at 25°[67]

$$HA + H_2O \longrightarrow A^- + H_3O^+; \quad \Delta\bar{V}$$
$$B + H_2O \longrightarrow BH^+ + OH^-; \quad \Delta\bar{V}$$

Acid, HA	pK_A	$\Delta\bar{V}(cm^3\,mol^{-1})$
H_2O	14	-22.07
HF	3.19	-13.7
HCN	9.21	-7.0
H_2S	7.0	-16.3
H_2CO_3	6.37	-27
HCO_3^-	10.3	-26
H_3PO_4	2.15	-16.5
$H_2PO_4^-$	7.2	-24.0
H_3BO_3	9.2	-32
H_2SO_3	1.85	-23
HSO_3^-	7.23	-23
$HCOOH$	3.77	-8.5
CH_3COOH	4.77	-11.6
$EtCOOH$	4.89	-13.5
$nPrCOOH$	4.83	-14.2
$nPentCOOH$	4.82	-14.2
$isoPrCOOH$	4.82	-14.8

$CH_2)_n(COOH)_2$	pK_A	1st ionization	2nd ionization
$n = 0$	1.25, 4.28	-6.72	-11.91
1	2.85, 5.67	-10.06	-18.55
2	4.28, 5.67	-12.86	-13.58
3	4.31, 5.41	-13.17	-13.59
4	4.48, 5.42	-13.48	-13.84
Phthalic acid	2.95, 5.41	-12.5	-16.5
Maleic acid	1.96, 6.24	-6.7	-22.9
Fumaric acid	3.01, 4.39	-9.9	-10.9
Citric acid	3.13, 4.77, 6.4	-8.8	-12.7, -18.3 (3rd)
$HOCH_2COOH$	3.87		-11.9
$MeCHOHCOOH$	3.86		-13.4
$HOCH_2CH_2COOH$			-13.9

	pK_A	$\Delta \bar{V}^a$ (cm3 mol^{-1})
Benzoic acid, 2-nitro-	2.21	−10.2
Benzoic acid, 3-nitro-	3.49	−8.7
Benzoic acid, 4-nitro-	3.45	−8.8
Benzoic acid, 3-methoxy	4.09	−10.3
Benzoic acid, 4-methoxy-	4.51	−11.3
Benzoic acid, 3-fluoro-	3.85	−9.8
Benzoic acid, 4-fluoro-	4.15	−10.5
Benzoic acid, 2-hydroxy-	3.0	−7.2
Phenylacetic acid,	4.31	−12.7
Phenylacetic acid, 4-nitro-	3.92	−10.9
Phenylacetic acid, 3-chloro-	4.12	−11.8
Phenylacetic acid, -4-methoxy-	4.36	−12.4
Sulphanilic acid		+2
ArSO$_2$OH		ca. −35
Phenol	9.95	−18.4
Phenol, 2-nitro-	7.27	−12.4
(excited state)		−6
Phenol, 3-nitro-	8.38	−12.9
(excited state)		−3
Phenol, 4-nitro-	7.16	−10.0
(excited state)		+8
Phenol, s-trinitro-	0.42	−10.0
1, 2-Naphthols		−17.8
In methanol		
Phenol		−39
Phenol, 4-bromo-		−36

$$\text{Base } B + H_2O \longrightarrow BH^+ + OH^-$$

	pK_A	$\Delta \bar{V}^a$ (cm3 mol^{-1})
Amines		−22 to −28
NH$_3$	9.23	−28.6
MeNH$_2$	10	−26.3
Me$_2$NH	10.8	−26.0
Me$_3$N	9.80	−27.0
H$_2$NCH$_2$CH$_2$NH$_2$	9.92	−27

Table 3.9 (cont.)

Acid, HA	$HA + H_2O \longrightarrow A^- + H_3O^+$; $\Delta\bar{V}$ $B + H_2O \longrightarrow BH^+ + OH^-$; $\Delta\bar{V}$ pK_A	$\Delta\bar{V}$ (cm³ mol⁻¹)
$H_2NCH_2CH_2NH_3^+$	6.9	−33
$^-O_2CCH_2NH_2$	9.8	−23
$PhNH_2$	4.61	−26
$PhNH_2$, 4-nitro-	1.0	−28
$PhNH_2$, 4-methyl-3-nitro	2.8	−25
$\begin{array}{c} CH_2{-}CH_2 \\ NH \end{array}$	8.0	−27
Pyrrolidine	11.3	−26
Piperidine	11.1	−24
Pyridine	5.2	−27
(methanol)		−53
Imidazole	6.9	−23
2,6-Dimethylpyridine	6.6	−25
2,6-Di-t-butyl-pyridine	(Methanol)	−53
1,4-Diazabicyclo-[2,2,2]-octane	(Methanol)	−67
		−29
Phenol red		−11.6
Bromocresol green		−16.8
Morpholine		−28.6
Piperazine		−27, −58 ($\Delta\bar{V}_2$)
Triethylenediamine		−28, −60

[a] Values may be obtained for the dissociation volumes of the corresponding conjugate acids by subtracting from the ionization volume of water (−22.07); thus NH_4^+; $\Delta\bar{V} = -22.07 -(-28.6) = +6.5$.

Volumes of dissociation may be tabulated in this form in other works.

where $[H^+]$ stands for the hydrated proton. For a conjugate acid–base pair K_A and K_B are related by the self-dissociation constant for water, K_w:

$$2H_2O \rightleftharpoons H_3O^+ + OH^-, \quad K_w = [H^+][OH^-]$$

whence

$$K_A K_B = K_w \tag{3.44}$$

The self-ionization of water is an important quantity and K_w is found to increase rapidly under pressure, $\Delta \bar{V}^0 = -22\,cm^3\,mol^{-1}$, which follows from the partial molar volumes of the separate ions. The large reduction in volume on ionization is attributed to solvation of the ions by hydrogen bonding. The relationship has been examined to extreme conditions of temperature and pressure[118] and the relationship, eq. (3.45), found to apply:

$$\log K_w(\rho, T) = 2\left(7.2 + 2.5\frac{\rho}{\rho_0}\right)\log\frac{\rho}{\rho_0} + \log K_w^0(\rho_0, T) \tag{3.45}$$

where ρ_0 is the standard density $= 1\,g\,cm^{-3}$ and ρ is the density at other conditions of temperature, T and the appropriate pressure. Water therefore becomes highly conducting at high temperatures and pressures—at $1000°$ and a pressure giving a density of $1.5\,g\,cm^{-3}$, $pK_w = 2$. Dissociations of weak acids and bases under pressure have been intensively studied beginning with Gibson's early observations;[54] volume changes are shown in Table 3.9. With the exception of the earlier members, $\Delta \bar{V}^0$ for ionization of carboxylic acids is remarkably constant at about $-14\,cm^3\,mol^{-1}$ including dicarboxylic acids[55] and hydroxy acids.[56] The first members of a series often show anomalous values, for instance:

	RCOOH	RCOO$^-$ + H$^+$	$\Delta \bar{V}^0$
$R = H$	$\bar{V} = 31.9$	23.4	-8.5
	} 20.0	} 17.06	
Me	51.9	40.46	-11.3
	} 16.2	} 13.5	
Et	68.1	54.0	-14.1
	} 16.5	} 16.4	
nPr	84.6	70.4	-14.6

Allowing about $16.5\,cm^3\,mol^{-1}$ per CH_2 group, the figures suggest an abnormally low partial molar volume for formic acid and an abnormally high one for formate and acetate ions for reasons which must involve specific hydration. Høiland has shown that volumes of ionization correlate well with changes in the degree of hydration (the 'hydration number').[46, 57] A volume change of $-3.4\,cm^3\,mol^{-1}$ per coordinated water molecule was inferred. Formic and acetic acids have small hydration numbers and their volumes of ionization are correspondingly less negative. From eq. (3.44) it follows that $\Delta \bar{V}_A + \Delta \bar{V}_B - \Delta \bar{V}_w = -22\,cm^3\,mol^{-1}$ where $\Delta \bar{V}_A$ and $\Delta \bar{V}_B$ refer respectively to ionization volumes of an acid and its conjugate base. Hence,

$$MeNH_2 + H_3O^+ \rightleftharpoons MeNH_3^+ + H_2O; \quad \Delta \bar{V}_B = -26.3 + 22 = -4.3$$

160

From Table 3.9 it will be observed that the volumes of ionization of neutral bases are in general more negative and more variable than those of carboxylic acids. Most fall within the range, $\Delta\bar{V} = -20$ to -30 cm^3 mol^{-1}.

The increase in ionization of weak acids and bases by pressure has been suggested as due to a lowering of the free energies of solvation of the ions.[132,133] This may be calculated, subject to the usual approximations, by the Born equation which on differentation with respect to pressure gives the activation volume of ionization in terms of ion size, r, and dielectric constant, ε:

$$\underset{\text{(solvation)}}{\Delta G} = \frac{Ne^2}{2}\left(1 - \frac{1}{\varepsilon}\right)$$

$$\Delta\bar{V} = \left(\frac{\partial\Delta G}{\partial P}\right)_T = \frac{Ne^2}{2}\left(1 - \frac{1}{\varepsilon}\right)\frac{1}{r^2}\left(\frac{\partial r}{\partial P}\right) - \frac{Ne^2}{2r}\cdot\frac{1}{\varepsilon^2}\left(\frac{\partial\varepsilon}{\partial P}\right)$$

Application of this equation to dissociations of some bases in both water and methanol, fig. 3.9, gives a reasonable fit to theoretical curves based on the volume of solvation of a typical ion-pair, Cs^+F^-.[133]

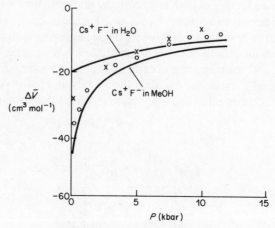

Fig. 3.9. Volume of solvation; full lines, Cs^+F^- in water and methanol as a function of pressure (calc.); experimental values for ionization of ammonia in water ($\times$) and piperidine in methanol ($\bigcirc$). (After Strauss[133])

It is likely that solvation is responsible for the differences observed in the acids shown in Table 3.9. This results in a complex interplay of thermodynamic properties but volume changes can give added insight into a difficult area. Thus, the acid dissociation constants of m- and p-nitrophenols differ considerably although the enthalpies of ionization are very similar (Table 3.10). Entropy differences are important here in determining acidity. The difference in partial molar volumes for the phenolate ions is very small and the observed difference in $\Delta\bar{V}^0$ resides in the undissociated phenols (99.71 cm^3 mol for m-nitrophenol and

98.23 for the para isomer). Volume measurements suggests, then, that the higher acidity of p-nitrophenol is due to its greater hydration and consequently the smaller degree of solvent reorganization needed upon dissociation.

Table 3.10. Thermodynamic parameters in nitro-
phenol dissociations[58]

	OH with NO₂ (meta)	OH with NO₂ (para)	
pK_A	8.39	7.15	M
ΔH	4.9	4.65	kcal mol^{-1}
ΔS	-21.8	-17.1	cal K^{-1} mol^{-1}
$\Delta \bar{V}$	-12.84	-11.32	cm^3 mol^{-1}

Dissociations of cationic acids, ammonium ions for example, should be associated with an increase in volume since the charge on nitrogen is neutralized. This is indeed the case, but values of $\Delta \bar{V}$, typically around $+5\,\text{cm}^3\,\text{mol}^{-1}$, are much less in absolute magnitude than those for neutral acids. That is, there is a much smaller volume difference between cationic acid and neutral base than between neutral acid and anionic base:

$$\text{Me}_3\text{NH}^+ \rightleftharpoons \text{Me}_3\text{N} + \text{H}_{aq}^+, \quad \Delta \bar{V} = +5.6$$
$$\bar{V}\ 36.11 \qquad\qquad 41.71 \quad 0$$

This is paralleled in partial molar volumes of cations and anions, the latter being by far the more negative. It would seem that volumes of electrostriction depend upon the sign of the charge and that these are greater for anions than for cations. This is probably due to the fact that in water, hydrogen bonds can form between the solvent and anionic solutes whereas cations are solvated by electrostatic interactions with the oxygen, Hydrogen bonding evidently is accompanied by a considerable decrease in volume.

In the pyridine series, values of $\Delta \bar{V}$ are more positive but these relate to methanol as a solvent.[13] However, it is notable that substituents in the 2,6 positions bring about a large increase (2,6-di-tert.butylpyridinium, 1, $\Delta \bar{V} = +22\,\text{cm}^3\,\text{mol}^{-1}$) and this has been suggested as due to steric hindrance to solvation of the free base (which presumably requires the solvent to approach the nitrogen closely in order to form a hydrogen bond). The pyridinium ion is still strongly solvated, however, since solvation depends upon the charge present which, furthermore, is to some extent delocalized around the ring.

(1)

Tris(hydroxymethyl)methylammonium $((HOCH_2)_3C-NH_3^+)$ has a very small volume of ionization, $\Delta \bar{V} = +1\,cm^3\,mol^{-1}$. The reason for this is not entirely clear but could be due to the cooperative effects of the amino and hydroxyl groups on hydrogen bonding around both the free base and the ammonium ion. The results is that the buffer solution made from this acid–base pair ('tris' buffer) has a pH almost pressure independent and therefore would be recommended for pressure studies on biological systems.

Anionic acids such as $H_2PO_4^-$ ($\Delta \bar{V} = -25\,cm^3\,mol^{-1}$) have much more negative ionization volumes than the neutral acids since the conjugate base must be a di-anion and the electrostriction volume, following the square of the charge, should be four times as much. Values for carbonic[59] and boric acids[60] are also very negative ($\Delta \bar{V} = -27.6$, $-35\,cm^3\,mol^{-1}$ respectively). The former presumably involves the net change on two equilibria,

$$H_2CO_3 \rightleftharpoons H_{aq}^+ + HCO_3^-$$

$$CO_2 + H_2O \rightleftharpoons H_2CO_3$$

and both would be expected to contribute $ca.\ -12$. Boric acid is also more complex in that a molecule of water is removed in addition to the loss of a proton,

$$B(OH_3) + H_2O \rightleftharpoons B(OH)_4^- + H_{aq}^+$$

The self-dissociation of water is very pressure dependent, $\Delta V = -22.1\,cm^3\,mol^{-1}$, since two ions are produced from neutral molecules and, particularly, that the partial molal volume of OH^- is very small ($-4\,cm^3$) compared to those of, for example, carboxylate ions. Measurements over a wide range of pressure and temperature have been made using conductivity.[61,62]

By assuming that the electronic excitation energy of an acid or its conjugate base is transformed into enthalpy of the excited state[63,64] an estimate may be made of the acidity of the excited state. This principle has been applied to some phenols under pressure[65] and the volumes of ionization of their excited singlet states measured. In all cases (Table 3.9) values are less negative by 7–$18\,cm^3\,mol^{-1}$ than for the ground states. This is most probably due to the fact that the excited states are all highly dipolar, (2), so that dissociation would resemble that of a zwitterion such as sulphanilic acid, (3), $\Delta \bar{V} = +2$.

3.2.2 Solvent Effects of Volumes of Ionization

Since the partial molar volumes of ions are more affected by solvent transfer than those of neutral substrates on the whole it would be expected that volumes of ionization should be quite solvent dependent. Relatively little information is available since ionizations of weak acids and bases will only occur at all in highly

$$\Delta \bar{V}^6 - 10 \cdot 9 \quad (cm^3\ mol^{-1})$$

(1) $\quad hv \rightarrow \quad$ +8 (2) $\quad$ +2 (3)

'polar' solvents, water, carboxylic acids, alcohols, and even then there are difficulties in interpretation due to ion-pairing and other changes in the nature of the solute species. Volumes of ionization of pyridine and piperidine in water and methanol have been compared,[29] Table 3.11, and shown to be far more negative in methanol due mainly to the smaller partial molar volumes of the ions in the less polar solvent (a consequence of its lower dielectric constant).

Table 3.11. Ionization volumes of base dissociations $(cm^3\ mol^{-1})$

B	$\bar{V}^0$ BOH	$\bar{V}^0$ B$^+$OH$^-$ (water)	$\Delta \bar{V}^0$	$\bar{V}^0$ BOMe	$\bar{V}^0$ B$^+$Me$^-$ (methanol)	$\Delta \bar{V}^0$
Pyridine	95.2	67.1	−28.1	118.9	65.5	−53.4
Piperidine	108.2	83.9	−24.3	134.9	85.4	−49.5

However, the precise identity of the species 'pyridinium hydroxide' and 'pyridinium methoxide' is uncertain. Since such species as (4) and (5) cannot exist in non-ionic form, it may be that these volumes to ion-pair dissociation:

(4) $\quad$ (5)

This information is now potentially available by way of partial molar volume measurements is view of the fact that $\bar{V}^0$ for single ions may be measured conveniently by ultrasonic vibration potentials in a variety of solvents.[53] Since values for the undissociated acids and bases are obtained from density measurements one may infer volumes of ionization for dissociations which occur hardly if at all.

For example, in ethanol

$$NH_4^+ \longrightarrow NH_3 + H_{solv}^+ \quad \Delta\bar{V}$$

$$\Delta\bar{V}^0 \quad 8.9 \qquad 28.8 \qquad -15.5 \qquad 4.4$$

compared with water,

$$12.2 \qquad 24.7 \qquad -5.7 \qquad 6.8$$

The indicator method may be used to determine $\Delta\bar{V}$ for dissociations in non-aqueous solvents such as methanol.[13] Values for some pyridines are available in this solvent but not at present in water for comparison.

3.2.3 Ion-Pairs Dissociation Volumes

Electrostatic attractions between oppositely charged ions result in pairing or higher aggregation of electrolytes, the extent of which depends upon the nature and charge of the ions and the solvent and concentration.[66] This association is characterized by an equilibrium constant with the usual thermodynamic properties. The extent of pairing may be determined by conductivity since only free ions are capable of contributing to the conductance of a solution. Partial molar volume measurements, however, can equally well be used. Volumes of formation of many ion-pairs are given in the recent review of Asano and LeNoble;[67] some representative values are given in Table 3.12.

$$C^+ + A^- \underset{\longleftarrow}{\overset{K}{\longrightarrow}} C^+A^-$$

The association of oppositely charged ions always results in an increase of volume since solvation of the aggregate is less complete by virtue of exclusion of solvent around the area of contact of the ions and the partial or even complete neutralization of charge which may result. Strong electrolytes such as Rb^+ and NO_3^- show a small value of $\Delta\bar{V}$, 5 cm^3 mol^{-1}, whereas weak electrolytes which on association form a bond with a great deal of covalent character have much larger values (e.g. Cu^{2+} OAc$^-$, 13; Fe^{3+} OH$^-$, 24).

The situation is more complex than this, however. Different types of ion-pair are well established. There are 'loose' or solvent-separated pairs in which one or more solvent molecules exist in the space between the ions which are still experiencing mutual electrostatic attraction; there are also 'tight' or contact ion-pairs which lack this intermediate solvation. Thus, one explanation for the difference in values of $\Delta\bar{V}$ for pairing of $RbNO_3$ and $TlNO_3$, both 1 : 1 salts with similarly sized cations, is that the former tends to associate to solvent separated

Table 3.12. Volumes of dissociation, $\Delta \bar{V}$, for ion-pairs[a]

$$[C^+A^-]_{solv} \xrightleftharpoons{K_m} C^+_{solv} + A^-_{solv}$$

Cation–anion pair		$K_m(\text{mol kg}^{-1})$ (1 bar)	$\Delta \bar{V}^a$ ($\text{cm}^3 \text{mol}^{-1}$)
Rb^+	NO_3^-		-5.2
Tl^+	NO_3^-		-12.2
Na^+	$SO_4^=$	0.029	-15.8
Mg^{++}	$SO_4^=$	4.7×10^{-3}	-8.0
Ca^{++}	$SO_4^=$	4.9×10^{-3}	-11.0
Zn^{++}	$SO_4^=$		-8.0
Mn^{++}	$SO_4^=$	4.4×10^{-3}	-7.4
N_i^{++}	$SO_4^=$	3.8×10^{-3}	$-10.2, -7.4(40)$
Co^{++}	$SO_4^=$	2.5×10^{-3}	-10.5
Cu^{++}	OAc^-		-13
Cu^{++}	HCO_2^- (30°)		-7
$Co(NH_3)_5NO_2^{++}$	$SO_4^=$		-10.5
La^{3+}	$SO_4^=$	2×10^{-13}	-22
Cr^{3+}	OH^-	1.4×10^{-10}	-13
Fe^{3+}	OH^-		-24
Fe^{3+}	Cl^-		-4.6
Fe^{3+}	SCN^-	6.9×10^{-3}	-15
Ce^{3+}	NO_3^-	0.034	-3.4
Ca^{++}	$CO_3^=$ (Solid)		-57
In acetone			
K^+	I^-	8.4×10^{-3}	-20
NMe_4^+	I^-	3.6×10^{-3}	-27
NE_4^+	I^-	7.7×10^{-3}	-17
MeN^+〈ring〉$-CO_2Me$ I^-		2.1×10^{-3}	-20
Co^{++}	$2Br^-$		-36
In ethanol			
K^+	I^-	0.001	-15
K^+	OEt^- (45°)		-39
NMe_4^+	Br^-	6.4×10^{-3}	-9
NBu_4^+	Br^-	7.5×10^{-3}	-8
Co^{++}	$2Cl^-$		-154
Co^{++}	$2Br^-$		-230.4
In 1-propanol			
NMe_4^+	Br^- (30°)		-16.2
Co^{++}	$2Cl^-$		-396
Co^{++}	$2Br^-$		-247.3
In 2-propanol			
NMe_4^+	Br^-		-20.7
	(0°)		-26
NBu_4^+	Br^- (3°)		-18
Co^{++}	$2Cl^-$		-65
Co^{++}	$2Br^-$		-36

Table 3.12. (*cont.*)

$$[C^+A^-]_{\text{solv}} \xrightleftharpoons{K_m} C^+_{\text{solv}} + A^-_{\text{solv}}$$

Cation–anion pair		$K_m(\text{mol kg}^{-1})$ (1 bar)	$\Delta \bar{V}^a$ $(\text{cm}^3\,\text{mol}^{-1})$
In isobutanol			
Na^+	I^- (30°)	1.4×10^{-3}	-15
Co^{++}	$2Cl^-$		-425
Co^{++}	$2Br^-$		-330
In ether solvents			
Na^+		Me_2O	-21
		THF	$-16\,(-24)$
(tight → solvent separated)			
Li^+		THF	$-10\,(-16)$
		Tetra-	-11
		hydropyran	
(tight → solvent separated)			

^{a} In water at 25° unless otherwise stated.

pairs ($\Delta \bar{V} = 5$) whereas the latter to tight pairs ($\Delta \bar{V} = 12\,\text{cm}^3\,\text{mol}^{-1}$).[68] This may be the result of the additional 'inert pair' of electrons which thallium possesses and which is available for covalency formation. Indeed distinct spectra of loose and tight ion pairs of the alkali metal salts of fluorene, in tetrahydrofuran, are observed. The effect of increasing pressure is to promote the transformation from tight to loose which is accompanied by a large volume diminution ($-24.5\,(Na^+)$ and $-15.6\,(Li^+)\,\text{cm}^3\,\text{mol}^{-1}$).[69] Solvent-separated ion-pairs are more capable of solvation and in this latter case a negative entropy effect is also observed supporting the notion of the immobilization of solvent molecules by the application of pressure.

$$\Delta V = -16\,\text{cm}^3\,\text{mol}^{-1}$$

Parallel with this observation is the conclusion of Symons[70] that an increase in dielectric constant favours a transformation from tight to loose ion aggregates and, as has been shown, the dielectric constant increases with pressure (section 2.1.14). Bjerrum[71] gave an equation relating the ion-pairing equilibrium constant with the average separation of the ions, a eq. (3.46)

$$K = \frac{10^3}{4\pi N Q(b) \rho} \left(\frac{\varepsilon k T}{4e^2} \right)^{1/2} \tag{3.46}$$

where

$$Q(b) = \int_2^b x^{-4} c^x \, dx \quad \text{and} \quad b = 4e^2/a\varepsilon kT$$

Some values of a for $ZnSO_4$ have been reported[72] and are found to increase with pressure ($a = 4.3$ Å (1 bar); 4.5 (1 kbar); 5.5 (3 kbar)). This is consistent with the gradual insertion of solvent under pressure between cation and anion or, better, the increasing difficulty of desolvation of the immediate solvation shell. A further complication may arise if conductivity is used to deduce ion-pair equilibria of acids. In addition to the increase in dissociation which may occur, the mobility of the proton also increases under pressure. This is due to normal ionic conductance of cationic proton carriers and also to the special mechanism of proton transfer in

aqueous and related solutions by the 'proton-jump' (excess conductance). Excess conductance can increase fourfold at 3 kbar (propanol as solvent).[73]

The volume change for ligand exchange is quite small but a very large effect of pressure accompanies an increase in coordination number, especially when charge separation occurs also [124] for example:

$$CoBr_2S_2 + 4S \longrightarrow Co\overset{++}{S_6} + 2Br^-, \quad \Delta\bar{V} = -109 \, cm^3 \, mol^{-1}$$

where S = acetone, but

$$CoBr_3S^- + Br^- \longrightarrow CoBr_4^= + S \qquad -0.8$$

Further complications exist when the cation is a transition metal since it will often have a number of molecules of water (usually four or six) coordinated and pairing with an anion such as halide may involve displacement of a water molecule and covalent binding of halide in the coordination sphere.[121] The pink octahedrally hydrated cobalt(II) ion, $Co(H_2O)_6^{++}$, is well known to be converted to a tetrahedrally substituted blue chloro species, $CoCl_4^=$ or an aquo analogue in the presence of chloride ion.[74, 75]

$$Co(H_2O)_6^{++} + nCl^- \rightleftharpoons Co(H_2O)_{4-n}Cl_n + (2+n)H_2O$$

There are presumably a series of such displacement equilibria but the octahedral hydrate is favoured over the tetrahedral species by pressure, $\Delta\bar{V}(t \to 0) = -16 \, cm^3 \, mol^{-1}$ (in aqueous sodium chloride as medium).[76] These changes mirror the increase in free energy of solvation of the octahedral ion over the tetrahedral. Tetrahedral nickel(II) tends also to be converted under pressure into an octahedral complex, $\Delta\bar{V} = -25 \, cm^3 \, mol^{-1}$. Possibly changes in the inner and outer solvation sphere are observed in manganous chloride.[77] Differences in molar volumes of the sulphates of Mn, Cu, and Zn are attributable to the successive tendency to form outer sphere rather than inner complexes.[125] In acetone, the tetrahedral solvate of the cobalt halides undergoes pressure-induced conversion to the octahedral solvate with concurrent charge separation and the

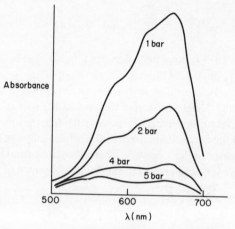

Fig. 3.10. Absorption spectrum of $CoCl_2$ in 1-butanol

halide ions released bring about a tetrahedral substitution reaction, almost pressure independent:[78]

$$CoCl_2ac_2 + 4ac \rightleftarrows Coac_6^{++} + 2Cl^-; \quad \Delta V^0 = -34$$

$$CoCl_2ac_2 + Cl^- \rightleftarrows CoCl_3ac^- + ac; \quad \Delta V^0 = -2$$

Similar transformations take place in alcohol solvents[79] in which volume changes of 150–400 $cm^3\,mol^{-1}$ are claimed, fig. 3.10, for primary alcohols, while secondary show a much smaller effect presumably due to a lesser ability to solvate.

Copper(II) in aqueous solution coordinates carboxylate ions with an overall increase in volume indicating covalency and charge neutralization in the complex. The volume changes are especially large for dibasic acids by which chelation can occur.[80]

The paramagnetic naphthalene anion is a useful probe for ion-pairing since the proximity of a $^{23}Na^+$ counterion (nuclear spin = 3/2) brings about nuclear-electron spin coupling observable in the e.s.r. spectrum. The magnitude of the

$$\Delta \bar{V} = +15\ cm^3\ mol^{-1}$$

coupling differs according to whether the ion-pair is intimate or solvent-separated and is pressure-dependent. These salts also show two peaks in the near ultraviolet spectrum assigned to tight and solvent-separated ion-pairs whose relative intensities may be used as a measure of the equilibrium constant. In tetrahydrofuran, the solvent-separated pairs exert more electrostriction and occupy 15 $cm^3\ mol^{-1}$ less space than the intimate pairs.[81] Similar results are inferred for sodium fluorenide in the same solvent, whose low dielectric constant is responsible for the large magnitude of the electrostrictive effect.[82,83] The volume change for fluorenyl lithium association in tetrahydrofuran is a even larger $-35\,cm^3\,mol^{-1}$.[131]

3.3 MISCELLANEOUS EQUILIBRIA

3.3.1 Hydrogen Bonding

Carboxylic acids in solution often exist in equilibrium with their hydrogen-bonded dimers,

$$2\ R-C\underset{O-H}{\overset{O}{\diagdown}}\ \rightleftharpoons\ R-C\underset{O-H\cdots O}{\overset{O\cdots H-O}{\diagup}}C-R$$

Association should be favoured by pressure although in hydrogen-bonding solvents the volume change may be offset by a reduction in the solvation of the carboxylic acid upon dimerization. In water, several such equilibria are indeed associated with a negative value of $\Delta\bar{V}$, as determined by conductivity (Table 3.12). For formic acid K is reported to increase steadily.[84] Other lower alkane carboxylic acids appear to show a maximum for association around 2–3 kbar. It is suggested that pressure, while increasing the hydrogen bonding interaction, decreases association between the alkyl groups, R, of the dimer when these are present which is a little strange since pressure usually increases association. It would be desirable to have confirmation of this maximum from physical measurements other than conductivity, e.g. infrared or Raman spectra.

Several studies of hydrogen bonding between alcohols in 'inert' solvents such as carbon tetrachloride and carbon disulphide, have been made.[84,86]

$$n\text{ROH} \rightleftharpoons (\text{ROH})_n$$

A large increase in intensity of the H-bonded OH stretching band in the infrared occurs under pressure, associated with $\Delta V^0 \sim -4\,cm^3\,mol^{-1}$ although this value depends on the number assigned to n. There is also a shift of the 'free' OH band, a sharp absorption at $ca.$ 3600 cm^{-1} towards lower energy as the pressure increases. This shift is solvent dependent and varies with both density and polarizability showing that its origin is in an increase in the attractive interactions between monomeric alcohol and the solvent. The ultraviolet spectrum of phenol in dioxan (absorbing at 280 nm) shows a small pressure-dependent shift to longer wavelength and this is attributed to hydrogen bonding between phenol and solvent as an oxygen donor,[87] $\Delta V^0 = -3.2\,cm^3\,mol^{-1}$.

$$\text{PhO-H} \quad \bigcirc \quad \rightleftharpoons \quad \text{PhO-H}\cdots\bigcirc$$

3.3.2 Organic Charge-transfer Complexes

Many molecules with high-lying filled molecular orbitals ('donors') are known to interact in solution with molecules having low-lying vacant orbitals ('acceptors'), the interaction being described as orbital overlap.[88] The complex so formed has only a transient existence (and indeed may be no more than a somewhat prolonged or 'sticky' collision between the two) as shown by the fact that no detectable change in the colligative properties of the solutions occur on mixing.[89] Nonetheless, the mixture of donor and acceptor species may result in an intense visible absorption since the interaction of the molecular orbitals gives rise to a new pair with a lower separation than between those in either component. Absorption of light may occur on a very short time-scale (10^{-16} s), much less than the lifetime. Typical donors are: aromatic molecules, especially with donor substituents, (methyl, —OR, —NR$_2$), polyenes, ether, amines. Typical acceptors are: tetracyanoethylene (1), chloranil (2), tetracyanoquinodimethane (3), polynitrobenzenes (e.g. 4), and generally π-systems flanked by numerous electron-withdrawing groups (—NO$_2$, —CN, —C—R), the halogens.

(1)   (2)   (3)   (4)

Colourless   ($\delta+$)   ($\delta-$)

Red

The usual method of evaluating the equilibrium constant is by the method of Benesi and Hilebrand.[88] Since optical absorption is the only means of measuring the concentration of the complex and since molar absorbance of the complex cannot be independently determined it is necessary to measure the absorbance at the maximum of the charge transfer band of a series of solutions in which the concentration of one component remains constant while that of the other varies. Then:

$$\frac{c_A l}{A_{AB}} = \frac{1}{K \varepsilon_{AB}} \frac{1}{c_B} + \frac{1}{\varepsilon_{AB}}$$

Table 3.13. Miscellaneous volumes of reaction, $\Delta \bar{V}$ (cm^3 mol^{-1})

	$\Delta \bar{V}$	Reference
(a) Hydrogen bonding		
HCOOH dimerization, 30°, H$_2$O	−14	40
CH$_3$COOH dimerization, 30°, H$_2$O	−13	
C$_2$H$_5$COOH dimerization, 30°, H$_2$O	−8.8	
C$_3$H$_7$COOH dimerization, 30°, H$_2$O	−6.2	
nC$_4$H$_9$OH dimerization, 25°, CS$_2$	−2.2, −4.6	84, 86
Phenol-dioxan, 30°, hexane	−3.2	87
(b) Charge transfer complexation		
1 + Hexamethylbenzene, 30°, CH$_2$Cl$_2$	−12 (−14.1)	89, 127, 128
1 + Hexamethylbenzene (dichloroethane)	−11.4	129
(chloroform)	−9.0	
(hexane)	−4.3	
1 + naphthalene	−4	
1 + Mesithylene	−7.1	
1 + Benzene	−3	
1 + Toluene	−4.9	
2 + Hexamethylbenzene	−11	
2 + Naphthalene	−5	
4 + Anthracene	−5	
4 + Naphthalene	−3	
N-Methyl-4-methoxycarbonylpyridinium iodide (5)	+16	95
Trinitrobenzene-iodine	0	
Tropyllium cation-hexamethylbenzene	0	
Dimerization		

2 NO$_2$ $\rightleftharpoons$ O$_2$N—NO$_2$ −23   100

2   $\rightleftharpoons$ −34   101

(c) Crown ether complexation, water, 25	A	B	134, 135
A = '18-crown-6' NaCl	+12	+8	
B = '15-crown-5' KCl	+13	+8	
RbCl	+9	+4	
CsCl	+9	+1.5	
CaCl$_2$	+24	+0.1	
BaCl$_2$	+19		

where C_A is the concentration of the constant component and C_B that of the variable one, A_{AB} is the absorbance of the complex, ε_{AB} its molar absorbance, and l, the path length. By plotting the left-hand side against $1/c_B$ both K and ε_{AB} may be evaluated from slope and intercept.

Spectra of charge-transfer complexes between neutral molecules increase in intensity indicating an increase in the stability constants with pressure (Table 3.12).[89-93] Values of $\Delta\bar{V}(-3$ to $-14\,cm^3\,mol^{-1})$ are consistent with this association and with charge separation which is the more pronounced the better

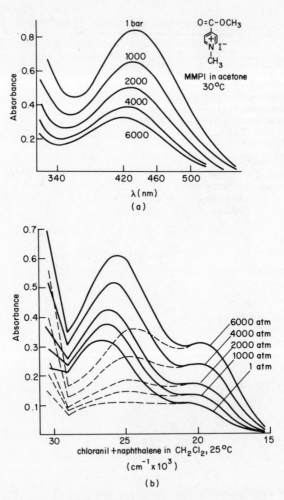

Fig. 3.11. (a) Charge-transfer absorption band of N-methyl-4-methoxycarbonylpyridinium iodide in acetone, 30.[95] (b) Absorption spectrum of the chloranil-naphthalene charge-transfer complex in CH_2Cl_2. The broken curves are of absorption corrected for chloranil absorption[89]

the donor or acceptor. By carrying out the measurements over a range of temperature in addition it was shown that both ΔH and ΔS for complex formation become more negative with pressure which is consistent with increased solvation of the complex, perhaps reflecting a greater degree of charge-transfer under pressure. This has the further effect of causing an increase in ε and a red-shift in the spectrum, i.e. excitation of the complex requires less energy under pressure, about $6\,\mathrm{nm\,kbar^{-1}}$, and an increase in the oscillator strength of the band. If the charge-transfer interaction is between oppositely charged ions and involves partial neutralization of charge, more completely upon excitation in the charge-transfer band, then the effect of pressure will be to diminish K. Such behaviour is found in pyridinium iodides.[5] such as the solvatochromic dye,[95] 5. Spectra are reduced in intensity and shifted towards the blue on pressurization, fig. 3.11a,[96] and a substantial positive value of $\Delta \bar{V} = +16\,\mathrm{cm^3\,mol^{-1}}$ is inferred. Finally, complexes between charged and neutral components show no pressure effect on K, $\Delta \bar{V} = 0$, since both complexation and excitation involve charge dispersion. The clear-cut difference in behaviour of these three types leaves little doubt that solvation affects behaviour and makes these systems useful probes of intermolecular interactions.

Many charge-transfer complexes exist in the solid state and may have well-defined crystal structures. The effect of pressure may be tested to hundreds of kbar[97] and it is found that electrical and optical properties are affected. The 1:1 complex of TCNE with perylene shows an 80% drop in resistance by 100 kbar; those of TCNQ-triethylammonium and TCNE-iodine first decrease then increase rapidly. Frequently at these high pressures, irreversible changes and chemical reactions occur[98, 99] (section 3.1.6).

5

3.3.3 Conformational Equilibria

Cyclohexanes exist in two 'chair' conformations interchangeable by rotation about carbon–carbon bonds. The two conformers of a substituted cyclohexane may not be equivalent either in energy or in volume, in which case pressure should affect the equilibrium.

1,4-Dibromocyclohexane exists as an equilibrium mixture of the diaxial (aa) and diequatorial (ee) conformers the former being the more stable by a small amount. The conformers are differentiated by their infrared spectra (ee, 738 and aa, $728\,\mathrm{cm^{-1}}$). Under pressure, the diaxial conformer increases at the expense of

$$\Delta \bar{V}(\text{ee} \rightarrow \text{aa}) = -3 \cdot 8 \text{ cm}^3 \text{ mol}^{-1}$$

the diequatorial by 15% at 1 kbar. Similar results are obtained with the dichloro analogue. The diequatorial isomer appears to occupy a larger volume than the diaxial by $3.8 \text{ cm}^3 \text{ mol}^{-1}$ (2.8 for the dichloro).[102] Chlorocyclohexane also shifts in favour of the axial conformer under pressure, $\Delta \bar{V}(\text{eq} \rightarrow \text{ax}) = -1.87 \text{ cm}^3 \text{ mol}^{-1}$.

Among substituted ethanes, trichloroethane has been studied by infrared spectroscopy under pressure. Two conformers are present, the gauche and the 'pseudo-trans':

$$\Delta \bar{V}(\text{t} \rightarrow \text{g}) = -3 \cdot 8$$

gauche   pseudo-trans

The gauche form is favoured under pressure, $\Delta \bar{V} = -3.8 \text{ cm}^3 \text{ mol}^{-1}$. It is possible that the reason for this is the higher dipole moment possessed by the gauche conformer which, unlike the dihalocyclohexanes, is not centrosymmetric. The isomer with higher dipole moment should show greater electrostriction and hence $\Delta \bar{V}$ should be more negative in solvents of low dielectric constant.

Little information is available concerning geometrical isomers although isomerization of alkenes may occur at sufficiently high temperature. *Cis*-dichloroethylene is favoured over *trans* at 185°, $\Delta \bar{V} = 2.4 \text{ cm}^3 \text{ mol}^{-1}$.[105]

3.3.4 The Reduction of Potassium Amide

Hydrogen reacts with amide ion in liquid ammonia with the equilibrium formation of solvated electrons giving rise to a blue colour and strong absorption in the near infrared:

$$K^+NH_2^- + 1/2H_2 \rightleftharpoons NH_3 + e^-_{\text{solv}} + K^+$$

At 25°, $K = 2 \times 10^{-5}$, $\Delta H = 67 \text{ kJ mol}^{-1}$. No change in the number of charges is involved but the equilibrium constant is very pressure dependent and shifted to the left with $\Delta \bar{V} = +64 \text{ cm}^3 \text{ mol}^{-1}$. This very large value appears to be associated with a large partial molar volume for the solvated electron, 98 cm^3 mol^{-1} at 100 bar. This must be due to the involvement of a considerable number of ammonia molecules involved in the solvation and their arrangement in a packing of low density. If the electron were shared by several solvent molecules, mutual repulsion of the charge would ensure this.

3.3.5 Micelle Formation in Detergents

Molecules with an ionic group such as —NR$_3^+$ or —SO$_3^-$ attached to a long hydrocarbon 'tail' are surface-active in water and act as detergents, emulsifying oils which are entrained within a molecular aggregate known as micelle. Micelles form spontaneously above a minimum concentration (critical micellar concentration, CMC) and have the polar groups oriented outwards into the water while the hydrocarbon ends are located in the centre. Micelle formation is accompanied by the loss of much hydrophobic interaction with the solvent and a discontinuity in the partial molar volume.[107]

Micelle formation is important both for surface activity and also for catalysis of many reactions involving polar transition states.[108] Several studies under pressure have shown that the CMC increases with pressure, that micelle formation is associated with a positive volume change, $\Delta \bar{V} = RT \, d(CMC)/dP$. This amounts to $+11$ cm^3 mol^{-1} for sodium dodecyl sulphate[109] and $+10$ for a cationic micelle[126] but may be more complicated as the value decreases with pressure and eventually becomes negative.[110-112] This is because the micelle is more compressible than the separate dissolved molecules.[113] The partial molar volumes of both cationic and anionic surfactants show a sharp increase at the CMC in agreement with this[114] (Figures 3.12 and 3.13).

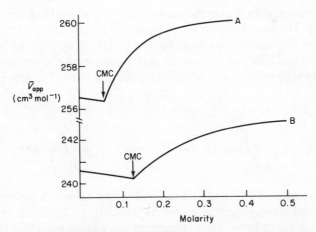

Fig. 3.12. Apparent partial molar volumes of some detergents: (a) decyltrimethylammonium bromide, (b) nonyltrimethylammonium bromide. CMC = critical micellar concentration

176

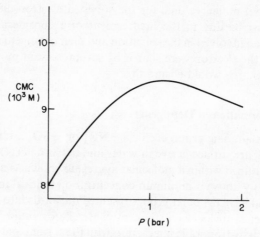

Fig. 3.13. Variation of critical micellar concentration of sodium dodecyl sulphate as a function of pressure

3.3.6 Keto–enol Tautomerism

The proton-transfer

is a frequent pre-equilibrium step in reactions which lead to substitution α- to a carbonyl group. The equilibrium concentration of the enol form is normally very small though in β-dicarbonyl compounds it is stabilized by internal hydrogen bonding and infrared spectroscopy can be used to determine K. Pressure appears to favour the keto form and the volumes of enolization of ethyl acetoacetate and

acetylactone are around $+5$ and $+3\,\mathrm{cm^3\,mol^{-1}}$ but solvent dependent in an apparently capricious way (acetylacetone; $\Delta V(\mathrm{MeOH})$, $+12$; (EtOH), 0).[115, 116] It would be desirable to know more concerning enolization equilibria of simple ketones. It is predicted that pressure favours the most solvated species which, for a mono-carbonyl compound, and in a protic solvent would be expected to be the enol.[117]

3.3.7 Crown Ether Complexation

A variety of macrocyclic polyethers are known which complex alkali metal cations resulting in solubilization of inorganic salts in a variety of solvents:

'18-crown-6' $\Delta \bar{V} = +12 \text{ cm}^3 \text{ mol}^{-1}$

The association constant depends upon the ring size and the diameter of the cation, requiring a good fit between the two for strong binding. In water, the volume change for complexation is positive (Table 3.13) as a result of the release of electrostricted solvent. The partial molar compressibility of the complex is also greater than that of the components which may indicate an increase in solvation upon compression.[134–135] In the example above, for instance, it is concluded that some 5 molecules of water are released on complexation, and 7–8 for complexation of a doubly charged cation such as Ca^{++}. The volume changes are smaller when the fit is less good such as between K^+ and the smaller crown ether, '15-crown-5' in which it is presumed that residual hydration of the potassium in the complex is possible and only about three molecules of water are released.

REFERENCES

1. E. A. Guggenheim, *Thermodynamics*, North Holland, Amsterdam (1959).
2. M. L. McGlashan, *Chemical Thermodynamics*, Amer. Chem. Soc. (1977).
3. G. J. Hills, *Soc. Exptl Biol. Symp. XXVI.* Chap. 1, Cambridge University Press (1972).
4. S. D. Hamann, *Mod. Aspects of Electrochem.*, **9**, B. Conway and J. Bockris (Eds.), Plenum, New York (1974).
5. T. W. Weber, *Chem. Eng. N.Y.*, **81**, (24), 153, 162 (1974).
6. M. Planck, *Ann. Phys.*, **32**, 462 (1887).
7. D. A. Lown and Lord Wynne-Jones, *J. Sci. Instr.*, **6**, 694 (1973).
8. D. A. Lown, H. R. Thirsk, and Lord Wynne-Jones, *Trans. Farad. Soc.*, **66**, 51 (1970).
9. E. M. Wooley and R. E. George, *J. Sol. Chem.*, **3**, 119 (1974).
10. A. Disteche and S. Disteche, *J. Electrochem. Soc.*, **112**, 350 (1965); **114**, 330 (1967).
11. S. D. Haman and M. Linton, *J. Chem. Soc., Farad. Trans. I*, **70**, 2239 (1974).
12. H. P. Hopkins, W. C. Duer, and F. J. Millero, *J. Sol. Chem.*, **5**, 263 (1976).
13. W. J. LeNoble and T. Asano, *J. Org. Chem.*, **40**, 1179 (1975).
14. R. C. Neumann, E. Kauzmann, and A. Zipp, *J. Phys. Chem.*, **77**, 2687 (1973).
15. M. Tsuda, I. Shirotani, S. Minomura, and Y. Terayama, *Bull. Chem. Soc. Jap.*, **49**, 2952 (1976).
16. F. J. Millero, *Water, Aqueous Solutions: Structure, Thermodynamics, and Transport Processes*, R. A. Horne (Ed.), 519, Interscience (1972).
17. F. T. Millero, *Chem. Rev.*, **71**, 147 (1971).
18. J. R. McCabe, R. A. Grieger, and C. A. Eckert, *Ind. Eng. Chem. Fundam.*, **9**, 156 (1970).
19. M. J. R. Dack, K. Bird, and A. J. Parker, *Austr. J. Chem.*, **28**, 955 (1975).
20. R. A. Grieger, C. Chandair, and C. E. Eckert, *Ind. Eng. Chem. Fundam.*, **10**, 24 (1971).

21. B. E. Conway, R. E. Verrall, and J. E. Desnoyers, *Can. J. Chem.*, 2738 (1966).
22. F. Kawaizumi and R. Zana, *J. Phys. Chem.*, **78**, 627 (1974).
23. F. Kawaizumi and R. Zana, *J. Phys. Chem.*, **78**, 1099 (1974).
24. F. J. Millero, *Rev. Sci. Instr.*, **38**, 1441 (1967).
25. M. R. J. Dack, K. J. Bird, and A. J. Parker, *Austr. J. Chem.*, **28**, 955 (1975).
26. R. Picker, E. Trenblay, and C. Jolicoeur, *J. Sol. Chem.*, **6**, 733 (1977).
27. O. Kratky, H. Leopold, and H. Stabinger, *Z. Angew. Phys.*, **27**, 273 (1969).
28. R. Zana and E. Yeager, *J. Phys. Chem.*, **70**, 594 (1966); **71**, 521 (1967).
29. S. D. Hamann and S. C. Lim, *Austr. J. Chem.*, **7**, 329 (1954).
30. J. R. Partington, *Advanced Treatise on Physical Chemistry*, Vol. II, 23, Longmans, London (1951).
31. H. Høiland and E. Vikingstad, *Proc. 4th Conf. Chem. Thermodyn.*, Marseilles.
32. S. Cabani, V. Mallica, L. Lepori, and S. T. Lobo, *J. Phys. Chem.*, **81**, 982 (1977).
33. J. Buchanan and S. D. Hamann, *Trans. Farad. Soc.*, **49**, 1425 (1953).
34. M. E. Friedman and H. A. Scheraga, *J. Phys. Chem.*, **69**, 3799 (1965).
35. H. Høiland, *Acta. Chem. Scand. A.*, 699 (1974).
36. A. Missenard, *Chem. Ind. Genie Chim.*, **95**, 632 (1966).
37. W-Y Wen, A. LoSurdo, C. Jolicoeur, and J. Boileau, *J. Phys. Chem.*, **80**, 466 (1976).
38. S. Cabani, G. Conti, L. Lepori, and G. Leva, *J. Phys. Chem.*, **76**, 1338 (1972).
39. H. Høiland, *J. Chem. Soc. Farad. Trans. I*, **70**, 1180 (1974).
40. K. Susuki, Y. Taniguchi, and T. Watanabe, *J. Phys. Chem.*, **77**, 1918 (1973).
41. M. Nakahara, *Rev. Phys. Chem. Jap.*, **44**, 57 (1973).
42. B. B. Owen and S. R. Brinkley, *Chem. Rev.*, **29**, 461 (1941).
43. R. E. Gibson, *Amer. Sc. J.*, **35A**, 49 (1938).
44. B. S. El'yanov and M. G. Gonikberg, *Izv. Akad. Nauk. SSSR, Ser, Khim. Nauk.*, 1044 (1967).
45. N. A. North, *J. Phys. Chem.*, **77**, 931 (1973).
46. H. Høiland, *J. Chem. Soc. Farad. I*, 1180 (1974).
47. L. G. Hepler, *J. Phys. Chem.*, **69**, 965 (1965).
48. S. D. Hamann and M. Linton, *J. Chem. Soc. Farad. I.*, 485 (1975).
49. S. D. Hamann and M. Linton, *J. Chem. Soc. Farad. I*, 70, 2239 (1974).
50. J. J. Christianson, R. M. Izatt, D. P. Wrathall, and L. D. Hansen, *J. Chem. Soc. A*, 1212 (1969).
51. C. L. Liotta, E. M. Perdue, and H. P. Hopkins, *J. Amer. Chem. Soc.*, **95**, 2439 (1975).
52. S. Katz and J. E. Miller, *J. Phys. Chem.*, **75**, 1120 (1971).
53. F. Kawaizumi and R. Zana, *J. Phys. Chem.*, **78**, 1099 (1974).
54. J. G. Mathieson and B. E. Conway, *J. Chem. Soc. Farad. I*, **70**, 752 (1974).
55. H. Høiland, *J. Chem. Soc., Farad. I.*, **71**, 797 (1975).
56. H. Høiland and E. Vikingstad, *J. Chem. Soc. Farad.*, **171**, 2007 (1975).
57. H. Høiland, E. Vikingstad, and T. Brun, *Proc. 4th. Int. Conf. Chem. Thermodynam.*, Marseilles; J. Rouquerol and R. Sabbah (Eds.), 73 (1975).
58. C. L. Liotta, A. Abidaud, and H. P. Hopkins, *J. Amer. Chem. Soc.*, **94**, 8624 (1972).
59. A. J. Read, *J. Sol. Chem.*, **4**, 53 (1975).
60. G. K. Ward and F. J. Millero, *J. Sol. Chem.*, **3**, 417 (1974).
61. G. Oloffson and L. G. Hepler, *J. Sol. Chem.*, **4**, 127 (1975).
62. F. J. Millero, E. V. Hoff, and L. Cahn, *J. Sol. Chem.*, **1**, 309 (1972).
63. T. Förster, *Z. Elektrochem.*, **54**, 44 (1950).
64. A. Weller, *Z. Elektrochem.*, **56**, 662 (1952).
65. S. D. Hamann, *Austr. J. Chem.*, **28**, 701 (1975).
66. A. K. Covington and T. Dickinson, *Physical Chemistry of Organic Solvent Systems*, Plenum, London (1973).
67. T. Asano and W. J. LeNoble, *Chem. Rev.* **78**, 407 (1978).
68. W. L. Masterton, H. Welles, J. H. Knox, and F. J. Millero, *J. Sol. Chem.*, **3**, 91 (1974).
69. S. Claesson, B. Lundgren, and M. Szwarc, *Trans. Farad. Soc.*, **66**, 3053 (1970).

70. T. R. Griffiths and M. C. R. Symons, *Mol. Phys.*, **13**, 90 (1960).
71. N. Bjerrum, *Kgl. Danske. Videnskab. Selskab.*, **4**, 26 (1906); **7**, 9 (1926).
72. Y. Tamiguchi, T. Watanabe, and K. Susuki, *Bull. Chem. Soc. Jap.*, **48**, 3032 (1976).
73. J. F. Cukerins and W. Strauss, *Austr. J. Chem.*, **29**, 249 (1976).
74. H-D. Ludemann and E. U. Franck, *Ber. Bunsenges. Phys. Chem.*, **71**, 455 (1967); **72**, 514 (1968).
75. A. H. Ewald and D. S. D. Hamann, *Austr. J. Chem.*, **9**, 54 (1956).
76. S. Rodriguez and H. Offen, *Inorg. Chem.*, **10**, 2086 (1971).
77. F. H. Fisher and D. F. Davies, *J. Phys. Chem.*, **69**, 2595 (1965).
78. I. Ishihara, K. Hara, and J. Osugi, *Rev. Phys. Chem. Jap.*, **44**, 11 (1974).
79. Y. Kitamura, *Rev. Phys. Chem. Jap.*, **39**, 1 (1969).
80. S. Katz, M. P. Donovan, and L. C. Robertson, *J. Phys. Chem.*, **79**, 1930 (1975).
81. W. J. LeNoble and P. Staub, *J. Organometall. Chem.*, **156**, 25 (1978).
82. S. Claesson, B. Lundgren, and M. Szwarc, *Trans. Farad. Soc.*, **63**, 3053 (1970).
83. B. Lundren, S. Claesson, and M. Szwarc, *Chem. Scripta*, **3**, 49 (1973).
84. E. Fishmann and H. Drickamer, *J. Chem. Phys.*, **24**, 548 (1956).
85. N. I. Shishkin and I. I. Norak, *Zhur. Tekh Fiz.*, **23**, 1485 (1953).
86. J. Osugi and Y. Kitamura, *Rev. Phys. Chem. Jap.*, **35**, 25 (1965).
87. K. Susuki and M. Tsuchina, *Bull. Chem. Soc. Jap.*, **48**, 1701 (1975).
88. H. A. Benesi and J. H. Hildebrand, *J. Amer. Chem. Soc.*, **71**, 2703 (1949).
89. A. H. Ewald, *Trans. Farad. Soc.*, **64**, 733 (1968).
90. T. Nakayama and J. Osugi, *Rev. Phys. Chem. Jap.*, **45**, 79 (1975).
91. T. Nakayama, M. Sasaki, and J. Osugi, *Rev. Phys. Chem. Jap.*, **46**, 57 (1976).
92. T. Nakayama and J. Osugi, *Rev. Phys. Chem. Jap.*, **45**, 79 (1975).
93. J. v. Jouanne, D. A. Palmer, and H. Kelm, *Bull. Chem. Soc. Jap.*, **51**, 463 (1978).
94. J. Osugi *et al.*, *Proc. 6th AIRAPT Conf. (Boulder)*, Plenum 651 (1979).
95. A. H. Ewald and J. A. Scudder, *J. Phys. Chem.*, **76**, 249 (1972).
96. J. Ham, *J. Amer. Chem. Soc.*, **76**, 3875 (1954).
97. H. G. Drickamer and C. W. Frank, *Electronic Transitions and the High Pressure Chemistry and Physics of Solids*, Chapman and Hall, London (1973).
98. T. Sakata, A. Onodera, H. Tsubomura, and N. Kawai, *J. Amer. Chem. Soc.*, **96**, 3365 (1974).
99. A. Onodera, T. Sakata, H. Tsubomura, and N. Kawai, *Proc. 4th ARIAPT Conf. Kyoto*, 713 (1974).
100. A. H. Ewald, *Disc. Farad. Soc.*, **22**, 138 (1956).
101. M. G. Gonikberg and L. F. Vereshchagin, *Z. Fiz. Khim.*, **23**, 1447 (1949).
102. S. D. Christian, J. Grundnes, and P. Klaboe, *J. Amer. Chem. Soc.* **97**, 3864 (1975).
103. S. D. Christian, J. Grundnes, P. Klaboe, C. J. Nielson, and T. Wildbaek, *J. Mol. Str.*, **43**, 33 (1976).
104. S. D. Christian, J. Grundnes, and P. Klaboe, *J. Chem. Phys.*, **65**, 496 (1976).
105. A. H. Ewald, S. D. Hamann, and J. E. Stutchbury, *Trans. Farad. Soc.*, **53**, 991 (1957).
106. W. Schindewolf, R. Vogelgesang, and K. W. Böddeker, *Angew. Chem. Int. Ed.*, **6**, 1076 (1967).
107. G. M. Musbally, G. Perron, and J. E. Desnoyes, *J. Coll. Interfac. Sc.*, **48**, 494 (1974).
108. E. J. Fendler and J. M. Fendler, *Adv. Phys. Org. Chem.*, **8**, (1970).
109. S. D. Hamann, *J. Phys. Chem.*, **66**, 1359 (1962).
110. R. F. Tuddenham and A. E. Alexander, *J. Phys. Chem.*, **66**, 1839 (1962).
111. S. Kaneshina, *Hyomen*, **12**, 197 (1974).
112. T. S. Brun, H. Høiland, and E. Vikingstad, *J. Coll. Interfac. Sc.*, **63**, 89 (1978).
113. S. D. Hamann, *Rev. Phys. Chem. Jap.*, **48**, 60 (1978).
114. J. E. Desnoyers and M. Arel, *Can. J. Chem.*, **45**, 3598 (1967).
115. J. Osugi, T. Mizukami, and T. Tachibana, *Rev. Phys. Chem. Jap.*, **37**, 72 (1968).
116. J. von Jouanne and J. Heidbig, *J. Magnet. Res.*, **7**, 1 (1972).
117. M. I. Kabachnik, *Izvest. Akad. Nauk. SSSR, Odtel Khim. Nauk.*, 98, (1955).

180

118. W. Holtzapfel, *J. Chem. Phys.*, **50**, 4424 (1969).
119. O. Redlich and D. M. Mayer, *Chem. Rev.*, **64**, 221 (1964).
120. R. E. Gibson and J. F. Kincaid, *J. Amer. Chem. Soc.* **59**, 579 (1937).
121. C. A. Angell and M. L. Abkemeier, *Inorg. Chem.*, **12**, 1462 (1973).
122. B. S. El'yanov and S. D. Hamann, *Austr. J. Chem.*, **28**, 945 (1975).
123. T. Matsuura, N. Kuraki and K. Konishi, *Sen-i Gakkaishi*, **21**, 598 (1965).
124. I. Ishihara, *Rev. Phys. Chem. Jap.*, **48**, 27 (1978).
125. K. Shimizu, N. Tsuchihashi, and Y. Kondo, *Rev. Phys. Chem. Jap.*, **47**, 80 (1977).
126. M. Ueno, M. Nakahara, and J. Osugi, *Rev. Phys. Chem. Jap.*, **47**, 25 (1977).
127 T. Nakayama, M. Susuki, and J. Osugi, *Rev. Phys. Chem. Jap.*, **46**, 57 (1976).
128. T. Nakayama and J. Osugi, *Rev. Phys. Chem. Jap.*, **45**, 79 (1975).
129. T, Nakayama, *Rev. Phys. Chem. Jap.*, **44**, 26 (1974).
130. K. Shimizu, N. Tsuchihashi, and Y. Furumi, Rev. Phys., Chem. Jap., **46**, 30 (1976).
131. W. J. LeNoble and A. R. Das, *J. Phys. Chem.*, **74**, 3429 (1970).
132. S. D. Hamann *et al.*, *Trans. Farad. Soc.*, **49**, 142 (1953); **51**, 1684 (1955); *Disc. Farad. Soc.*, **22**, 70 (1956).
133. W. Strauss, *Aust. J. Chem.*, **10**, 359 (1957).
134. H. Høiland, J. A. Ringseth, and E. Vikingstad, *J. Solution Chem.*, **7**, 515 (1978).
135. H. Høiland, J. Ringseth, and T. S. Brun, *J. Solution Chem.*, **8**, 779 (1979).

Chapter 4

Effects of Pressure on Rate Processes

4.1 THE EFFECT OF PRESSURE ON REACTION RATES

4.1.1 Transition State Theory

The empirical dependence of reaction rates upon appropriate concentrations of reagents has been recognized since the pioneering experiments of Bodenstein on the combination of hydrogen and iodine[1] (1887). Thermodynamic concepts were introduced into the theory of rate processes in 1935 by Polanyi[2] and by Eyring[3] by means of the transition state theory. According to their ideas, a reaction proceeds by a smooth and gradual rearrangement of atomic positions between those of reagents and those of products. The energy of the system, however, increases initially to a maximum before falling to that of products. The maximum corresponds to an intermediate structure known as the activated complex or transition state and it is the necessity to pass through this high energy species which explains the increase in rate with temperature. The transition state, although possessing only metastable existence, is ascribed all the thermodynamic properties of a normal species, energy, entropy, volume, etc. and is assumed to exist in equilibrium with reagents. The rate of reaction will be the rate at which the transition state passes along the reaction coordinate to products. This will depend upon the equilibrium constant for its formation, K^+, and a 'transmission coefficient', κ, expressing the probability of the transition state going forward to products against returning to reagents:

$$\text{Reagents} \quad \underset{\xrightarrow{\hspace{1cm}}}{\overset{K^+}{\rightleftharpoons}} \quad \begin{array}{c}\text{transition}\\\text{state}\end{array} \quad \xrightarrow{\hspace{1cm}} \quad \text{Products}$$

the specific rate coefficient, k, is given by

$$k = \frac{\mathbf{k}T}{h} K^+$$

$$= \frac{\mathbf{k}T}{h} \exp\left(\frac{-\Delta G^+}{RT}\right) \cdot \frac{\prod^{\text{rgts}} \gamma}{\gamma^+} \tag{4.1}$$

where $\mathbf{k}$, h are the Boltzmann and Planck constants and ΔG^+ the standard free energy change between reagents and transition state ('free energy of activation'). The activity coefficient term is usually assumed, for dilute solution and because of lack of knowledge, to be unity and will not be explicitly included subsequently. Partial differentiation gives, analogous to eq. (3.29),[227]

$$\left(\frac{\partial \ln k}{\partial P}\right)_T = -\frac{\partial}{\partial P} \cdot \frac{\Delta G^+}{RT} = \frac{-\Delta V^+}{RT} \tag{4.2}$$

where ΔV^+ is the difference in partial molar volumes between reagents and transition state, known as the 'volume of activation'. Since $\Delta V^+ = (\Sigma_{\text{rgts}} \bar{V} - \bar{V}^+)$ and since it is possible to measure partial molar volumes of reagents with ease, this equation permits the determination of the partial molar volume of the transition state, one of the few absolute properties of the transition state to be accessible. Equation (4.2) predicts an increase in rate if the transition state occupies a smaller volume than the total volume of reagents and conversely and, since rates may be measured to a high degree of precision, provides an extremely useful diagnostic probe into reaction mechanisms. The volume of activation is the most important parameter to be obtained from pressure studies on reaction rates and several thorough reviews have been devoted to the subject.[4-9, 220-223, 231]

4.1.2 The Measurement of Activation Volumes

The temperature coefficient of reaction rate, $\partial \ln k/\partial T^{-1} = E_a/RT^2$ gives the 'activation energy' E_a and is almost constant over a large range of temperature since the heat capacities of reagents and transition state are rather temperature invariant. This is not so for the pressure derivative which yields ΔV^+, since volumes of both reagents and transition states change with pressure and there is no *a priori* reason for their having similar compressibilities. In general, plots of $\ln k$ against pressure may have slopes ranging from positive to negative but they invariably curve towards the pressure axis, that is, the slope and consequently $\langle \Delta V^+ \rangle$ tends to diminish with pressure. It is purely a convention that the slope at $P \to 0$ is the quantity recorded for mechanistic studies and comparison and should properly be denoted, ΔV_0^+ although the subscript is usually understood. Fig. 4.1 shows typical rate plots. The evaluation of ΔV_0^+ requires a measurement of the slope of the curve as it intercepts the y-axis and this is fraught with some difficulties since there is no theoretical function to describe the curve and the part of interest lies on an extrapolated sector. It is clear, though, that the experimental points should be made at frequent intervals at the low pressure end. Several procedures are available for the determination of ΔV_0^+:

(a) *Curve-fitting to a polynomial* This is the procedure adopted today by most workers. The data are fitted by means of a suitable computer program to a polynomial of second order or, if it is of sufficient precision third order, using a least squares criterion:

$$\ln k_p = A + BP + CP^2(+DP^3) \tag{4.3}$$

in which $A = \ln k_0$ and may or may not be constrained to this value. A plot of $\log(k_p/k_0)$ against pressure is more linear than of $\log k_p$ and more precisely extrapolated to $P = 0$.[224] It has been suggested that a cubic equation,

$$\ln k_p = A \text{ (or } \ln k_0) + BP + CP^3 \tag{4.4}$$

may better fit data, especially over a large pressure range, since it would show an inflection but no turning point as does the parabola, and this latter feature is absent from real data.[452]

(b) *A graphical method* has been proposed by Whalley[9] in which a smooth curve is drawn through the experimental points and the average slopes (chords) at a

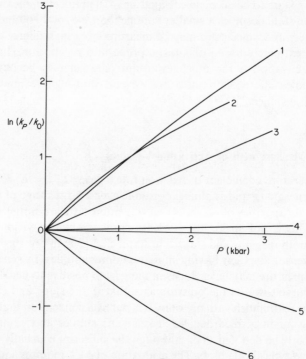

Fig. 4.1. Typical rate-pressure plots: (1) dimerization of cyclo-pentadiene at 20°; (2) nitration of benzene, 0°; (3) nitration of toluene, 0°; (4) hydrolysis of s-trioxan; (5) rearrangement of N-chloroacetanilide, 25°; (6) hydrolysis of $Co(NH_3)_5OSO_3^+$ by OH^-

fixed pressure interval plotted against mean pressures. Extrapolation gives an intercept taken as $\partial \ln k/\partial P_{p\to 0}$.

(c) *Benson and Berson*[225] attempted to calculate the curvature of a plot of $\ln k$ against pressure allowing for compressibility differences by means of the Tait equation (eq. (2.14)). A function of the type, eq. (4.5), was proposed:

$$\frac{\ln (k_p/k_0)}{P} = A'' + B'' P^{0.523} \tag{4.5}$$

A critical comparison of these different methods has been made by Hyne and coworkers who concluded that the second-order polynomial with unconstrained intercept (i.e. *a*) was most satisfactory.[226] The problem has also been discussed by Kelm[359] and others[497]. Extensive compilations of activation volumes have been published.[482, 8, 9, 178, 179, 221, 222]

Complex rate expressions

The treatment above applies directly to unit reactions, i.e. those unimolecular or bimolecular processes occurring in a single concerted process. Frequently, however, reactions of interest are the result of a multistep sequence so that a measured rate constant, and therefore also the activation volume, is a composite quantity. In order to be able to interpret the latter, it is important to dissect it into its components. It is difficult to treat this topic generally and a few specific examples follow:

(a) $$A \xrightarrow[\text{slow}]{k_1} B \xrightarrow[\text{fast}]{k_2} \text{Products}$$

The terms 'slow' and 'fast' are relative, the point here being that the intermediate, B, is removed almost as fast as it is formed so that its concentration always remains very low. The first step is rate-determining and any subsequent fast steps are not relevant to the kinetics, therefore

$$k_{obs} = k_1 \quad \text{and} \quad \Delta V_{obs}^{\ddagger} = \Delta V_1^{\ddagger}$$

This situation simplifies the treatment although no information concerning k_2 may be obtained from studies of the forward rate alone:

(b) $$A \underset{k_{-1}}{\overset{k_1}{\rightleftharpoons}} B \xrightarrow{k_2} C$$

The formation of the intermediate is now reversible but several situations may arise depending upon the relative rates of each step.

(1) $k_1 \ll k_{-1}$ and $k_2 \ll k_{-1}$; the pre-equilibrium lies heavily to the left and k_2 is rate-determining, as in many examples of acid and base catalysis (section 4.3.6):

$$\text{rate} = k_2[B] \quad \text{but } [B] = \text{constant since } d[B]/dt = 0$$

(steady-state assumption)

$$\text{and } k_1[A] = k_{-1}[B] + k_2[B]$$

$$\text{hence } [B] = \frac{k_1[A]}{(k_{-1}+k_2)}$$

therefore

$$\text{rate} = \frac{k_1 k_2[A]}{(k_{-1}+k_2)} \approx \frac{k_1}{k_{-1}} \cdot k_2[A] \quad \text{and} \quad \underline{k_{obs} = Kk_2} \tag{4.6}$$

where

$$K = k_1/k_{-1} \quad \text{hence} \quad \underline{\Delta V_{obs}^{\ddagger} = \Delta \bar{V}_1 + \Delta V_2^{\ddagger}}$$

that is, the activation volume is composed of two terms, the volume *change* for the pre-equilibrium and the volume of *activation* for the rate determining step k_2.

(2) $k_1 \ll k_{-1}$ and $k_2 \gg k_1$; the pre-equilibrium again lies to the left but now there is little return of B to reagents but instead, a rapid progress to products. This scheme is followed by many solvolytic reactions (section 4.3.7). Equation (4.6) applies but as $(k_{-1}+k_2) \approx k_2$, $k_{obs} \approx k_1$ and $\Delta V_{obs}^{\ddagger} = \Delta V_1^{\ddagger}$. Essentially the same conclusion will result if $k_1 > k_{-1}$.

(3) $k_1 \approx k_2$. Steady-state conditions now no longer apply and the complete analysis of such a sequential reaction scheme is complex. This often implies that

the intermediate, B, is stable and its reaction to products may be separately studied to obtain ΔV_2^+: or a reagent may be added to scavenge B rapidly as it is formed from A and thus to give ΔV_1^+:

$$A \underset{k_{-1}}{\overset{k_1}{\rightleftharpoons}} B \underset{k_{-2}}{\overset{k_2}{\rightleftharpoons}} C \overset{k_3}{\longrightarrow} \text{Products}$$

More than one pre-equilibrium step is sometimes observed (e.g. in nitration of aromatic hydrocarbons, section 4.2.11). If $k_1 < k_{-1}$, $k_2 < k_{-2}$ and $k_3 < k_1, k_2$ the rate-determining step is k_3 and

$$k_{obs} = k_1 k_2 k_3 \quad \text{and} \quad \Delta V_{obs}^+ = \Delta \bar{V}_1 + \Delta \bar{V}_2 + \Delta V_3^+ \tag{4.7}$$

(d) Even more complex schemes are sometimes observed, for example peroxide decomposition (section 4.3.2), polymerization (section 4.2.4) or general acid/base catalysis (section 4.3.6). No detailed analyses of volumes of activation of these have been attempted but the principle remains that a rate constant which is itself a complex quantity will yield a 'volume of activation' which is the sum of a number of components and thereby make interpretation difficult unless they can be disentangled.

(e) Product branching, in which an observed reagent is consumed by two concurrent pathways, is fairly common. A good instance is the formation of olefin and alcohol from an alkyl halide by E2 and S_N2 processes

$$RCH_2CH_2Br + OH^- \begin{array}{c} \overset{k_E}{\nearrow} RCH{=}CH_2 \quad (= P_E) \\ \underset{k_s}{\searrow} RCH_2CH_2OH \quad (= P_s) \end{array}$$

$$[= S]$$

hence

$$\text{rate} = k_{obs}[S][OH^-] = (k_E + k_s)[S][OH^-]$$

$$\Delta V_{obs}^+ = \frac{k_E}{k_{obs}} \cdot \Delta V_E^+ + \frac{k_s}{k_{obs}} \cdot \Delta V_s^+ \tag{4.8}$$

The individual rate constants, k_E and k_s, can be obtained by partitioning k_{obs} in the proportions of the products;

$$k_E = \frac{P_E}{(P_E + P_s)} \cdot k_{obs} \quad \text{and} \quad k_s = \frac{P_s}{(P_E + P_s)} \cdot k_{obs}$$

hence individual activation volumes for the two reactions may be calculated.

4.1.3 The Problem of Concentration Units

Specific rate constants express the amount of substance reacting in unit time and the amount may be expressed in a variety of units. A problem arises since some are pressure dependent while others are not. Pressure independent units of concentration are molality (moles solute per kg of solvent) or mole fraction (dimensionless, the relative number of moles of solute to the combined number of moles of solute(s) and solvent). No problems arise in the application of eq. (4.2) if such units are used to define the rate constants which may be evaluated by the usual integrated rate equations. However, it has become traditional to express concentrations in molarity units (moles per unit volume) and, since the volume diminishes with pressure by an amount determined by the compressibility of the solvent, the molarity apparently increases with pressure. This has led some workers to correct concentrations measured at 1 bar for use at higher pressures, a procedure which has been pointed out by Hamann[6, 8, 353] as being incorrect. A second-order process involving the collision of two molecules is clearly not made any more probable by compression of the solution since the same numbers of molecules of all types are still present. By contrast, increasing the molarity of the solution at 1 bar by addition of solute *will* result in a higher frequency of collision of solute molecules. Concentrations expressed as molarities ($mol\,dm^{-3}$) *at 1 bar* constitute a pressure-independent unit and therefore it is not necessary to correct concentrations initially determined at this pressure.

Corrections for compression should be applied when the concentration in molarity units is actually measured at an elevated pressure and the value at 1 bar should be used in rate equations. This is so when spectrophotometric measurements are used to obtain concentration and a fixed path-length cell is used, e.g. fig. 1.32(a), since the number of absorbing molecules in the light path increases with pressure. A sliding cell of the type shown in fig. 1.32(b) does not require such a correction.

4.1.4 Factors Influencing the Volume of Activation

Just as the partial molar volume of a stable molecule may be considered to be made up of two parts, the intrinsic part or Van der Waals volume, $\bar{V}_1$, and the solvation part, $\bar{V}_2$, related to changes in the volume of solvating medium brought about by the solute,

$$\bar{V} = \bar{V}_1 + \bar{V}_2 \qquad (4.9)$$

the same division may be made to the volume of activation and it is convenient to consider factors affecting each in turn.

(i) *Contributions to the intrinsic term, $V_1^{\ddagger}$*

The formation of a bond between two molecules is accompaned by a reduction in volume and, conversely, bond breaking brings about an increase. Order of magnitude calculations may be carried out to ascertain how reasonable are estimates made from pressure effects on rates. A model often used for bond extension is of two touching spheres representing the bound atoms, being converted to a cylindrical volume on bond extension to the transition state. Since the cleavage of di-*t*-butyl peroxide involves O—O bond fission the activation volume observed, $+6\,cm^3\,mol^{-1}$, may reasonably be interpreted as implying a 25% increase in the bond which seems very plausible.[229]

A similar volume contraction is associated with neutral bond formation. No volume change is involved in bond deformation but cyclization is sometimes associated with a small positive contribution due to the exclusion of solvent from the centre of the ring.[190]

Steric crowding is usually associated with lower partial volumes as compared to non-crowded isomers, and it is often assumed that sterically crowded transition states are likewise of relatively low volume,[176] possibly on account of reduced solvation. Pressure effects on rates of sterically hindered reactions may therefore be different from those on less hindered examples which has practical advantages since the application of pressure sometimes permits reactions to occur which are otherwise prevented by steric factors (section 3.3.3). An example of this effect is found in quaternizations of crowded and uncrowded pyridines,[176]

$$\Delta V^{\ddagger}\ -50 \qquad\qquad -22\ cm^3\ mol^{-1}$$

These changes, involving both components, have been summarized by Le Noble[221] as follows:

	contribution to $\Delta V^{\ddagger}$ $(cm^3\ mol^{-1})$	
bond cleavage	$+10$	
bond formation	-10	
bond deformation	~ 0	
cyclization	~ 0	
displacement	-5	
cyclization	~ 0	
ionic dissociation	-20	(solvent dependent)
ion recombination	$+20$	
charge dispersal	$+5$	
charge concentration	-5	
steric hindrance	$(-)$	
diffusion control	$> +20$	

(ii) *Contributions to the solvation term, $\Delta V_s^{\ddagger}$*

The partial volumes of ions are highly solvent dependent, compared with those of neutral, non-polar substances (Tables 3.1–4 above) and this is due to a

---

*The reason for this difference is to be found in the position of the transition state; that for the hindered, less exothermic reaction is later on the reaction coordinate and $\Delta V^{\ddagger} \neq \Delta \bar{V}$, an application of the Hammond Principle.

substantial contribution from $\bar{V}_2$. The origin of this effect must lie in changes in the structure and ordering of the solvent molecules in the immediate vicinity of the solute, the solvation sphere, hence there is potentially available information concerning this region. During an activation process, there is frequently a change in the charge distribution of the substrate; ions may be created or neutralized and transition states may possess dipolar character. Volumes of activation are then highly solvent sensitive and $\Delta V_2^{\ddagger}$ may be the dominant term. Several semi-quantitative approaches have been made for the estimation of $\Delta V_2^{\ddagger}$.

The electrostatic model The simplifying assumption is made of a spherical ion considered to be surrounded by a continuous medium which possesses the same dielectric constant as the bulk solvent. From electrostatic theory it may be shown that the work of transferring a conducting sphere, radius r, bearing charge q from a vacuum to a medium of dielectric constant ε is given by

$$W = \frac{q^2}{2r} \cdot \left(1 - \frac{1}{\varepsilon}\right) \tag{4.10}$$

Assuming an ion can be regarded as a conducting sphere and the medium as a continuous dielectric, Born[232] then gave the molar work of transfer of ion from vacuum to dielectric medium at high dilution, that is, the free energy of solvation, ΔG_s, as

$$\Delta G_s = -\frac{N_0 z^2 e^2}{2r}\left(1 - \frac{1}{\varepsilon}\right) \tag{4.11}$$

where N_0 is the Avagadro number, z the charge on the ion and e the electronic charge ($ze = q$).

Since $\Delta V = \partial \Delta G / \partial P$, we may differentiate (4.11) with respect to each of the variables namely ε and also r since the ion in principle will be compressible:

$$\Delta V_2 = \frac{N_0 e^2 z^2}{r} \cdot \frac{\partial(1/\varepsilon)}{\partial P} - N_0 e^2 z^2 \cdot \left(1 - \frac{1}{\varepsilon}\right) \cdot \frac{\partial(1/r)}{\partial P} \tag{4.12}$$

The function $\partial(1/\varepsilon)/\partial P\,(= 1/\varepsilon^2 \cdot \partial\varepsilon/\partial P)$ has been measured for many bulky solvents (though the assumption that this is the same as that of the microscopic environment of the ion is open to question) eq. (2.61), Table 2.15. The second term, related to the compressibility of the ion itself, is less easy to estimate, but as will be shown may usually be ignored. An order of magnitude calculation assuming a univalent ion of radius $10\,\text{Å}$ in ethanol ($1/\varepsilon^2\, \partial\varepsilon/\partial P = -8.2 \times 10^{-6}\,\text{bar}^{-1}$) may be made as follows; the dielectric term:

$$\Delta V_2 = \frac{N_0[\text{mol}^{-1}] \cdot e^2[\text{C}^2] \cdot (1/\varepsilon^2\, \partial\varepsilon/\partial P)[\text{N}^{-1}\,\text{m}^2]}{r[\text{m}] \cdot 4\pi\mu_0[\text{J}^{-1}\,\text{C}^2\,\text{m}^{-1}] = [\text{N}^{-1}\,\text{m}^{-1}\,\text{C}^2\,\text{m}^{-1}]} \tag{4.13}$$

where μ_0 is the permittivity of free space

$$= \frac{(6.02 \times 10^{23})(1.602 \times 10^{-19})^2 \cdot (-8.2 \times 10^{-6} \times 10^{-5})}{(10 \times 10^{-10}) \cdot 4\pi \cdot (8.854 \times 10^{-12})}$$

$$= 1.11 \times 10^{-5}\,\text{m}^3\,\text{mol}^{-1} = \underline{-11\,\text{cm}^3\,\text{mol}^{-1}}$$

In hexane $(1/\varepsilon^2 \, \partial\varepsilon/\partial P = 120 \times 10^{-6}\,\text{bar}^{-1})$ the volume contraction would amount to $-160\,\text{cm}^3\,\text{mol}^{-1}$, the compressive term: no information on the compressibility of an isolated solute molecule is available but if, instead, we use the bulk compressibility of the pure substance the effect will be overestimated since compression, at least at low pressures, brings about changes in packing rather than in molecular deformation. Assuming $(1/v \,.\, dv/dP) \sim 10^{-4}\,\text{bar}^{-1}$ for many organic compounds (Table 2.1–6), then $(1/r \,.\, dr/dP) \sim 10^{-12}\,\text{bar}^{-1}$ $\sim 10^{-17}\,\text{N}^{-1}\,\text{m}^2$. Then for ethanol $(\varepsilon = 24, (1 - 1/\varepsilon) = 0.9583)$

$$\Delta V = \frac{(6.02 \times 10^{23})(1.602 \times 10^{-19})^2 (0.9583)(10^{-17})}{4\pi(8.854 \times 10^{-12})}$$

$$\sim 10^{-21}, \quad \text{i.e. } 10^{-15}\,\text{cm}^3\,\text{mol}^{-1}$$

which is negligible. This may not be the case for metal ions in water for which a similar type of calculation suggests the compressive term to be twice the dielectric term for sodium chloride.[6]

It is usual for organic chemists to express solvent electrostriction as

$$-\frac{q^2}{r} \cdot \left(\frac{1}{\varepsilon^2} \frac{\partial \varepsilon}{\partial P} \right) = -\frac{q^2}{r} \cdot \Phi \tag{4.14}$$

section 2.1.15 (the Drude–Nernst equation).[234] The use of this equation is, at best, only semiquantitative for the following reasons:

(a) The dielectric constant function is necessarily measured on bulk solvent whereas a solute molecule experiences a 'microscopic' property which is moreover discontinuous.

(b) Ions other than those of single atoms are rarely spherical and it is difficult to assign a radius. The charge is not usually dispersed evenly within an organic ion but is localized on a few atoms.

(c) Since the desired objective is to estimate the solvent effect upon a rate of reaction, it is necessary to consider the charge q associated with the transition

state. This is usually fractional—in fact, a dipole, e.g. $\overset{\delta+}{\underset{}{\diagdown}}\text{C} \cdots \text{Br}^{\,\delta-}$,[235] the use of δ

here hides our ignorance of the magnitude of the dipole moment, which, if the value were known, could be inserted into the Drude–Nernst equation. In fact, as $\Delta V^{\ddagger}$ is an experimental quantity q may be estimated from solvent-dependence of $\Delta V^{\ddagger}$.

(d) The value of $\bar{V}_2$ calculated is very sensitive to the magnitude of q and when the same reaction is studied in a series of solvents there is no guarantee that the transition state is identical in each nor that changes in charge distribution are the same.

(e) Although the electrostatic term may be predominant in the solvation of ions, it is by no means the only interaction between solute and solvent. Donor–acceptor forces, the interactions between filled and vacant orbitals which may be thought of as a sort of incipient covalency, can be of great importance. This molar volume of the iodide ion in acetone, for instance, is abnormally small, probably due to this type of solvation. Hydrogen-bonding may be present and polarizability of solute or solvent can enhance the electrostatic forces. These terms may all be included under the heading of specific solvation and are not taken into account by the electrostatic model.[236]

Despite these shortcomings in the quantitative application of the electrostatic theory, qualitative predictions justify its continued use. Volumes of activation of ion-forming reactions are more negative the less polar the solvent (e.g. Table 4.10). The r^{-1} relationship seems to be justified when tested by measuring partial volumes of alkali metal ions but the radius is less important for di- and tri-valent ions. Thus, electrostriction volumes in water are given as $-8.0/r$, -32.5 and $-58.5 \, cm^3 \, mol^{-1}$ for mono, di- and tri-valent cations.[233]

Heydtmann and coworkers[166, 167] have developed an electrostatic model in which the solute is a dipole (dipole moment μ) and which is transferred from vacuum to a spherical cavity (radius r) in a continuous dielectric medium (dielectric constant, ε). Kirkwood[237] gave for the molar free energy of this process,

$$\Delta G_t = -N_0 \cdot \frac{\mu^2}{r^3} \left(\frac{\varepsilon - 1}{2\varepsilon + 1} \right)^{'} [+\phi] \tag{4.15}$$

from which the volume change is given by

$$\Delta \bar{V}_t = -\frac{\partial \Delta G_t}{\partial P} = \frac{-N_0 \mu^2}{r^3} \cdot \frac{\partial}{\partial P} \left(\frac{\varepsilon - 1}{2\varepsilon + 1} \right) = \frac{-N_0 \mu^2}{r^3} \cdot q_p + \frac{\partial \phi}{\partial P} \tag{4.16}$$

where

$$q_p = \partial \left(\frac{\varepsilon - 1}{2\varepsilon + 1} \right) \bigg/ \partial P = \frac{3}{(2\varepsilon + 1)^2} \left(\frac{\partial \varepsilon}{\partial P} \right)_T$$

The term ϕ was introduced to take account of non-electrostatic terms, geometrical deformation and specific solvation.

The volume of activation is then given by

$$\Delta V^{\ddagger} = \Delta V_0^{\ddagger} + \Delta V_1^{\ddagger} + \Delta V_2^{\ddagger}$$

$$= \Delta V_0^{\ddagger} + \frac{\partial \sum \phi}{\partial P} - N_0 \left(\frac{\mu_{ts}^2}{r_{ts}^3} - \frac{\sum \mu_{rgt}^2}{\sum r_{rgt}^3} \right) \cdot q_p \tag{4.17}$$

$\Delta V_0^{\ddagger}$ is the volume of activation in a hypothetical liquid $\varepsilon = 1$; $\Delta V_1^{\ddagger}$ represents the contribution from specific terms and $\Delta V_2^{\ddagger}$ from electrostatic terms: μ_{ts}, r_{ts} refer to the transition state and μ_{rgt}, r_{rgt} refer to all reagents taking part in the activation process. For a related series of solvents, the specific terms for a given reaction are more or less constant and a plot of $\Delta V^{\ddagger}$ against q_p is reasonably linear (fig. 4.2a) while, when a more varied series of solvents is included, the variation in specific interactions becomes apparent, fig. 4.2(b). Values of Φ are given in Table 2.15. This type of electrostatic model has also been applied to formally neutral reactions such as the Diels–Alder reaction[237] in which some trend between the dielectric constant function and the volume of the transition state can be discerned.

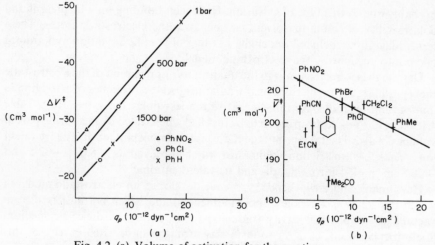

Fig. 4.2. (a) Volume of activation for the reaction:

$$PhCOCH_2Br + \underset{Me}{\overset{\displaystyle N}{\bigcirc}} \longrightarrow PhCOCH_2\underset{Me}{\overset{\displaystyle N}{\bigcirc}}{}^+ \quad Br^-$$

as a function of the solvent parameter, q_p.[166] (b)

$$Et_3N + EtI \longrightarrow Et_4N^+ I^-$$

An empirical solvation parameter, E_T, which is used by many workers to assess the 'polarity' of a solvent is defined as the energy, hv, for excitation of the solvatochromic dye,

$$E_T = hv = \frac{hc}{\lambda} \Big/ \text{kcal mol}^{-1}$$

E_T changes with pressure although by very little, 4–8 nm kbar^{-1}, the absorption maximum shifting towards the blue. This apparently indicates an increase in polarity of all solvents tested as pressure is increased. Although a correlation between $\ln k_p$ and $\Delta\lambda/\Delta P$ for some Menshutkin reactions[488] was noted, other workers have remarked this as being coincidental[489] and that the parameter $\Delta\lambda/\Delta P$ in no way correlates with dielectric constant or any of its derived functions. It would seem that solvation of the ground state, which is highly dipolar, is accomplished by means of specific as well as electrostatic interactions but little information can be deduced concerning changes of solvation energy in the excited state, assumed to be governed by the Franck–Condon principle.

4.1.5   The Dipole Moment of the Transition State

Adaptation of the Kirkwood solvation model to rates leads to the expression, eq. (4.18)

$$\ln \frac{k}{k_0} = \frac{N_0}{RT} \cdot \left[\frac{\mu_{\pm}^2}{r_{\pm}^3} - \Sigma \frac{\mu_{rgt}^2}{r_{rgt}^3} \right] \left(\frac{\varepsilon - 1}{2\varepsilon + 1} \right)$$

$$= \frac{N_0 [\Delta \mu^2 / r^3]}{RT} \cdot \frac{(\varepsilon - 1)}{(2\varepsilon + 1)} \tag{4.18}$$

All of these quantities with the exception of $\mu_{\pm}$, the transition state dipole moment, are accessible experimentally. Because of the approximate nature of the theory no single value calculated from rates in one solvent is likely to be reliable and therefore the estimation of $\mu_{\pm}$ is made from a plot of $\ln k/k_0$ (k_0 refers to the

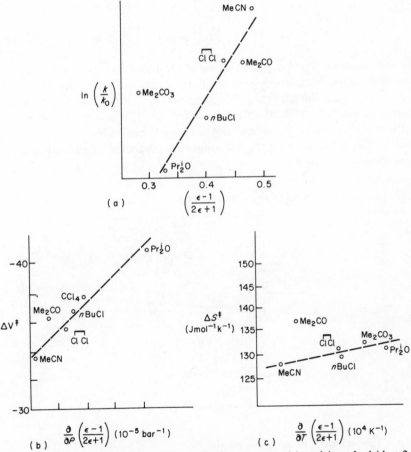

Fig. 4.3. Evaluation of transition state dipole moments, (a) maleic anhydride + 2-chlorobutadiene, eq. (4.18), (b) maleic anhydride + 2-methoxybutadiene, eq. (4.19), (c) maleic anhydride + 2-chlorobutadiene, eq. (4.20). (After McCabe and Eckert)

rate in a solvent $\varepsilon = 1$, in practice something like a paraffin would need to be used or it can be regarded as a constant and appear as intercept) against $(\varepsilon - 1)/(2\varepsilon + 1)$, avoiding solvents known to interact specifically. A typical example of this application is shown in fig. 4.3. The slope gives the difference in μ^2/r^3 terms in which $\mu_{\ddagger}$ is the only unknown. In principle, any function involving $[\Delta\mu^2/r^3]$ can be used for this purpose. The following are applicable:

$$\Delta V^{\ddagger} = \Delta V_0^{\ddagger} - N_0[\Delta\mu^2/r^3]\cdot\frac{\partial}{\partial P}\left(\frac{\varepsilon - 1}{2\varepsilon + 1}\right) = \Delta V_0^{\ddagger} - N_0[\Delta\mu^2/r^3]\,q_p \quad (4.19)$$

where $\Delta V_0^{\ddagger}$ refers to the volume of activation in a non-electrostrictive solvent, i.e. $\Delta V_1^{\ddagger}$:

$$\Delta S^{\ddagger} = \Delta S_0^{\ddagger} + N_0[\Delta\mu^2/r^3]\cdot\frac{\partial}{\partial T}\left(\frac{\varepsilon - 1}{2\varepsilon + 1}\right)$$
$$= \Delta S^{\ddagger} + N_0[\Delta\mu^2/r^3]\cdot q_T \quad (4.20)$$
$$\bar{V}^{\ddagger} = \bar{V}_0^{\ddagger} - \left(\frac{\mu^2}{r^3}\right)^{\ddagger}\frac{(\varepsilon - 1)}{P(2\varepsilon + 1)} = \bar{V}_0 - \left(\frac{\mu^2}{r^3}\right)^{\ddagger}q_p \quad (4.21)$$

The last equation, involving the volume of the transition state as a function of solvent (obtained by measuring partial molar volumes of reagent and the volume of activation), has the advantage of giving $(\mu^2/r^3)^{\ddagger}$ as the slope in a plot of $\bar{V}^{\ddagger}$ against q_p directly rather than as the difference of two large numbers as in methods giving $[\Delta\mu^2/r^3]$. Some values of $\mu^{\ddagger}$ for some Diels–Alder reactions are given in Table 4.1, showing the variation to be expected by the application of these methods, Table 4.1.

Table 4.1. Dipole moments of the transition state, $\mu^{\ddagger}$ for Diels–Alder reactions[237]

	$\mu^{\ddagger}$ (Debye)		
Method	Cl	OMe	Me
$\ln k \left/ \frac{(\varepsilon - 1)}{(2\varepsilon + 1)}\right.$	11	11	10
$\Delta V^{\ddagger}/q_p$	6	8	8
$\Delta S^{\ddagger}/q_T$	7	11	7
$\bar{V}^{\ddagger}/q_p$	3	4	4

4.1.6 The Solvent Compression Model

Kondo and coworkers have developed a model of electrostriction in which a solute molecule is considered surrounded by a solvation shell of solvent which is under an internal pressure and is therefore compressed.[238,173,175] If n moles solute are dissolved in N moles of solvent, the total volume of solution, V_t, is given by

$$V_t = V_0(N - nz) + V_s n + V_e nz \qquad (4.22)$$

Each solute molecule is considered surrounded by z solvating molecules (z = solvation number) and V_0, V_s, and V_e represent molar volumes of bulk solvent, pure solute, and solvent in the solvation shell, respectively. The three terms in eq. (4.22) represent residual bulk solvent, van der Waals volume of solute, and volume of solvation shells. The volume change due to electrostriction is given by $(nz(V_e - V_0))$ and corresponds to $(\Delta V_1 + \Delta V_2)$ in eq. (4.9)

The partial molar volume of solute, $\bar{V}_s^0$, will depend upon V_e which is in turn considered due to an internal pressure operating within the solvation sphere, P_s and is made up of the van der Waals volume, V, and the volume change due to electrostriction:

$$\bar{V}_s^0 = V + z(V_e - V_0) = \left(\frac{\partial V_e}{\partial P_s}\right)_{N,T} \qquad \text{on a molar basis} \qquad (4.23)$$

It is further considered that the volume change is a result of the pressure P_s operating upon normal compressibility of solvent for which

$$\beta_T = -\left(\frac{1}{V} \cdot \frac{\partial V}{\partial P}\right)_{T,N} = -\left(\frac{1}{V_e} \cdot \frac{\partial V_e}{\partial P_s}\right)_{T,N} \qquad (4.24)$$

hence the volume change on solvation, $\Delta V = (V_e - V_0)$, is given by

$$\Delta V = -\beta_T V_0 P_s \qquad (4.25)$$

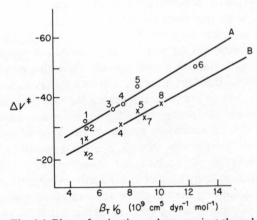

Fig. 4.4. Plots of activation volume against the solvent parameter $\beta_T V_0$; (a) $Et_3N + EtI \rightarrow Et_4N^+ I^-$. (b) Pyridine + MeI $\rightarrow$ N-Methylpyridinium iodide. (1) methanol; (2) nitrobenzene; (3) bromobenzene; (4) chlorobenzene; (5) benzene; (6) p-xylene; (7) acetone; (8) carbon tetrachloride

hence

$$\bar{V}_s^0 = V_s - z\beta_T V_0 P_s \tag{4.26}$$

with the further assumption that external pressure will be experienced by the solvation sphere so that $P_s(P) = (P_s(1\,\text{bar}) + P)$. Similarly, for a transition state,

$$\bar{V}^{\pm} = V^{\pm} - z\beta_T V_0 P_s \tag{4.27}$$

and

$$\Delta V^{\pm} = (\bar{V}^{\pm} - \sum \bar{V}_{\text{rgts}}) - z\beta_T V_0 P_s \tag{4.28}$$

This predicts a linear relationship between $-\Delta V^{\pm}$ and $\beta_T V_0$, a solvent property, with slope $= zP_s$. A reasonably linear plot is found for several Menshutkin reactions, fig. 4.4, and the term zP_s can be evaluated but the individual values of z and P_s cannot at this stage be separated.

4.1.7 Volume Profiles

A comparison of the partial molar volumes of both transition state and products with that of reagents often gives further insight into the nature of the transition state than a value of $\Delta V^{\pm}$ alone. The volume change for a reaction $\Delta \bar{V}$ can be measured either by separately measuring partial molar volumes of reactants and of products when

$$\Delta \bar{V} = (\sum_{\text{reactants}} \bar{V}^0 - \sum_{\text{products}} \bar{V}^0) \tag{4.29}$$

or can be measured by applying dilatometry to the reaction directly. A schematic diagram of 'reaction coordinate' against volume can then be drawn, fig. 4.5.

The alkylation of the thiazolidine, fig. 4.5(a), with volume change largely due to electrostriction, suggests that charge-development in the transition state is about half complete (rather more because of the q^2 relationship). The profile for the electrophilic substitution, (b), for which both reagents and products are neutral reveals a highly dipolar transition state presumably with a maximum of charge separation. The Diels–Alder reaction, (c), is shown to have a smaller transition state than product. As this is a neutral reaction, the rate being not very solvent dependent, it is more likely that the additional volume contraction in the transition state is due to 'secondary attractive interactions' within the system than to solvent electrostriction though the latter possibility remains. By comparison, the volumes of reaction and of activation for the Diels–Alder reaction between maleic anhydride of and 1-methoxybutadiene[38] are identical $(-32\,\text{cm}^3\,\text{mol}^{-1})$ suggesting the transition state to be very similar to product unless some kind of compensating effect is taking place, for instance, a small degree of charge separation in the transition state compensating by electrostriction for incomplete volume contraction due to bond formation. It must always be borne in mind that these indirect probes into the nature of the transition state frequently lead to ambiguity when applied singly. Volume measurements are a supplement to other data not a replacement.

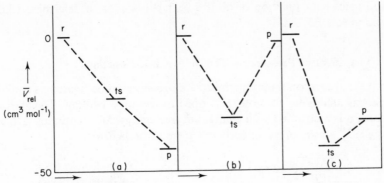

Fig. 4.5. Volume profiles for some typical reactions;
(a) Menschutkin reaction[239]

$$\Delta V^{\ddagger} = -25 \text{ cm}^3 \text{ mol}^{-1}; \Delta V = -45 \text{ cm}^3 \text{ mol}^{-1}.$$

(b) Neutral electrophilic displacements,[240]

$$\Delta V^{\ddagger} = -51; \Delta V = -3.$$

(c) Diels–Alder reaction[38]

$$\Delta V^{\ddagger} = -44 \cdot 7; \Delta V = -33 \cdot 3.$$

4.1.8 The Relationship of Activation Volume to other Activation Parameters

The theory of absolute rates, based on transition state concepts, gives as the rate of a chemical reaction,

$$k = \frac{\mathbf{k}T}{h} e^{\Delta S^{\ddagger}/R} e^{-\Delta H^{\ddagger}/RT} \qquad (4.30)$$

where the pre-exponential factor can be regarded as constant, $\Delta S^{\ddagger}$ is the entropy of activation, and $\Delta H^{\ddagger}$ the enthalpy of activation. Thus, these two quantities should embody all effects upon the reaction rate whether from structural change, solvent effects, or pressure. Both $\Delta S^{\ddagger}$ and $\Delta H^{\ddagger}$ are obtained from the effects of temperature change upon the rate constant so that it is necessary to measure reaction rates as a function of both T and P in order to determine whether pressure affects ΔS^+, ΔH^+, or both.

4.1.9 The Effect of Pressure on Thermodynamic Functions

Since the transition theory permits rate processes to be treated as equilibria, expressions analogous to those developed in section 3.1 follow. The activation equilibrium is associated with changes in free energy ΔG^+, enthalpy ΔH^+, and entropy ΔS^+ whose pressure derivatives are given below:

$$\left(\frac{\partial \Delta G^+}{\partial P}\right)_T = \Delta V^+ \tag{4.31}$$

$$\left(\frac{\partial \Delta S^+}{\partial P}\right)_T = -\left(\frac{\partial \Delta V^+}{\partial T}\right)_P \quad \text{as } \Delta S = \partial \Delta G / \partial T \tag{4.32}$$

$$\left(\frac{\partial \Delta H^+}{\partial P}\right)_T = \Delta V^+ - T\left(\frac{\partial \Delta V^+}{\partial T}\right)_P \quad \text{as } \Delta H = \Delta G + T \Delta S \tag{4.33}$$

$$\left(-\frac{\partial \Delta V^+}{\partial P}\right)_T = \Delta \beta^+ = 2RT.C$$

(C = the quadratic coefficient in eq. (4.3): $\Delta \beta^+$ is known as the compressibility coefficient of activation). These higher derivatives carry information relating to the transition state and are therefore in principle of interest in interpretation of mechanisms. At present, little use is made of the pressure derivatives of entropy and enthalpy since most of the available data are of insufficient precision and their interpretation is uncertain besides. In later sections, examples of the pressure- and temperature dependence of ΔV^+ will be found.[497] It appears that ΔV^+ and $\partial \Delta V^+/\partial T$ are of similar sign. This implies that for those reactions whose volume of activation is negative (accelerated by pressure) the entropy of the transition state is becoming less negative with pressure than that of reagents which in turn means that the transition state is less compressible. Frequently, parallel trends in ΔV^+ and ΔS^+ are noted in a related series of reactions.

Activation parameters may be expressed at constant pressure, the usual quantities, or at constant volume using eqs. (4.35), (4.36):

$$\Delta E_v^+ = \Delta H_p^+ - \frac{T\alpha \Delta V^+}{\beta_T} \tag{4.35}$$

$$T\Delta S_v^+ = T\Delta S_p^+ - \frac{T\alpha \Delta V^+}{\beta_T} \tag{4.36}$$

where α is the coefficient of thermal expansivity. It is possible that constant volume quantities may be more easily interpreted than those at constant pressure, especially in mixed solvents of high thermal expansivity,[198] fig. 4.15 below. The quantity $(T\alpha \Delta V^+/\beta_T)$ has the dimensions of energy and denotes the extra energy needed to keep the system at constant volume.

4.1.10 Viscosity Effects and Diffusion Control

For a bimolecular reaction to occur, it is clear that the two reacting molecules must first diffuse together during their random wanderings, and then react. The rate at which physical encounters occur under normal conditions is extremely high, about $10^9 \, M^{-1} s^{-1}$, so it follows that for most organic reactions at least, the probability of an encounter being followed by reaction is very low indeed. This is because covalent bond reorganization requires the reactants to approach in the correct orientation and with a minimum of energy in the appropriate modes so that, usually diffusion is not a critical aspect of the reaction. If the rate of reaction is comparable with the collisional rate, the reaction is said to be under diffusion control which will occur under the following circumstances:

(a) The reaction is intrinsically very fast, each encounter leading to reaction. This is found for acid–base reactions in water, or many inorganic anion–cation reactions, also the disappearance of highly reactive intermediates.[399]

(b) The encounter rate is extraordinarily slow so that encounters are prolonged and normally slow reactions are able to take place with high probability before the species diffuse apart. This condition is met in highly viscous solvents since the encounter rate is given by

$$k_{en} = \frac{2RT}{3\eta} \cdot \frac{(r_A + r_B)}{r_A r_B} \tag{4.34}$$

where r_A, r_B are the radii of the two molecules and η the viscosity of the medium. Viscosity effects may be important in the termination of radical polymerization as

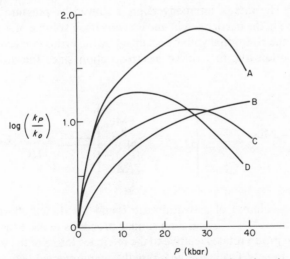

Fig. 4.6. Effects on rate, attributed to increasing viscosity for bimolecular reactions of EtBr with OR^-; (a) eugenoxide in isopropanol; (b) methoxide in methanol; (c) isopropoxide in isopropanol; (d) eugenoxide in eugenol–isopropanol 1 : 1. (After Hamann[397])

polymerization proceeds (section 4.3.2) and also at very high pressures. It would be necessary to increase the viscosity by a factor around 10,000 to reduce encounter rates to that of the slow reactions usually studied by chemical means so that the media would be almost vitreous but Hamann has shown that, at pressures in excess of 10 kbar, rates of some bimolecular displacements can reach a maximum, fig. 4.6, and then fall with pressure.[397] The maximum occurs at lowest pressure for those solvents which have the highest pressure coefficient of viscosity, such as eugenol, suggesting that diffusion control sets in at viscosities around 10^4 poise. A further example may be the reaction of metal coordination compounds in glycerol.[398] If a reaction is under diffusion control, an increase in pressure, invariably increasing the viscosity of the medium, will be accompanied by a large decrease in rate and hence a high positive contribution to $\Delta V^{\ddagger}$.

4.2 EXAMPLES OF PRESSURE EFFECTS ON RATE PROCESSES

4.2.1 Conformational Change

It has been mentioned that volumes of distinct conformers of molecules may be different (section 3.2.6) and this leads to pressure effects on their rates of interconversion. Such processes are among the simplest of unimolecular reactions occurring in solution and may be studied by temperature-dependent NMR which may be carried out up to about 2 kbar[10]. The rotation about the C—N bond in N,N-dimethylacetamide is retarded by a free energy barrier of 87 kJ mol^{-1}.[12] The rate of interconversion is slowed by pressure[11] associated with an increase in the barrier height and an activation volume of $+9$ cm^3 mol^{-1} which must be the additional volume occupied by the orthogonal conformer and also a contribution due to reduced electrostriction since the dipole moment decreases.

Very similar volumes of activation are found for dimethylformamide and dimethylbenzamide[444,445] suggesting similar processes in each case that is, the dimethylamino group rotates relative to the reference frame of the solvent rather than the RCO-moiety. Cyclohexane is rapidly interconverted between two chair forms, the intermediate, 'transition state' probably being the twist form.

Pressure causes an increase in rate as shown by the change in line width of the resonance due to averaging of the axial and equatorial protons, for which $\Delta V^{\ddagger} = -1.9$ cm^3 mol^{-1}.[11]

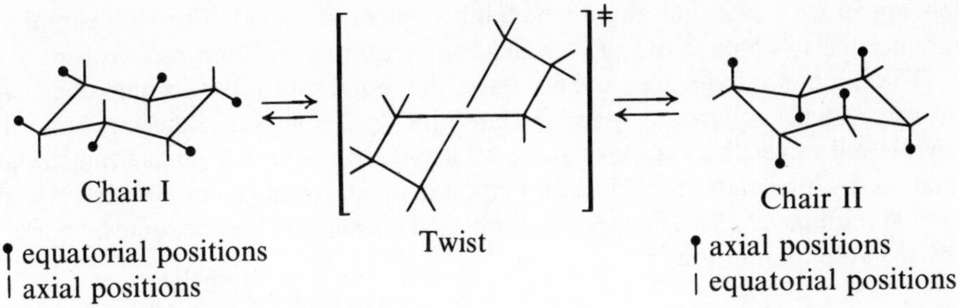

Chair I

↑ equatorial positions
⏐ axial positions

Twist

Chair II

↑ axial positions
⏐ equatorial positions

4.2.2 Thermal Racemizations

A relatively high energy barrier to conformational change exists in certain molecules on account of steric hindrance to bond rotation resulting in the stable existence of enantiomeric forms. These may be racemized at a suitably high temperature and rates measured. Typical examples are biphenyls with 2-4 ortho substituents,[12]

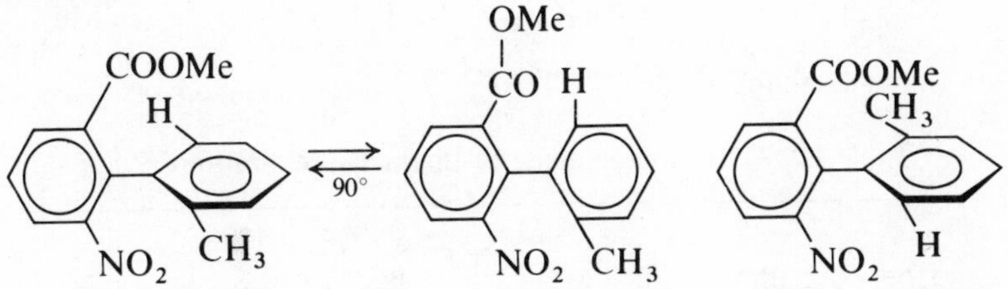

$$\Delta V^\ddagger = -26 \text{ cm}^3 \text{ mol}^{-1}$$

and sulphonanilides,[14]

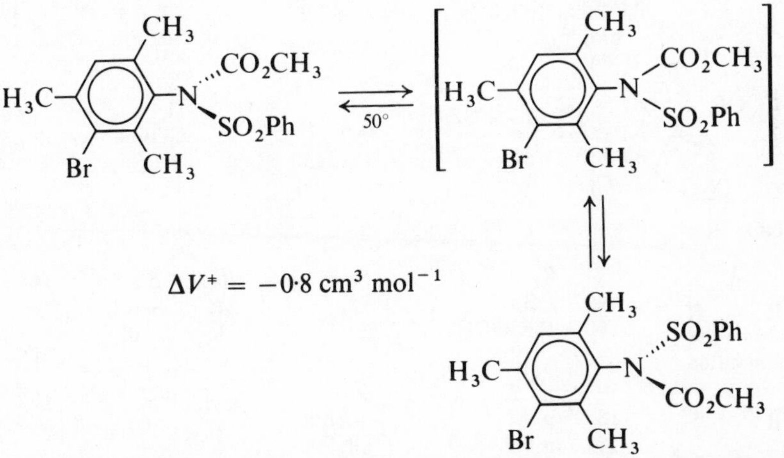

$$\Delta V^\ddagger = -0.8 \text{ cm}^3 \text{ mol}^{-1}$$

The biphenyl racemization is undoubtedly accelerated by pressure which would seem to rule out stretching of the central bond in the transition state though it is uncertain how bond bending, which is supposed to occur to permit passage of the ortho groups,[13] can require a volume contraction. However, there is a sharp change in $\Delta V^{\ddagger}$ at 2 kbar above which it becomes quite small. Further examples are needed to confirm the effect which may require a different mechanism.

The sulphonanilides are more normal in that bond rotation is accompanied by a very small volume change, $\pm 2\,\mathrm{cm^3\,mol^{-1}}$, Table 4.2. Racemization of pyramidal molecules may take place by inversion through a planar transition state or by dissociation and recombination of an attached group. The first route would require very little change of volume and is undoubtedly the route favoured by the sulphonium ion,[15]

$$\mathrm{PhCOCH_2}\overset{+}{\underset{\mathrm{Et}}{\mathrm{S}}}\text{-}\mathrm{Me} \rightleftharpoons \left[\begin{array}{c} \mathrm{CH_2}\text{-}\overset{+}{\mathrm{S}}\text{-}\mathrm{Me} \\ \mathrm{PhCO} \qquad \mathrm{Et} \end{array} \right] \rightleftharpoons \mathrm{PhCOCH_2}\overset{\mathrm{Et}}{\underset{+\text{-}\mathrm{Me}}{\mathrm{S}}}$$

$$\Delta V^{\ddagger} = 0$$

With a better cationic leaving group (t-butyl) is held to racemize by dissociation:

$$\mathrm{Me_3C}\overset{+}{\underset{\mathrm{Et}}{\mathrm{S}}}\text{-}\mathrm{Me} \rightleftharpoons \left[\mathrm{Me_3C}\overset{+}{} \quad :\mathrm{S}\overset{\mathrm{Me}}{\underset{\mathrm{Et}}{}} \right] \rightleftharpoons \mathrm{Me_3C}\overset{\mathrm{Et}}{\underset{+\text{-}\mathrm{Me}}{\mathrm{S}}}$$

$$\Delta V^{\ddagger} = +6\cdot4\,\mathrm{cm^3\,mol^{-1}}$$

Table 4.2. Activation volumes for thermal racemizations

$\mathrm{Ar-N}\begin{smallmatrix}\mathrm{SO_2Ph}\\\mathrm{CO_2Me}\end{smallmatrix}$; Ar =	2-Me-4-Br-phenyl	3,5-Me₂-nitrophenyl	2-Me-naphthyl	Reference
$\Delta V^{\ddagger}$ (cm³ mol⁻¹)	-0.8	$+1.9$	$+0.5$	14
$T(°C)$	50			

biphenyl COOR₁ / NO₂ R₂	$R_1 = \mathrm{Me}$ $R_2 = \mathrm{Me}$	H Me	H COO⁻	
	$\Delta V^{\ddagger}$ -32	(-283^a)	$+1.6$	12, 16
	$T(°C)$ 90	90	52	

$R_1\overset{+}{\underset{R_2}{\mathrm{S}}}R_3$ in water	T	R₁	R₂	R₃	$\Delta V^{\ddagger}$	15
	40	t. Bu	Et	Me	$+6.4$	
	60	PhCOCH₂	Et	Me	0	

$R^1\overset{\mathrm{S}}{\underset{R^2}{}}\text{=}\mathrm{O}$ in toluene		R₁	R₂		$\Delta V^{\ddagger}$	15
	192	p. tol	Ph		-2	
	187	p. tol	α-naph		0	
	43	p. tol	allyl		$+1$	
		p. tol	benzyl		$+26$	

^{a} This compound shows an enormously steep rise of $\ln k$ with pressure up to ca. 100 bar after which the curve levels sharply. The reason for this is uncertain and the data does not fit the quadratic expression. Further investigation would be desirable.

The volume change is not large (although it is solvent-dependent) considering the delocalization of positive charge which is occurring, so the transition state must be quite early along the reaction coordinate if this is indeed the mechanism. Among sulphoxides (Table 4.2) several examples of planar inversion are reported but $\Delta V^{\ddagger}$ for the benzyl compound is so different, $+26$ cm^3, that dissociation must be occurring, though this time a geminate radical pair is postulated.

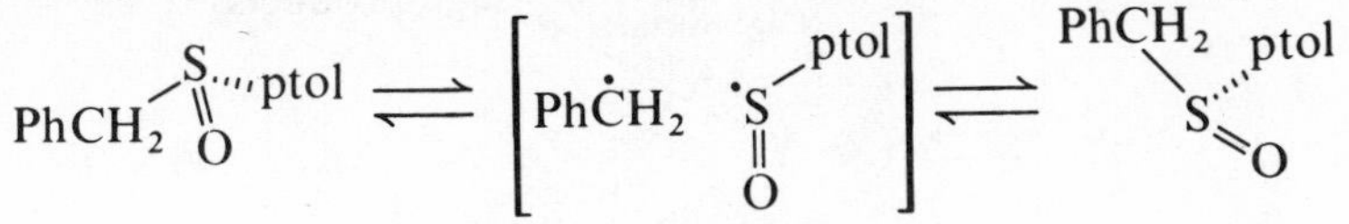

$$\Delta V^{\ddagger} = +26 \text{ cm}^3 \text{ mol}^{-1}$$

4.2.3 Thermal Homolytic Processes

Homolysis of a covalent bond is one of the simplest of elementary reactions and occurs to all types of bonds at a sufficiently high temperature. Among compounds which homolyse below 100°, and are important as initiators of radical chain reactions, are many peroxy compounds which may undergo scission into two radicals or into two radicals and CO_2 while azo compounds also cleave two bonds simultaneously forming two radicals and nitrogen. The kinetics

Mechanism A
one bond
fission

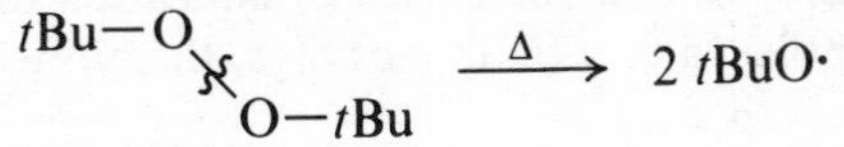

di-t-butyl peroxide

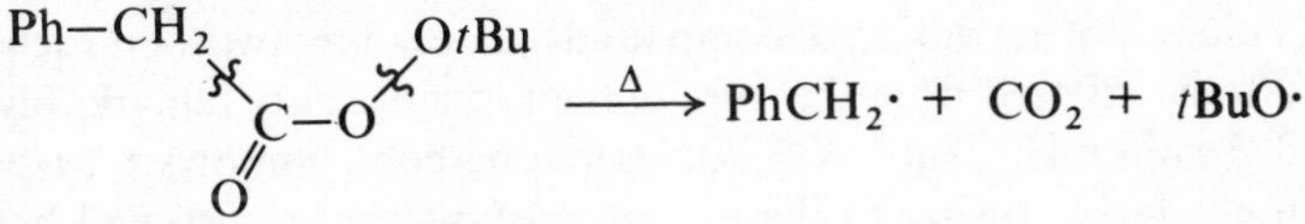

Mechanism B
two bond
fission

t-butyl peroxyphenylacetate

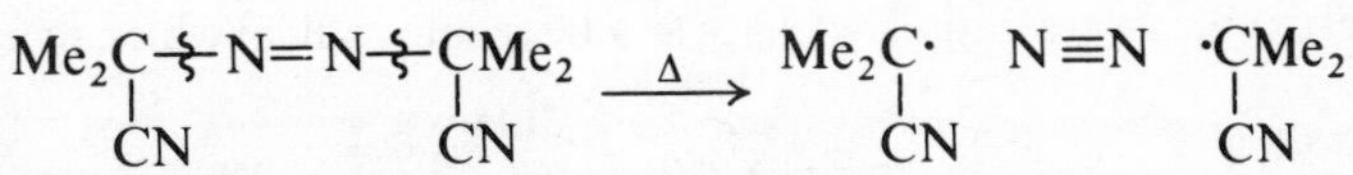

azobis(isobutyronitrile)
AIBN

of these reactions have long been studied,[32] many examples at high pressure.[33] Table 4.3 summarizes apparent activation volumes for homolyses, measured from the rates of disappearance of the substrates and it will be seen that a considerable range of values occurs although all are positive. At first sight it might be supposed that the volume change for stretching one bond would be less than that associated with stretching two yet the opposite appears to be the case.[36]

Neuman and coworkers have amply demonstrated that the observed activation volume, $\Delta V_{obs}^{\ddagger}$, is not necessarily a measure of the homolytic process alone yet can provide considerable insight into the complexities of these reactions. A more complete reaction scheme shows that the initially formed radicals reside for a time in a solvent 'cage' and may diffuse apart to form products or may undergo various cage reactions. These include return to the starting material by coupling and disproportionation. If there is a significant proportion of return, the

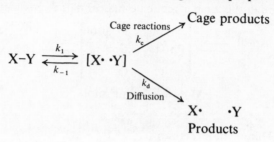

measured rate will be less than k_1 and the value of $\Delta V_{obs}^{\ddagger}$ greater than $\Delta V_1^{\ddagger}$ because of the pressure effect on modes of reaction of the radical pair:

$$\Delta V_{obs}^{\ddagger} = \Delta V_1^{\ddagger} + RT\partial \ln\left[1 + k_{-1}/(k_c + k_d)\right]/\partial P \qquad (4.36)$$

One may assume that return will not occur in decompositions of azo compounds since the radicals are not likely to react with molecular nitrogen and this is probably true also of peroxides reacting by mechanism B since, additionally, the entropy of the process, which now involves the organization of three species, would be highly unfavourable. In two-bond cleavage, therefore, we can ignore k_{-1} so that

$$k_{obs} \approx k_1 \quad \text{and} \quad \Delta V_{obs}^{\ddagger} \approx \Delta V_1^{\ddagger}$$

Volumes of activation of compounds undergoing two-bond scission (which may also be inferred from the form of their kinetics), are remarkably similar, around $3\text{--}5 \text{ cm}^3 \text{ mol}^{-1}$ (Table 4.3). Values for one-bond homolysis, $10\text{--}20 \text{ cm}^3 \text{ mol}^{-1}$, are much larger owing to the incidence of internal return and hence the need for diffusion ($\Delta V_d^{\ddagger} + \text{ve}$) to lead to reaction. The two are clearly distinguishable and this provides a useful mechanistic criterion; it may be inferred that t-butyl peroxybenzoate initially cleaves to a benzoate radical which in a subsequent step

$$\text{Ph} \cdot \text{C} \overset{\displaystyle O}{\underset{\displaystyle O-Ot\cdot Bu}{\diagdown}} \xrightarrow[\;(\leftarrow)\;]{\text{slow}} \text{Ph} \cdot \text{C} \overset{\displaystyle O}{\underset{\displaystyle O\cdot}{\diagdown}} \xrightarrow{\text{fast}} \text{Ph} \cdot + CO_2 + \cdot Ot \cdot Bu$$

$$\Delta V^{\ddagger} = 10 \text{ cm}^3 \text{ mol}^{-1}$$

loses CO_2. The peroxypivalate, on the other hand, undergoes synchronous cleavage of butoxy and phenyl radicals,[21]

$$Me_3C-C \overset{\displaystyle O}{\underset{\displaystyle O-Ot. Bu}{\diagdown}} \xrightarrow{\text{slow}} Me_3C\cdot + CO_2 + \cdot Ot.Bu$$

$$\Delta V^{\ddagger} = 0\cdot3 \text{ cm}^3 \text{ mol}^{-1}$$

Pressure increases return where this is possible and, indeed, increases the extent of all cage reactions apparently by making diffusion from the solvent cage slower

Table 4.3. Volumes of activation, $\Delta V^{\ddagger}_{obs}$, for thermal homolyses

	Solvent	Temperature (°C)	Mechanism	$\Delta V^{\ddagger}_{obs}$ (cm³ mol⁻¹)	Reference
Peroxides					
Di-*t*-butyl peroxide	heptane	140–200	A	10.1	495
	benzene	120	A	13	17
	toluene		A	5	
	CCl₄		A	13	
	cyclohexene			7	
Dibenzoyl peroxide	CCl₄	60	A	10	18
	PhCOMe		A	4.8	19
t-Butyl peroxybenzoate	cumene	80	A	10	20
	PhCl		A	13	20
t-Butyl peroxypivalate	cumene	90	B	0.3	21
t-Butyl peroxyisobutyrate	cumene	90	B	1.6	21
t-Butyl peroxycyclohexane-carboxylate	cumene	80	B	3.9	20
t-Butyl peroxyphenylacetate	cumene	80	B	1, 0.4[a]	22, 23
m-Chloroperoxyphenyl acetate	cumene	80	B	1.6[a]	
p-Chloroperoxyphenyl acetate	cumene	80	B	1.2[a]	
p-Methylperoxyphenyl acetate	cumene	80	B	0.2[a]	
p-Methoxyperoxyphenyl acetate	cumene	80	B	0.2[a]	
p-nitroperoxyphenylacetate	cumene	80	B	3	25
Et⧹ ⧸CO.OO*t*.Bu H⧸ ⧹nPr	cumene	100	A	6.8	24
Et⧹ ⧸nPr H⧸ ⧹CO.OO*t*.Bu	cumene	100	A	9	24

206

Table 4.3. (cont.)

	Solvent	Temperature (°C)	Mechanism	$\Delta V_{obs}^{\ddagger}$ (cm³ mol⁻¹)	Reference
Azo compounds					
Azocumene	cumene	55	B	5.0	26
	PhCl		B	4.3	
4'4-Dimethylazocumene	cumene		B	4	
Azoisobutyronitrile	toluene	70	B	3.8	27
Di-t-butyl hyponitrite	octane	55	B	2.8	28, 29
p-Nitrophenylazotriphenyl-methane	octane	60	A	18	26
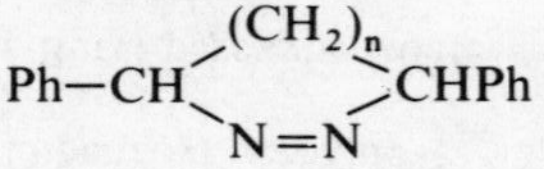 Ph—CH(CH₂)ₙCHPh / N=N n = 2	cumene		A	20	
n = 3	t.Bu-Ph		A	21	
n = 4	PhMe	60	B	5.3	30, 32
	PhMe	60	B	5.3	
	PhMe	60	B	5.5	
Others					
Pentaphenylethane	toluene	70	A	13	
C₆H₁₀=C=N-(1-CN-C₆H₁₀)	PhCl	100	A	13	

[a] At 2 kbar.

$(\Delta V_d^{\ddagger} + ve)$.[23] The effect of pressure on the cage effect, the tendency of a radical pair to return rather than to diffuse apart, can lead to further information.[23,29] The fate of the primary radicals from the decomposition of t-butyl peroxycyclohexanecarboxylate, 1, may react in the solvent cage by coupling, giving the ether, 2, or disproportionation giving cyclohexene and t-butanol. If diffusion occurs, the separate radicals abstract hydrogen from the solvent to form cyclohexane and t-butanol. The relative yield of these products gives a measure of their relative rates of formation[20,23]

The yield of cyclohexane diminishes greatly with pressure while ether and cyclohexene increase proportionately. In terms of volumes of activation for the constituent reactions, $\Delta V_c^{\ddagger} \sim \Delta V_{c'}^{\ddagger}$ (i.e. there is little difference in the ratio of ether and cyclohexane with pressure) but $\Delta V_d^{\ddagger} - (\Delta V_c^{\ddagger} + \Delta V_{c'}^{\ddagger})] = \Delta V_{cage} \sim 12\,cm^3\,mol^{-1}$. This has been called an activation volume for the cage effect and may be even larger $(+31\,cm^3\,mol^{-1}$ for AIBN decomposition in toluene).[35] Frequently the measured volumes of activation are somewhat solvent dependent, for example dibenzoyl peroxide, Table 4.3. This reflects a change in the cage effect, i.e. ease of diffusion in different media and is related to viscosity.[34,323] In fact, the volume of activation for the cage effect is correlated with the volume of activation for viscous flow (section 2.2.1)[3] within several hydrocarbon solvents.

A homolytic process of interest in polyethylene production is β-scission which has the effect of reducing the chain length. This is the reverse of the propagation

$$\text{ᶰCH}_2-\text{CH}_2-\underset{\underset{\text{H}}{|}}{\dot{\text{C}}}-\text{CH}_2\text{ᶰ} \longrightarrow \text{ᶰ}\dot{\text{C}}\text{H}_2 + \text{CH}_2{=}\underset{\underset{\text{H}}{|}}{\text{C}}-\text{CH}_2\text{ᶰ}$$

$$\Delta V^{\ddagger} > 10\,cm^3\,mol^{-1}$$

step in polymerization (section 4.2.3) and so should have a large positive volume of activation. This undesirable process is minimized by working at high pressures and low temperatures. The collision of two ethyl radicals in solution gives products of dimerization and of disproportionation. Both are presumably cage

$$C_2H_5-N{=}N-C_2H_5 \xrightarrow{h\nu} [2\,C_2H_5{\cdot}] \begin{array}{c} \nearrow CH_2{=}CH_2 + C_2H_6 \\ \\ \searrow C_4H_{10} \end{array}$$

reactions but the amount of disproportionation increases with pressure with $\delta\Delta V^{\ddagger} = 2.7\,cm^3\,mol^{-1}$ which seems surprising in reactions involving no appreciable activation since butane should have the smaller volume.[484]

A further contribution to $\Delta V^{\ddagger}$ may come from solvent electrostriction since, although homolytic processes are formally neutral reactions, it is well known that a polar component is often present. Thus, the rates of many perester decompositions are affected by substituents present in an aromatic ring such as to correlate with substituent constants σ^{+}.[34] This indicates the generation of partial charges in the transition state, for example,

$$\Delta V_{obs}^{\ddagger} = 1\ cm^3\ mol^{-1}$$

(although for perbenzoate decomposition, $\rho = -0.35$ is too small to support such a feature[453]). Solvent electrostriction will be more important in solvents of low dielectric constant such as are commonly used and will make a negative contribution, $\Delta V_2^{\ddagger}$, to the volume of activation. The observed value is estimated to contain a solvent electrostriction component of $ca.$ $-3\,cm^3\,mol^{-1}$,[22] thus reducing $\Delta V_1^{\ddagger}$ from 4 to about 1 $cm^3\,mol^{-1}$. There are no doubt other anomalous features to volumes of activation for these reactions which may reveal hidden complexities. For example, the pressure dependence of decomposition of p-nitroperoxyphenylacetate shows a plateau and a rise at higher pressures. This may indicate the onset of some other reaction such as induced decomposition, the attack of radicals upon the substrate.[25] The addition of scavengers, species which react rapidly with radicals, has a considerable effect on $\Delta V^{\ddagger}$ by affecting cage lifetimes. The value for AIBN in the presence of iodine is 9.4 and in its absence 3.8 $cm^3\,mol^{-1}$.[27]

4.2.4 Radical Reactions

Organic radicals are for the most part extremely reactive and undergo the following processes usually at very high rates:[83]

(a) dimerization, two radicals combining, their unpaired electrons becoming a bonding pair:

$$CH_3{\cdot} \quad {\cdot}CH_3 \longrightarrow CH_3-CH_3$$

(b) addition to a π-system; this results in the retention of radical character:

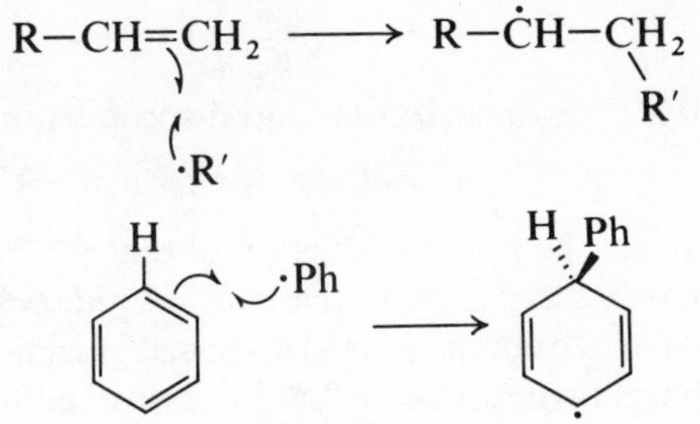

$$R-CH=CH_2 \longrightarrow R-\dot{C}H-CH_2$$
$$\qquad\qquad\qquad\qquad \underset{\cdot R'}{\big\langle} \qquad\qquad\qquad \underset{R'}{\big\backslash}$$

(c) abstraction of an atom, often hydrogen, from a molecule capable of yielding a more stable radical

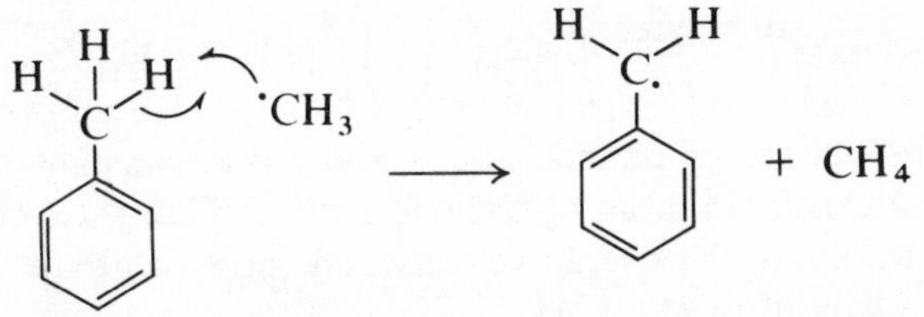

$$+ CH_4$$

If hydrogen abstraction occurs between two radicals the process is known as disproportionation and destroys the radical character of each:

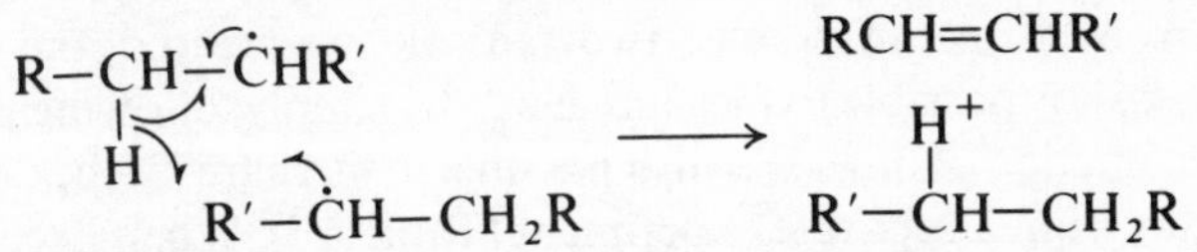

All such processes are of importance in radical polymerization of vinyl compounds and in which pressure may play a vital role.[424] The principal processes are summarized below.

(a) initiator radical production:

$$In-In \xrightarrow{\Delta\,or\,h\nu} 2\,In\cdot. \quad Rate = k_d[In-In], \quad \Delta V^+ \sim +5$$

(b) initiation of polymerization:

$$\underset{(M)}{R-CH=CH_2} + In\cdot \longrightarrow \underset{(M\cdot)}{R-\dot{C}H-CH_2In.}$$

$$Rate = k_i[M][In\cdot], \quad \Delta V^+ \sim -10$$

(c) propagation of polymerization:

$$\underset{(M\cdot)}{R\dot{C}H.CH_2In} + \underset{(M)}{RCH=CH_2} \longrightarrow RCH-CH_2In, etc$$
$$\qquad\qquad\qquad\qquad\qquad\qquad\qquad\qquad\quad |$$
$$\qquad\qquad\qquad\qquad\qquad\qquad\qquad CH_2-CH-R$$

$$Rate = k_p[M\cdot][M], \quad \Delta V^+ \sim -20$$

(d) termination:

$$2 \sim CH_2-\overset{\bullet}{C}HR \longrightarrow \text{dimerization and disproportionation products}$$

$$\text{(M·)} \qquad \text{Rate} = k_t[M·]^2 \qquad \Delta V^{+} \sim +15$$

Polymerization requires the prior production of radicals from an independent source, the initiator, which decomposes thermally or photochemically. Initiators are usually peroxides or azo compounds as discussed in section 4.2.2. The growth of the polymer chain then proceeds by successive additions of monomer molecules terminating in a radical-destruction process. The kinetics of such a complex sequence are involved but well understood.[310] The overall rate of polymerization often takes the form

$$\text{rate} \frac{d[\text{polymer}]}{dT} = k_p\left(\frac{k_d f}{k_t}\right)^{1/2}[M][\text{In–In}]^{1/2} \qquad (4.37)$$

where the symbols are as defined above and f is a constant expressing the efficiency of radicals In· ininitiating chain formation. Thus three elementary rate constants contribute to the overall velocity and consequently to the activation volume for the overall reaction, ΔV^{+}_{pol}.

$$\Delta V^{+}_{pol} = \Delta V^{+}_p + \Delta V^{+}_d/2 - \Delta V^{+}_t/2$$

From the approximate values above it is clear that a value ~ -20 is to be expected. The square root terms arise since two radicals are created and destroyed in processes k_d, k_t. It is possible to measure ΔV^{+}_{pol} by simply following the amount of polymer as a function of time. It is also possible to measure k_d, the decomposition of the initiator from a separate experiment (Table 4.3). When decomposition of the initiator is accomplished by light, as is possible for many azo-compounds, a technique involving rapid alternation of light and darkness (the rotating sector method) may be used to obtain individual values of k_p and k_t.[319] The theory is complex but it is a standard technique in polymerization studies[311] and may be applied at high pressure.[319,312-316] The propagation step involves bond formation and a neutral transition state and also the removal of bulky π-bonds which is accompanied by a volume contraction around $-20 \text{ cm}^3 \text{ mol}^{-1}$, so that k_p increases with pressure. On the other hand, k_t falls with pressure for reasons which are not entirely clear, $\Delta V^{+}_t \sim +20 \text{ cm}^3 \text{ mol}^{-1}$. Termination by dimerization should have a negative volume of activation although it is likely that these reactions do not need activation at all and are diffusion-controlled which causes a positive ΔV^{+}. Termination by hydrogen transfer—disproportionation—could have a larger transition state than reagents on account of C–H bond extension although an estimate of $\Delta V^{+} = -3$ to -6 has been made[317] for this process.* Increasing viscosity of the medium has also been implicated in determining this value,[400,401] in accordance with other diffusion-controlled reactions (section 4.1.10). The overall activation volume, ΔV^{+}_{pol} comes out at about $-20 \text{ cm}^3 \text{ mol}^{-1}$ and agrees with values based on the individual volumes of activation from Table

---

* And other hydrogen transfers have ΔV^{+} negative (Table 4.3).

4.4. Polymerization in the absence of initiator may occur though the process is not well understood. It is strongly accelerated by pressure.[318]

Since the propagation rate increases with pressure and the termination rate decreases, the average molecular weight of the polymer must increase—some four-fold by 3 kbar—a fact which could be of commercial significance. Before polymerization begins there is usually a period of inactivity although the initiator is decomposing and producing radicals. This is known as an induction period and is due to extraneous radical scavengers, oxygen, etc., removing initiator radicals faster than the monomer is able. The induction period may be greatly lengthened by pressure since initiator radicals are produced more slowly, the activation volume for peroxide or azo compound decomposition being positive.

Copolymerization is the incorporation into one polymer chain of two or more chemically different monomers. Suppose n_A and n_B moles of two monomers, A and B, are mixed and polymerization initiated. At each propagation step the radical end of the polymer may be an A or a B unit and each will be capable of adding either an A or a B monomer. Four rate processes then need to be considered, each with its volume of activation:

$$
\begin{array}{l}
\sim\!\!\text{A}\cdot \left\{
\begin{array}{l}
\xrightarrow[k_{AA}]{A} \sim\!\!\text{A--A}\cdot \\
\xrightarrow[k_{AB}]{B} \sim\!\!\text{A--B}\cdot
\end{array}
\right. \\[2em]
\sim\!\!\text{B}\cdot \left\{
\begin{array}{l}
\xrightarrow[k_{BA}]{A} \sim\!\!\text{B--A}\cdot \\
\xrightarrow[k_{BB}]{B} \sim\!\!\text{B--B}\cdot
\end{array}
\right.
\end{array}
$$

The ratios $r_A = k_{AA}/k_{AB}$ and $r_B = k_{BB}/k_{BA}$ are the radical selectivities expressing the degree of preference of a given radical end adding to a like or a different monomer. The instantaneous composition of the copolymer forming at any instant, $\mathrm{d}\,n_A/\mathrm{d}\,n_B$, depends upon the r-values and the proportions of the two monomers present:

$$\frac{\mathrm{d}\,n_A}{\mathrm{d}\,n_B} = \frac{r_A(n_A/n_B)+1}{r_B(n_B/n_A)+1} \tag{4.38}$$

This is the copolymerization equation;[321] if $n_A \sim n_B$ and if the r-values are ~ 1, then no preference will be exercised and the copolymer will be random. If the r-values are < 1 each monomer radical will prefer the other species and the copolymer will tend to be alternating. If the r-values are > 1, each monomer radical prefers its own species and separate homopolymers, not a copolymer, will result.[310] Pressure would affect the r-values if activation volumes for a given radical adding to two monomers were different:

$$\frac{\partial \ln r_A}{\partial P} = \frac{\partial(\ln k_{AA} - \ln k_{AB})}{\partial P} = -\frac{\Delta V_{AA}^{\ddagger} - \Delta V_{AB}^{\ddagger}}{RT} \tag{4.39}$$

and

$$\frac{\partial \ln r_B}{\partial P} = \frac{\partial(\ln k_{BB} - \ln k_{BA})}{\partial P} = -\frac{\Delta V_{BB}^{\ddagger} - \Delta V_{BA}^{\ddagger}}{RT} \tag{4.40}$$

Reactivity ratios, r, have been measured under pressure and found to change appreciably. For the copolymerization of ethylene (E) and vinyl acetate (V), the following values have been reported,[322]

	35 bar	600 bar
r_E	0.743	0.94
r_V	1.515	1.20

The increase in one value together with the decrease in the other, i.e. $\partial(\ln r_A r_B)/\partial P \sim 0$, is in accordance with a model which considers each activation volume term to be the sum of contributions from the reacting species, i.e.

$$\Delta V_{AA}^{\ddagger} = \Delta \bar{V}_{A\cdot} + \Delta \bar{V}_A.$$

The four activation volumes for this system have been calculated;[322] $\Delta V_{EV}^{\ddagger} = -4$; $\Delta V_{EE}^{\ddagger} = -16$; $\Delta V_{VV}^{\ddagger} = -9$; $\Delta V_{VE}^{\ddagger} = -21\ \mathrm{cm^3\ mol^{-1}}$. As the pressure is increased, therefore, the tendency will be for each terminating radical to favour progressively the addition of ethylene rather than vinyl acetate with consequent modification of the polymer properties. Numerous other binary polymer systems have been studied similarly.[439] Terpolymerization (three different monomers) gives six reactivity ratios. Estimates of their volumes of activation for the system: styrene–acrylonitrile–diethylfumarate have been made by Jenner.[440] Radical abstractions are concerted substitution reactions which may be compared to S_N2 processes from which they differ in the main by being essentially neutral throughout the exchange. Relatively slow reactions occur between stable radicals such as diphenylpicrylhydrazyl (DPPH) and thiols:

$$Ph_2N\!-\!\overset{\displaystyle \cdot}{N}\!-\!C_6H_2(NO_2)_3 \qquad\longrightarrow\qquad Ph_2N\!-\!\underset{\displaystyle H}{\overset{\displaystyle |}{N}}\!-\!C_6H_2(NO_2)_3$$
$$H\!-\!SR \qquad\qquad\qquad\qquad\qquad + \dot{S}R$$

These all have large and negative volumes of activation, $\Delta V^{\ddagger} \sim -12$ to $-20\ \mathrm{cm^3\ mol^{-1}}$ and the same range of values is found for abstractions by highly reactive radicals. These may be studied by incorporation of the abstracted group into a growing polymer (a termination step) (Table 4.4). A further example is the

$$\text{\textasciitilde CH}_2-\overset{Ph}{\underset{\cdot}{CH}} \xrightarrow{CCl_4} \text{\textasciitilde CH}_2-\overset{Ph}{\underset{Cl}{CH}} + C\dot{C}l_3$$

$$\xrightarrow{RCHO} \text{\textasciitilde CH}_2-\overset{Ph}{\underset{H}{CH}} + R\dot{C}O$$

$$\xrightarrow{NEt_3} \text{\textasciitilde} \quad \text{\textasciitilde CH}_2-\overset{Ph}{\underset{H}{CH}} + Et_2N-\dot{C}HCH_3$$

Table 4.4. Activation volumes of some radical reactions
Polymerization (no solvent, 30°)

Monomer	$\Delta V_{pol}^{\neq}$	$\Delta V_p^{\neq}$ /cm^3 mol^{-1}	$\Delta V_t^{\neq}$	Reference
Styrene	−17.1	−17.9	+13.1	312
		−14.3	+7.3	319
Methyl methacrylate	−15.6	−19.0	+25.0	313
Butyl acrylate	−26.3	−22.5	+20.5	314
Butyl methacrylate	−17.4	−23.2	+17.8	315
Octyl methacrylate	−19.2	−24.7	+20.8	316
Vinyl acetate	−17.2	−23.3	+16.3	320

Abstraction	Solvent	T /°C	$\Delta V^{\neq}$	Reference
R·^{a} + CCl$_4$ → R-Cl + CCl$_3$·	Styrene	40	−19	325
	Styrene-H$_2$O emulsion	50	−11	324
Ph· + (PhCO—O—)$_2$	CCl$_4$		−15	319
R· + NEt$_3$	Et$_3$N	60	−2	327
R· + H $\diagdown$ $\diagup$ C—isoPr $\diagup$ O	Styrene	50	−13.5	325
H $\diagdown$ $\diagup$ C—nPr $\diagup$ O				
DPPH· + t. BuSH	Toluene	35	−29.5	326
n. BuSH			−14.5	
nC$_6$H$_{13}$SH			−16.8	
t. C$_8$H$_{17}$SH			−20.7	
R· + MeCOOCH—CH=CH$_2$ $\vert$ H	Allyl acetate	60	−11.2	324

Table 4.4 (cont.)

Abstraction		Solvent	T /°C	$\Delta V^{\ddagger}$	Reference
DPPH· + LO⟨ring, t.Bu top, t.Bu bottom⟩R	L = H, R = H	Toluene 25		−13.3	356
	R = Me			−13.1	
	L = D, R = H			−13.3	
	R = Me			−13.2	
	R = tBu			−12.7	
DPPH· + HO⟨ring, R top, R bottom⟩R	R = Me			−13.5	
	R = Ph			−11.4	

β-scission

	Solvent	T /°C	$\Delta V^{\ddagger}$	Reference
$(tBu)_2C{=}\dot{N} \rightarrow tBu\cdot + tBuC{\equiv}N$	Isooctane	−25	+3.0	499
$tBuP(OEt)_3 \rightarrow tBu\cdot + P(OEt)_3$	Isopentane	−150	+0.2	499

Radical dimerization

	Solvent	T /°C	$\Delta V^{\ddagger}$	Reference
$2(tBu)_2C{=}\dot{N} \rightarrow [(tBu)_2C{=}N{-}]_2$	isooctane	−80	−18	499

[a] R· = growing polystyrene radical.

induced decomposition of a peroxide by the radicals which it produces thermally:

$$\text{Ph·} \underbrace{\qquad}_{O} \overset{O}{\underset{}{\overset{\parallel}{C}}}-Ph \quad \overset{C=O}{\underset{Ph}{}} \longrightarrow \quad \overset{Ph-O}{\underset{Ph}{}}C{=}O \ + \ CO_2 \ + \ Ph·$$

$$\Delta V^{\ddagger} = -15 \text{ cm}^3 \text{ mol}^{-1}$$

The values are considerably more negative than those for S_N2 reactions of the neutral charge-type. The difference must lie in the solvent effect. Charge dispersal in the nucleophilic displacement leads to release of solvating solvent to the extent of about $10 \text{ cm}^3 \text{ mol}^{-1}$.

Palmer and Kelm have measured isotope effects for hydrogen atom abstraction by stable radicals. All such reactions are 'normal' in that the PKIE's arise from differences in zero-point energies of the O—H, O—D bonds and no tunnelling is implicated (section 4.2.5). No effect by pressure was noted as expected since the volumes of the transition states should not depend upon isotopic substitution.[356]

One example of a hydrogen atom transfer is, however, a likely candidate for tunnelling. This is the intramolecular rearrangement of the tri-tert.butylphenyl radical,[498] followed by e.s.r., and for which isotope effects as large as 150 have

been reported. It is reported that here too the isotope effect is pressure dependent and the volume of activation isotope dependent, $\Delta V_H^{\ddagger} = 5.3 \pm 1.7$; $\Delta V_D^{\ddagger} = -1.2 \pm 2 \, \text{cm}^3 \, \text{mol}^{-1}$.[499]

The fate of reactive radicals may be greatly influenced by pressure. The t-butoxy radicals produced by decomposition of di-t-butyl peroxide may fragment to acetone and a methyl radical or may abstract hydrogen from the solvent, the former being unaffected by pressure and the latter rate favoured.[357]

Abstractions sometimes interfere with the normal course of radical polymerization in that the growing polymer may react with solvent, initiator, etc. to terminate chain growth but still retain a radical which may initiate chain growth.

$$\Delta V^{\ddagger} = -10 \text{ to } -20 \, \text{cm}^3 \, \text{mol}^{-1}$$

This is known as chain transfer and may result in very low molecular weight polymers being formed. If chain transfer, for which $\Delta V^{\ddagger}$ is strongly negative becomes the principal mode of termination, the velocity of polymerization becomes

$$v = k_d \cdot k_p / k_{tr} \quad \text{and} \quad \Delta V^{\ddagger} = \Delta V_d^{\ddagger} + \Delta V_p^{\ddagger} - \Delta V_{tr}^{\ddagger}$$

giving rise to values which are very much less negative than for polymerization terminated by dimerization and disproportionation. The polymerization of maleic anhydride is an example; short chains are formed and $\Delta V^{\ddagger} = -4.6 \, \text{cm}^3 \, \text{mol}^{-1}$. An internal abstraction is found to occur in ethylene polymerization which results in chain branching and affects physical properties. The driving force is the conversion of a primary to a more stable secondary radical. Chain branching has a far less negative activation volume than propagation so that its importance diminishes with pressure.

Zhulin and coworkers have noticed an empirical relationship between rates of hydrogen abstraction and their volumes of activation; the faster reactions have the most negative volumes. In polymerization this takes the form[448]

$$\log k_{tr} - \log k_p = \text{const.} \, (\Delta V_{tr}^{\ddagger} - \Delta V_p^{\ddagger})$$

Benzylic bromination by N-bromosuccinimide shows a similar relationship, relative values compared with toluene being compared[449]

R, R'	H, H	H, Ph	H, Me	Me, Me
k_{rel}(1 bar)	1	8	21	43
(5 kbar)	1	10	39	88
$\Delta\Delta V^{\ddagger}$/cm^3 mol^{-1}	0	-2.4	-4.8	-5.5

If differences can be ascribed to the Hammond Principle,[60] the faster reactions, being more exothermic, should have an earlier transition state.

4.2.5 Thermal Cycloadditions and other Pericyclic Reactions

The formation of a ring by bond formation between termini of two unsaturated systems is termed a cycloaddition,[45] the type being specified by the numbers of π-electrons which formally take part on each component, thus:

$$(2+2)\quad \begin{array}{c}Me_2C{=}C{=}O\\[4pt]EtO{-}CH{=}CH_2\end{array}\quad\longrightarrow\quad \begin{array}{c}Me_2C{-}\overset{O}{\overset{\|}{C}}\\ HC{-}CH_2\\ EtO\end{array}$$

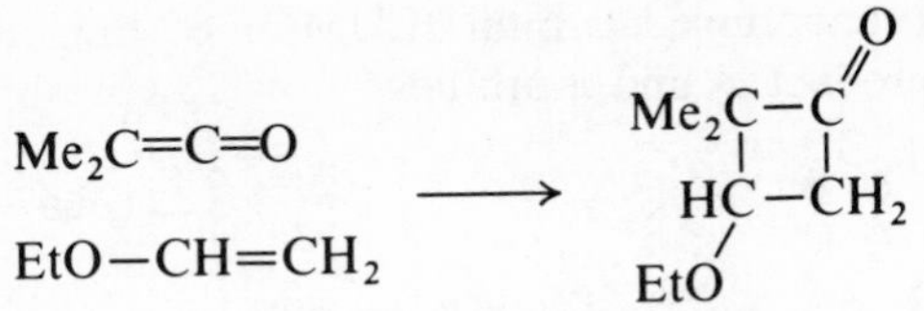

$(4+2)$ Diels–Alder reaction

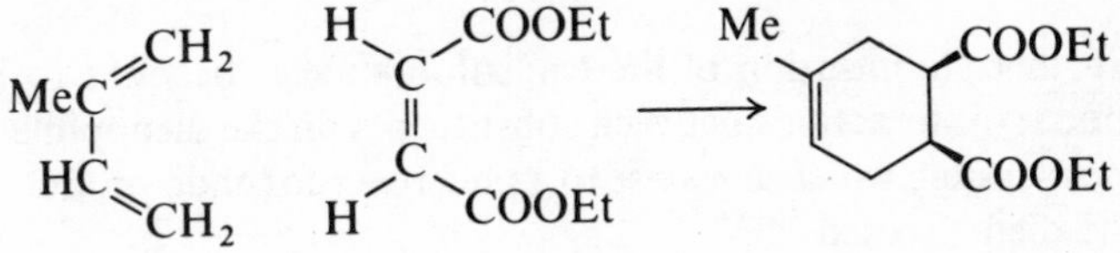

$$MeC\!\!\begin{array}{c}{=}CH_2\\ \end{array}\ HC\!\!\begin{array}{c}\\ {=}CH_2\end{array}\quad \begin{array}{c}H\quad COOEt\\ C\\ \|\\ C\\ H\quad COOEt\end{array}\ \longrightarrow\ \begin{array}{c}Me\quad COOEt\\ \\ COOEt\end{array}$$

$(6+4)$

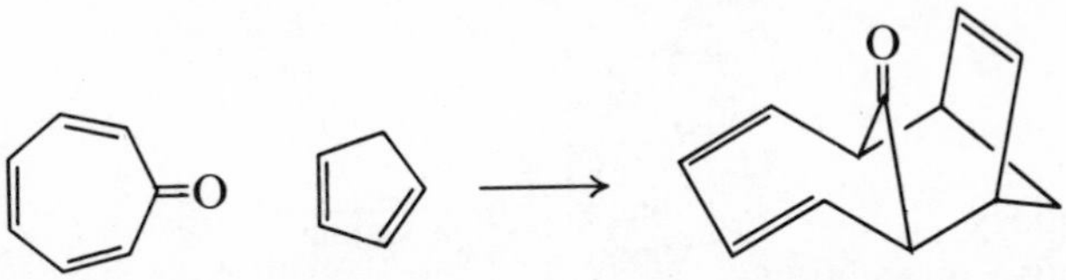

The Diels–Alder reaction, a $(4+2)$ cycloaddition, is the best known and most thoroughly studied. A 1,3-diene adds thermally to an olefinic component (the dienophile) to give a cyclohexene derivative. It is generally agreed[46] that the

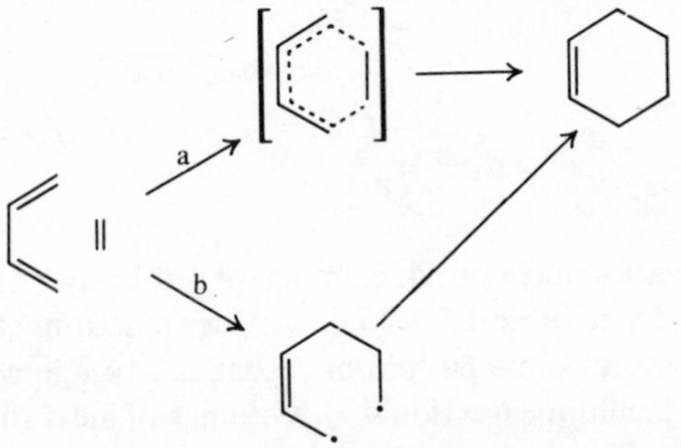

mechanism consists of a synchronous reorganization of three electron pairs through a quasi-aromatic transition state to form the product in a single step, a. The alternative diradical mechanism, b, was for some years considered[37] but must now be deemed incorrect. The evidence on which this conclusion is based includes retention of stereochemistry of substituents at the dienophile,[47,48] linear free energy correlations,[49] secondary isotope effects,[50] activation parameters—especially very large, negative entropies of activation[38] and smallness of kinetic solvent effects.[52] Furthermore, there are theoretical grounds for preferring the one-step mechanism. The Diels–Alder reaction is one member of the family of cycloadditions in which the symmetries of the interacting orbitals correlate with those of the products and is therefore an allowed thermal process[53] by the Woodward–Hoffmann rules. Thus, the reaction proceeds by the interaction of the highest filled molecular orbital (HOMO) of one component—normally the diene—with the lowest unoccupied orbital (LUMO) of the other, which transform smoothly to product σ- and π-orbitals.*

Further stabilization of the transition state is believed to be acquired from secondary interactions between substituents on the dienophile if present and the diene π-system which gives rise to a preference for endo products over exo when a cyclic diene is used.[53, 54]

Volumes of activation have provided much of our knowledge of the transition state; values are given in Table 4.5. All Diels–Alder reactions studied are strongly accelerated by pressure; some 60-fold at 5 kbar is average which makes this a useful preparative technique (section 4.4). Volumes of activation lie in the range -30 to $-40\,\mathrm{cm^3\,mol^{-1}}$ a value which must indicate a very compact transition state consistent with the formation of two bonds and quite incompatible with the diradical pathway. The reactions are in general very clean and well-behaved kinetically so that rate data at high pressures are among the most precise available. Mixing of reagents under pressure and accurate gas chromatographic analysis have been used to achieve this.[38, 39, 41, 55, 56] In many cases, volumes of reaction have also been measured and it is frequently but not invariably found that the transition state occupies a smaller volume than the product.[58] This difference must arise when secondary attractive forces in the transition state are important and cases where this is not possible, such as the addition of

---

*The notation for this type of interaction is $({}_{\pi}4_s + {}_{\pi}2_s)$. The subscripts s refer to 'suprafacial', meaning that bonding at both termini on the 4- and the 2-electron systems occurs on the same side of the plane of symmetry. The opposite would be a, 'antarafacial', in which bonding occurs on opposite sides of the π-system.[67, 80]

cyclopentadiene to acetylenedicarboxylic esters[38] (on account of geometrical restrictions), $\Delta \bar{V}$ is more negative than $\Delta V^{\ddagger}$. On the whole, cyclic dienes lead to a more compact transition state than acylic.[54] It must also be remembered that cyclic compounds have smaller partial molar volumes than acyclic analogues (Table 3.4) so that the nature of the product dictates the magnitude of $\Delta V^{\ddagger}$ to some extent. The extent of secondary interactions, as judged by the magnitude of $(\Delta V - \Delta V^{\ddagger})$, appears greater for 1-substituted butadienes than for 2-substituted. This is because in the former case the conjugating substituent may more strongly interact with the butadiene π-system and perturb the orbital electron distribution. Temperature dependence of the rates at high pressures have also been studied[57] which show that the increase of pressure which facilitates the reaction is accomplished by a reduction in the entropy of activation though the $\Delta H^{\ddagger}$ increases somewhat, Table 4.5.

Table 4.5. Activation parameters at high pressure

P(bar)	$\Delta H^{\ddagger}$ (kJ mol^{-1})	$-\Delta S^{\ddagger}$ (J K^{-1} mol^{-1})	$\Delta G^{\ddagger}$ (kJ mol^{-1})
1	79.0	171	138
1000	83.6	150	134
2000	84.8	140	131.5
3000	86.5	127	130
5000	87.3	117	127.5

Despite their being 'neutral' reactions, these cycloadditions are strongly dependent upon the presence of polar substituents on either moiety.[49] Conjugating donor substituents on the diene, especially at C1, raise the energy of the HOMO while acceptor substituents on the dienophile lower the LUMO, both trends facilitating reaction. From the limited data available, 2-substituted butadienes show a rate sequence MeO > Me > Cl while increasingly negative volumes of activation and values of $(\Delta \bar{V} - \Delta V^{\ddagger})$ fall in this order.[38,39,58] A possible explanation may lie in transition states for the more reactive dienes more closely resembling products (i.e. 'later' on the reaction coordinate) which would be predicted by the Hammond postulate[60] and requiring less from secondary attachments than do less nucleophilic dienes in order to stabilize the transition state. In several investigations the solvent has been varied widely (Table 4.6). There is a rough correlation of rates with solvent 'polarity' as measured by one or other semi-empirical scale[59] (k(MeCN)/k(isoPr$_2$O $\approx$ 10–20) but a moderate correlation of volumes of activation with the pressure coefficient of the Kirkwood dielectric constant function q_p has enabled Eckert to estimate the dipole moment of the transition state.[38] With maleic anhydride as the dienophile, this is 4D for penta-1,4-diene or 2-methoxybutadiene as diene and 3D for 2-chlorobuta-1,3-diene, values which, if they have no great precision, appear reasonable but are not available by any other method.

The importance of secondary interactions is also shown by the differential activation volumes, $\delta\Delta V^{\ddagger}$, for formation of exo or endo products in reactions with a cyclic diene. Strong secondary interactions are only possible in the endo orientation and lead to a lower volume than for exo product so that the endo/exo product ratio increases with pressure.[40]

The formation of heterocyclic rings by $(4+2)$ cycloaddition raises the further mechanistic possibility of a dipolar transition state or intermediate in addition to the diradical and concerted mechanisms. This may indeed by the case for certain reactions capable of stabilizing cationic or anionic centres,[63] e.g.

In two systems studied under pressure, acrylic aldehyde dimerization[61]

and diene additions to nitrosobenzenes[41]

values of $\Delta V^{\ddagger}$ lie between -30 and $-40\,\text{cm}^3\,\text{mol}^{-1}$ and the rates are not correlated strongly with solvent polarity so that here also a normal Diels–Alder reactions seems indicated. The formation of a 1,4-adduct from styrenes and tetracyanoethylene (in addition to the 1,2-adduct) is accelerated only moderately by pressure ($\Delta V^{\ddagger} = -17\,\text{cm}^3\,\text{mol}^{-1}$) and also by a polar solvent which could point to a two-step mechanism;[118] or at least to a dipolar transition state.

Product ratios As mentioned above, two products may result when a cyclic diene adds to an ethylene derivative which is not symmetrical about the longitudinal axis. Substituents in the adduct are designated exo or endo according to whether they are on the same or opposite side to the non-dienic bridge. Routes to these products are parallel and independent and, should the transition states to each differ in volume, the product ratio will be pressure-dependent. Substituents located endo in the transition state occupy a smaller volume than when exo[84] and the partial molar volumes of endo product are slightly smaller than endo.[55] The compactness of the endo transition state is ascribed to secondary interactions between substituent and diene moiety. For additions of mono-substituted ethylenes ($R = H$, $X = CN$, COOR, CHO, etc.), the endo product is even more favoured at high pressure although not by much. The difference in activation volumes, $\Delta\Delta V^{\ddagger} = (\Delta V^{\ddagger}_{endo} - \Delta V^{\ddagger}_{exo})$, only amounts to $\sim 0.5\, cm^3\, mol^{-1}$.[84] Additions of 1,1-disubstituted ethylenes lead to a competition for the endo position when the order prevails; $Br > Cl > COOMe \sim CN > Me > Et \sim Bu$ which points to a combination of steric and polar factors operating. An intriguing aspect of this reaction is the slight difference in activation volumes for the two diastereoisomeric transition states from butadiene and dimethyl fumarate[85] ($\Delta\Delta V^{\ddagger} = 0.9\, cm^3\, mol^{-1}$) which leads to some induced asymmetry in the cyclohexenedicarboxylic acid product.

Structural selectivity ('regioselectivity') is also affected by pressure in favour of the isomer originating from the transition state of lower volume. From isoprene and crotonaldehyde two adducts are formed (four if stereoisomers are counted),

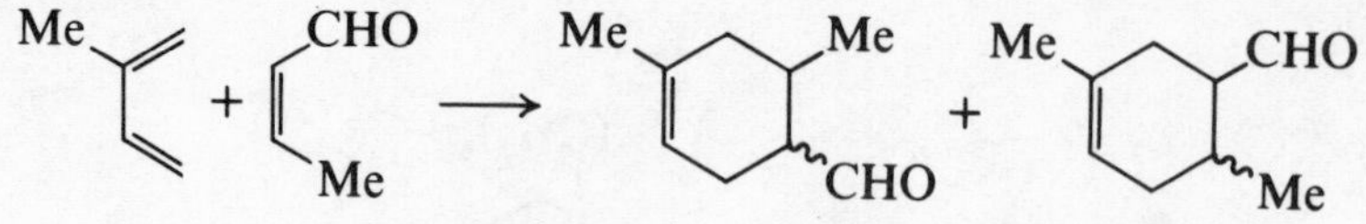

	6 kbar	60	40
	8	64	36
	10	72	28

and the variation in product ratio may be translated into a difference in activation volumes for the two reactions, $\Delta\Delta V^{\ddagger}$.

(6+4)-Cycloadditions Few examples of this thermally allowed concerted reaction are known. The reaction between tropone, and a reactive diene such as cyclopentadiene, occurs at moderate temperatures and has been studied under pressure.[62] The small value of $\Delta V^{\ddagger}$ is at first sight surprising but is due to the

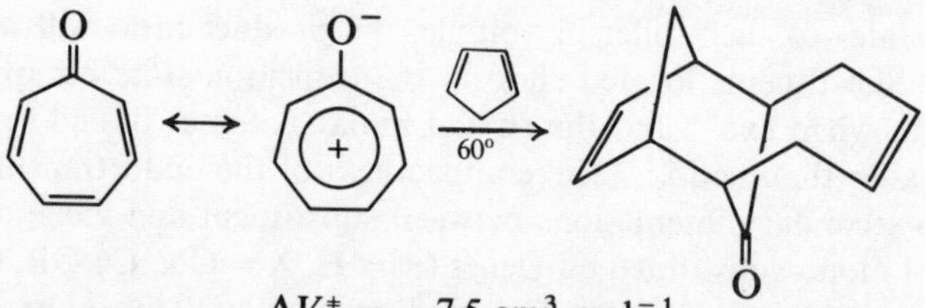

$$\Delta V^{\ddagger} = -7\cdot 5 \text{ cm}^3 \text{ mol}^{-1}$$

abnormally low partial molar volume of tropone which has a large dipole moment, 4.2D, due to contributions from the 'aromatic' canonical structure, absent in the product. There is no reason to suppose the result to be inconsistent with a concerted cycloaddition, and the volume of reaction is similar though even less negative, $\Delta V = -4.3 \text{ cm}^3 \text{ mol}^{-1}$.

A further class of $(4+2)$ cycloadditions are termed '1,3-dipolar' and take place between reactive olefins and molecules with a three-atom system of the general type, $-x = y - \ddot{z}^-$ which act as the 4π component.[81] Examples of 1,3-dipoles are: diazoalkanes, $R_2C{=}\overset{+}{N}{=}\overset{-}{N}$, cyanates $R{-}C{\equiv}\overset{+}{N}{-}\overset{-}{O}$, and azomethine ylides $R{-}C{\equiv}\overset{+}{N}{-}NR'$. Concerted cycloadditions occur with the formation of five-membered rings:

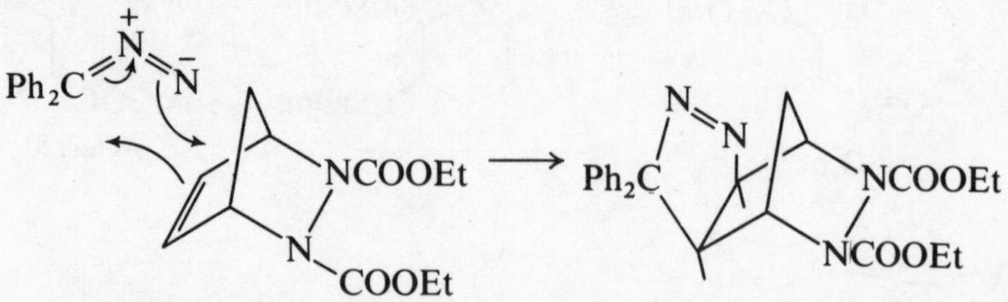

One might expect the volume of activation to be less negative than for neutral Diels–Alder reactions and especially less than those for dipole-forming cycloadditions since the polar nature of the reagent is lost in the product. That this is so may be judged from the one result in Table 4.6. That this value is as large as

Table 4.6. Volumes of activation and reaction for Diels–Alder reactions and other cycloadditions

Diene	Dienophile	Solvent	T (°C)	$\Delta V^{\neq}$ (cm³ mol⁻¹)	$\Delta \bar{V}$ (cm³ mol⁻¹)	Reference
Isoprene	Isoprene	None	60	−25		37
		nBuBr	40	−36 (−30)	−25.6	4 (6)
			50	−50	−41.9	44
		nBuBr	60	−35.3	−44.1	
			70	−40.0	−45.9	
				−40.0	−47.8	
				−41.3		
	Methyl acrylate	nBuBr	21	−30.8	−36.9	42
	n-Butyl acrylate	PhBr	80	−42	−23	43
	Acrylonitrile	nBuBr	21	−33.1	−37.0	42
	But-1-en-3-one	nBuBr	21	−36.9	−37.1	42
	Maleic anhydride	Acetone	35	−39.0	−35.9	38, 39
		MeCN		−37.5	−34.5	
		nBuCl		−38.0		
		CH₂Cl·CH₂Cl		−37.0	−35.5	
		CH₂Cl₂		−39.8	−33.4	
		(MeO)₂C=O		−39.3		
		EtOAc		−37.4	−36.8	
		(isoPr)₂O		−38.5	−38.3	
		MeNO₂		−32.5	−30.7	
Penta-1,3-diene	Maleic anhydride	Acetone	35	−47.3	−31.3	38
		MeCN		−43.1		
		nBuCl		−48.9	−33.3	
		CH₂Cl·CH₂Cl		−44.7		
		(MeO)₂C=O		−45.6		
		(isoPr)₂O		−51.4		

Table 4. 6. (cont.)

Diene	Dienophile	Solvent	T (°C)	$\Delta V^{\ddagger}$ (cm³ mol⁻¹)	$\Delta \bar{V}$ (cm³ mol⁻¹)	Reference
1-Methoxybuta-1,3-diene	Maleic anhydride	MeCN	35	−32.0	−32.4	38
		nBuCl		−45.4	−35.5	
		CH₂Cl·CH₂Cl		−43.7	−30.4	
		(MeO)₂C=O		−53.6	−32.2	
		MeNO₂		−43.0	−28.2	
2-Methoxybuta-1,3-diene	Maleic anhydride	Acetone	50	−36.2	−34.1	38
		MeCN		−33.5	−31.9	
		nBuCl		−36.7		
		CCl₄		−37.6		
		CH₂Cl·CH₂Cl		−35.5		
		(isoPr)₂O		−40.7		
2-Chlorobuta-1,3-diene	Maleic anhydride	Acetone	65	−48.6		38
		MeCN		−41.6	−36.9	
		nBuCl		−51.1		
		CH₂Cl·CH₂Cl		−48.2		
		(MeO)₂C=O		−42.9		
		(isoPr)₂O		−43.7		
2,3-Dimethylbuta-1,3-diene	Maleic anhydride	nBuCl	30	−41.3	−36.3	40
	Methyl acrylate	nBuCl	40	−30.2	−37.0	
	n-Butyl acrylate	nBuCl	40	−29.6	−37.0	
	Dimethyl fumarate	nBuCl	40	−32.9	−37.2	
	4-Chloronitrosobenzene	PhMe	25	−39.7	−32.5	41
		CCl₄		−29.3	−33.5	
		PhCl		−26.8	−30.6	
		CH₂Cl·CH₂Cl		−30.0	−32.6	
		CH₂Cl₂		−28.8	−30.9	
		EtOH		−37.0	−32.2	
		PhNO₂		−22.8	−31.2	

Diene	Reagent	Solvent				
	4-Ethoxynitrosobenzene	PhMe	25	−39.2	−38.2	
		CCl₄		−32.0	−40.5	
		PhCl		−26.7	−35.3	
		CH₂Cl·CH₂Cl		−27.5	−38.4	
		CH₂Cl₂		−27.8	−36.6	
		EtOH		−29.1	−40.8	
		PhNO₂		−26.6	−33.7	
Acrolein	Acrolein	heptane	70	−37		45
But-1-en-3-one	But-1-en-3-one			−41		
Acrolein	But-1-en-3-one			−36.5		
But-1-en-3-one	Acrolein			−35		
Cyclopentadiene	Cyclopentadiene	n-BuCl	40	−23.7	−33	40
			20	−31		44
	Dimethyl fumarate	n-BuCl	30	−32.7	−36.7	40
	Methyl arylate	n-BuCl	40	−30.1	−35.7	
	Dimethyl acetylene-dicarboxylate	n-BuCl	35	−30.2	−33.9	
	4-Chloronitroso-benzene	PhMe	25	−33.9	−41.6	41
		PhCl		−37.1	−50	
		CH₂Cl₂		−25.7	−35	
		EtOH		−39.5	−40	
		dioxan	60	−7.5	−4.3	
1,2,3,4-Tetrachloro-cyclopentadiene	Tropone (Cyclohepta-2,4,6-trienone) Methyl acrylate	n-BuCl	40	−24.6	−33.2	
Cyclohexa-1,3-diene	Maleic anhydride	n-BuCl	35	−37.2	−30.3	70
Diphenyldiazo-methane	(bicyclic structure, NCOOEt / NCOOEt)	PhCH₃	24.5	−32		

Table 4. 6. (cont.)

Diene	Dienophile	Solvent	T (°C)	$\Delta V^{\ddagger}$ (cm³ mol⁻¹)	$\Delta \bar{V}$ (cm³ mol⁻¹)	Reference

Activation volume differences

$\Delta\Delta V^{\ddagger} = 0$ (ref) $-2\cdot3$ $-1\cdot0$ $-3\cdot3$ 395

$\Delta\Delta V^{\ddagger} = 0$ $+0\cdot83$ 396

$-32\,\mathrm{cm^3\,mol^{-1}}$ points to the fact that diphenyldiazomethane has only a small dipole moment ($1.43\,\mathrm{D}$[82]) and the dipolar structure written above is only one of several contributing structures which render the molecule almost neutral. The addition of cyanate to HCN falls into this category and is a potentially useful source of oxadiazoles[446]

$$\Delta V^* = -20\,\mathrm{cm^3\,mol^{-1}}$$

(2+2)-*Cycloadditions* The formation of cyclobutane from two molecules of ethylene does not take place thermally. For such a cycloaddition to take place in a concerted fashion, the symmetries of reagent and product orbitals would not correlate[67] unless the sterically unfavourable orthogonal approach were adopted. These restrictions may be relaxed for highly strained olefins such as trans-cycloheptane[64] which dimerizes or for heteroatom derivatives or cumulenes; the following examples of (2+2) cycloadditions are typical:

(a) azo-esters with nucleophilic olefins:

(b) ketens with nucleophilic olefins:

(c) isocyanates with olefins:

(d) allene dimerizations:

Extremely electrophilic olefins such as tetracyanoethylene are prone to add to nucleophilic olefins such as vinyl ethers:

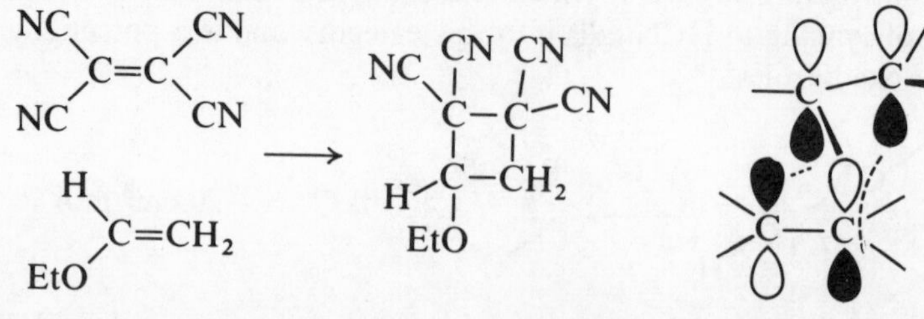

As with $(4+2)$ cycloadditions, the mechanistic possibilities of concerted, diradical, or dipolar routes arise and the latter becomes distinctly favourable when heteroatoms are present.[63] However, it would be surprising if such diverse reactions all partook of the same reaction pathway so that high pressure studies may provide a useful criterion. In general, the concerted mechanism should show $\Delta V^{\ddagger} \sim -30\,\mathrm{cm^3\,mol^{-1}}$ and a small solvent dependence; the diradical pathway should be accompanied by a much less negative value of $\Delta V^{\ddagger}$, -10 to -20 and also little solvent effects, while the dipolar route should show a very large negative value and rates strongly dependent on solvent polarity. Tetracyanoethylene additions to vinyl ethers have been studied by Kelm[68,69] and by the author.[70] The volumes of activation are very large and negative, even more so than for Diels–Alder reactions (Table 4.7). Volumes of reaction are much less negative. Further evidence concerning this reaction all points to a two-step pathway *via* a dipolar intermediate.

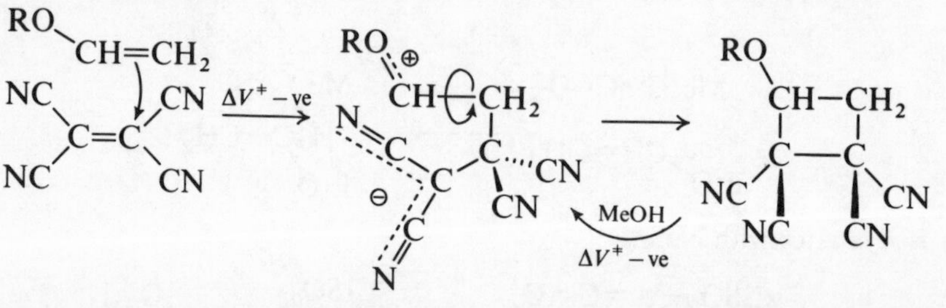

Moreover, the cycloadduct will open in methanol to the same dipolar intermediate with a large negative activation volume, $\Delta V^{\ddagger} = -16.7\,\mathrm{cm^3\,mol^{-1}}$.[389] Products from 2-substituted vinyl ethers are not stereospecific[65b,c] and unreacted vinyl ether undergoes some geometrical isomerization showing that the first step is reversible. Rates are highly solvent dependent and an estimate of the dipole moment of the transition state from this indicates it to be well towards a zwitterion. The entropy of activation is large and negative (-170 to $-120\,\mathrm{J\,K^{-1}}$) which must be a consequence of solvation although, strangely, this value is not solvent dependent. In view of this evidence, the activation volumes can be seen as a consequence of a very ordered and compact transition state necessary to accomplish bonding, and the electrostriction of solvent around the polar sites. Stereochemical evidence suggests that cycloadditions of isocyanates also partake of a two-step dipolar mechanism[71] although keten cycloadditions may be concerted:[72,76] this also may apply to reactions of allenes both with themselves and with electrophilic olefins. The evidence for concertedness in keten–olefin cycloadditions includes the existence of secondary isotope effects at either terminus of the olefin component[72,73,74] showing that bonding at both sites is occurring in the transition state, although an orthogonal (antarafacial) mode of approach to the keten may be adopted. There are similar findings from cycloadditions to azoesters.[75] For all these reactions, the most consistent picture is of a concerted cycloaddition in which the participants adopt the orthogonal, $(_{\pi}2_s + _{\pi}2_a)$ geometry,[77,78] although the extent of bond formation may not be the same at each terminus allowing the transition state to possess dipolar character since, in some instances at least, rates of these reactions are dependent upon solvent polarity.[79] Indeed, it is quite likely that the mechanism goes over to a two-step dipolar type when the donor molecule is one readily capable of accepting a full positive charge, such as an enamine.

Volumes of activation for additions of ketens are few but are very negative as is that for the addition of diethyl azodicarboxylate, Table 4.7. It must be concluded that it is not possible to distinguish between concerted and dipolar non-concerted mechanisms on the basis of a single activation volume, both giving rise to values in the region of $-40 \, cm^3 \, mol^{-1}$. The effect of changing solvent may permit a distinction to be made, since values of $\Delta V^{\ddagger}$ should be more sensitive to the dielectric constant function than is the Diels–Alder reaction. This appears to apply to the tetracyanoethylene-vinyl ether reaction, for which an estimate of the dipole moment of the transition state has been made (section 4.15) and a value of 15 D obtained.[69] By further partitioning $\Delta V^{\ddagger}_{exp}$ into components $\Delta V^{\ddagger}_1$ (intrinsic change due to bond making) and $\Delta V^{\ddagger}_2$ (due to solvent interaction) it was shown that the former contributes $-14 \, cm^3 \, mol^{-1}$, in accordance with the formation of only one bond, while $\Delta V^{\ddagger}_2$ ranges from -32 in carbon tetrachloride to -5 in acetonitrile and thus is larger the lower the dielectric constant of the solvent, as predicted. A further mechanistic point arises from the existence of charge-transfer complexes in mixtures of these donors and acceptors (section 4.2.3). While they are known to be favoured by pressure and their spectra are visible in the mixtures, it remains uncertain as to whether these species actually occur on the reaction pathway to cycloadducts. Present evidence favours the formation of charge-transfer complexes by a separate equilibrium, independent of the cycloaddition. Volume profiles for the cycloadditions are similar to those of Diels–Alder reactions (i.e. the transition state has a smaller volume than the products) and the difference is even more pronounced, $\Delta V^{\ddagger} - \Delta V = = -20\text{–}25 \, cm^3 \, mol^{-1}$.[487] Such compactness is consistent with a highly dipolar transition state *en route* to the intermediate.

Diradical $(2+2)$ cycloadditions are known in a few instances such as the side reaction occurring in the dimerization of chloroprene. The Diels–Alder product is favoured by pressure over the cyclobutane product with an activation volume which is $10 \, cm^3 \, mol^{-1}$ more negative in accordance with its smaller transition state.[388] By contrast, very little discrimination in product ratio by pressure is

Table 4.7. Volumes of activation and reaction for (2 + 2) cycloadditions

Acceptor	Donor	T (°C)	Solvent	$\Delta V^{\ddagger}$ (cm³ mol⁻¹)	$\Delta \bar{V}$ (cm³ mol⁻¹)	Reference
Tetracyanoethylene	EtO—C=CH₂ (H)	25	CH₂Cl₂	−38.0 (−55)	−31.9	71 487
	nBuO—C=CH₂ (H)	30 30	CCl₄ Benzene Acetone Acetonitrile	−34.5 −37 −43 −35 −29	−29.4	71, 69
	EtO—C=CMe₂ (H)		CH₂Cl₂	−29.2	−29.5	487
	EtO—C=CMe₂ (Ph)	25	CH₂Cl₂	−46.5 (−55) −45.5	−27.7	71
	CH₂=CH—CH₂—CH₂C₆H₄OMe			−34.0	−26.7	70
	(dihydropyran)	27.5	CH₂Cl₂	−43.0 (−36)	−36.1	487

Structure	Temp	Solvent		
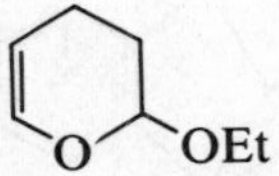 (dihydropyran, OEt)	25		−45.3	
Me–CH=C(OEt)H (EtO, H, Me, H)	25	CH$_2$Cl$_2$	−50	−30.3
	25	CH$_2$Cl$_2$	−55	−29.7
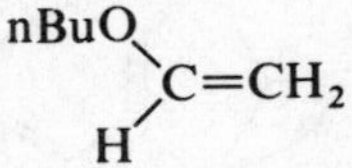 nBuO–C(=CH$_2$)H	25.5	PhCH$_3$	−51	
Me$_2$C=C=O, nBuO	25	PhCH$_3$	−30	
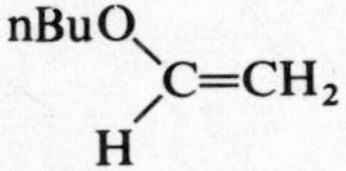 nBuO–C(=CH$_2$)H	24.5	PhCH$_3$	−46	

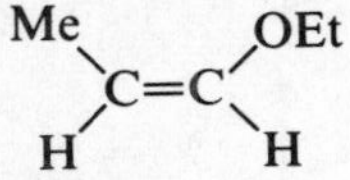

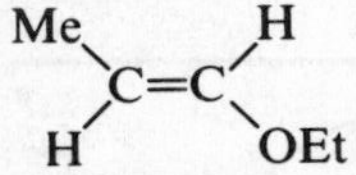

Tetracyanoethylene

Ph$_2$C=C=O

Me$_2$C=C=O

EtO$_2$C—N=N—CO$_2$Et

Diels–Alder, 4 + 2   1

$(2+2)$ $[1 : 2 = 0.4(1\,\text{bar}),$
$12(10\,\text{kbar})]$

observed between the $(2+4)$ and $(2+2)$ pathways for addition of tetrachloro-benzyne to norbornadiene, $\delta\Delta\bar{V} = 0.7\,\text{cm}^3$ (in favour of $(2+2)$). It is suggested

that in this case the transition state for $(2+2)$ addition bears a dipole which reduces its volume and compensates for the incomplete bond-formation.[483]

Claisen, Curtius, and Cope rearrangements are concerted neutral processes in which a formally bi-allyl system cleaves at the centre bond and rejoins at the termini. A rearrangement of 6-electrons is involved via a cyclic transition state. These pericyclic reactions are classed as [3,3]-sigmatropic shifts.[80] The Claisen rearrangement is the thermal conversion of an allyl aryl ether to the dienone, which subsequently isomerizes to o-allylphenol.[89, 90] The Cope rearrangement of hexa-1,5-dienes[91] can be seen to be stereospecific from the following example; the driving force is the production of a more stable hexa-1,5-diene, usually having a greater number of terminal substituents.[92] Finally, the Curtius rearrangement of allylic azides may be regarded as a nitrogen analogue of the Cope rearrangement.[93]

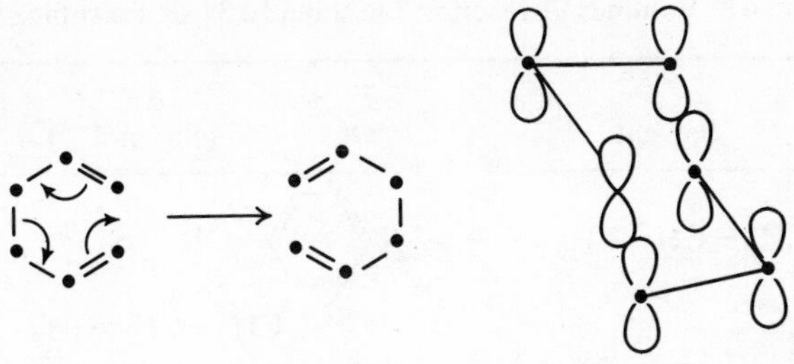

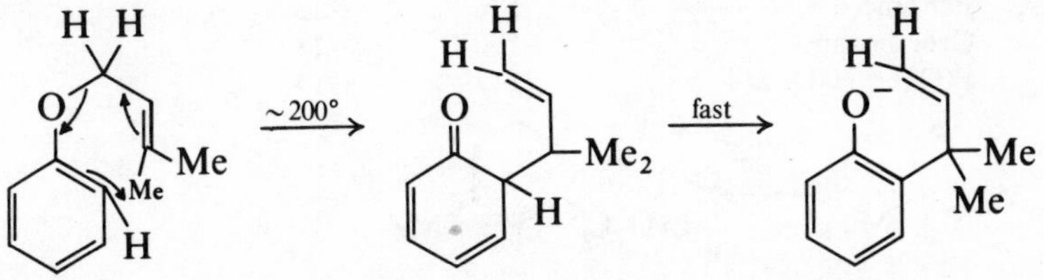

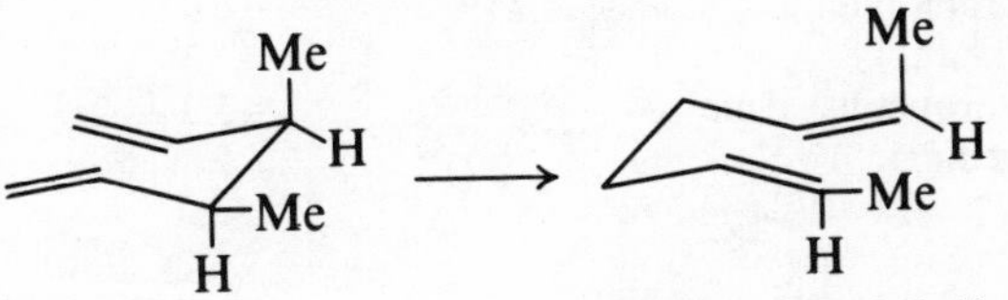

meso

Cis, trans octa-2,6-diene

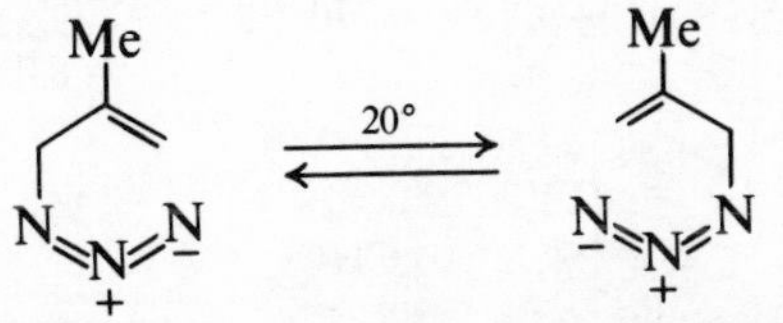

Volumes of activation are all remarkably similar for these examples, supporting the essentially similar pathways (Table 4.11). All show a slightly more compact transition state than either reagents or products consistent with the cyclic species proposed. There is very little effect either on rates or activation parameters due to solvent. The volumes of activation are consistent with the formation of a cyclic transition state with no charge development. Somewhat similar to the Claisen rearrangement is the thermal interconversion of a 2-alkoxypyridine-1-oxide, into 1-alkoxy-2-pyridone. For R = benzyl, the rearrangement is believed to be concerted but the benzhydryl analogue, R = —CHPh$_2$, certainly rearranges via a radical pair which results in nuclear polarization (CIDNP[148]), characteristic of this mechanism. It is highly gratifying to find a positive value of $\Delta V^{\neq}$ for benzhydryl ($\Delta V^{\neq} = +10 \, \text{cm}^3 \, \text{mol}^{-1}$) but a negative value for benzyl ($\Delta V^{\neq} = -30$) due to the cyclic transition state and despite the loss of dipolar character of the N-oxide group.[147]

Table 4.8. Volumes of reaction for some [3,3]-sigmatropic shifts

	Solvent	T (°C)	$\Delta V^{\ddagger}$ (cm³ mol⁻¹)	Reference
$X\langle\bigcirc\rangle$–OCH$_2$–CH=CH$_2$ $\longrightarrow$ $X\langle\bigcirc\rangle$–OH, CH$_2$–CH=CH$_2$				
X = Me	None	160	−10.3	86
	Octan-1-ol		−7.3	
	Ethanediol		−6.2	
	Decahydronaphthalene		−7.2	
X = Me	Benzene	186	−18	87
	Cyclohexane	176	−18	
	EtOH—H$_2$O, 2:1	147	−14	
(NC, EtO$_2$C, Et, Me) $\longrightarrow$ (NC, EtO$_2$C, Et, Me)				
	1,4-Dibromobutane	119	−6.0	86
	Octan-1-ol		−5.3	
	Decahydronaphthalene		−5.3	
	Propanediol		−5.8	
(Me, $^-$N=N=N$^+$) $\rightleftharpoons$ (Me, N=N=N$^-$ +)				88
	CH$_2$Cl$_2$	20	−9.5 (forward)	$\Delta V = -1.5$
			−8.0 (reverse)	
$\square\!\square \longrightarrow$ (hexamethylbenzene)	25 CHCl$_3$		−34.6	392
(Ph, Ph, CO$_2$Et, CO$_2$Et) $\longrightarrow$ (Ph, Ph, CO$_2$Et, CO$_2$Et) (E,Z) → (Z,Z)	90 C$_4$Cl$_4$		−1	393
			−10	

The butadiene–cyclobutene interconversion Thermal ring-opening of cyclobutenes is found to occur, the stereochemistry being regulated by orbital symmetry considerations.[67] Rotation in the same sense ('conrotation') at C3, C4 is required so that a *cis*-3,4-disubstituted cyclobutene is converted to the (E, Z) butadiene and the *trans*- to (E, E) or (Z, Z) isomers, the reverse reactions similarly.

Despite bond-breaking and ring-opening in this neutral reaction, a slightly negative activation volume is found, approximately −1 cm³ mol⁻¹. Subsequent isomerization of the product to the more stable (Z, Z) isomer is found to be even

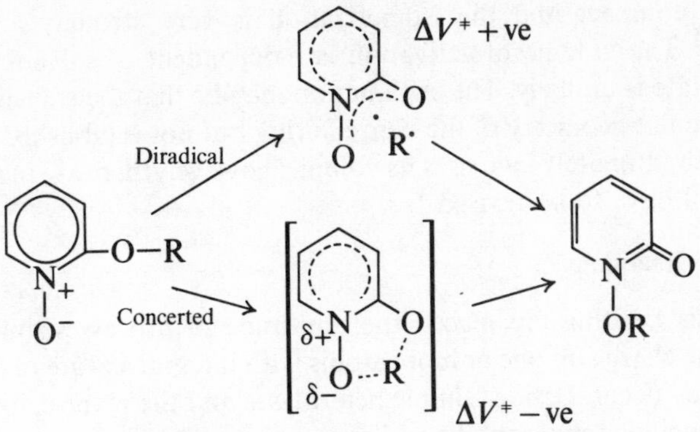

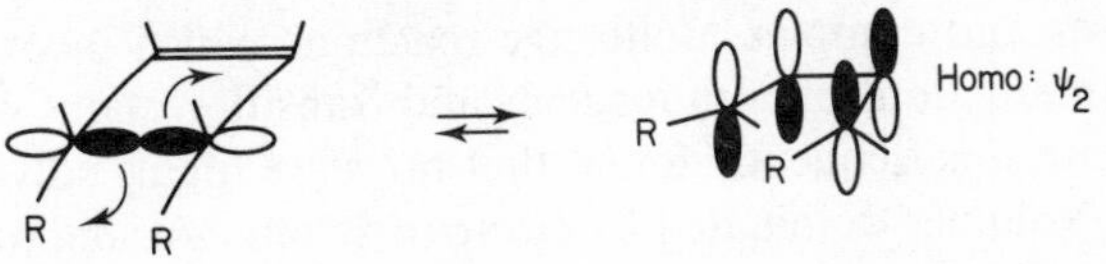

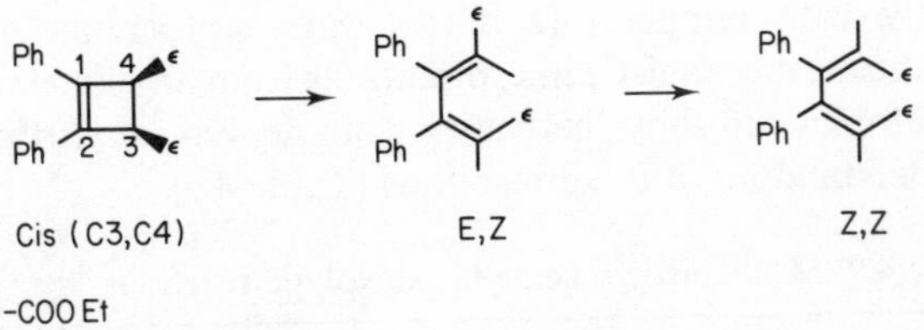

$\Delta V^{\ddagger} + \text{ve}$

Diradical

Concerted

$\Delta V^{\ddagger} - \text{ve}$

Homo: ψ_2

Cis (C3,C4)   E,Z   Z,Z

$\epsilon = -COO\,Et$

more pressure-accelerated, $\Delta V^{\ddagger} \sim -10\,\text{cm}^3\,\text{mol}^{-1}$ which may be due to a dipolar transition state.[393]

An example of a formally 'forbidden' pericyclic reaction is the transformation of 'Dewar benzene' into benzene, a process which does take place (with the hexamethyl derivative) at temperatures around $100°$ with the high activation energy of $155\,\text{kJ}\,\text{mol}^{-1}$. It is assumed that breaking of the central bond gives rise initially to a diradical.

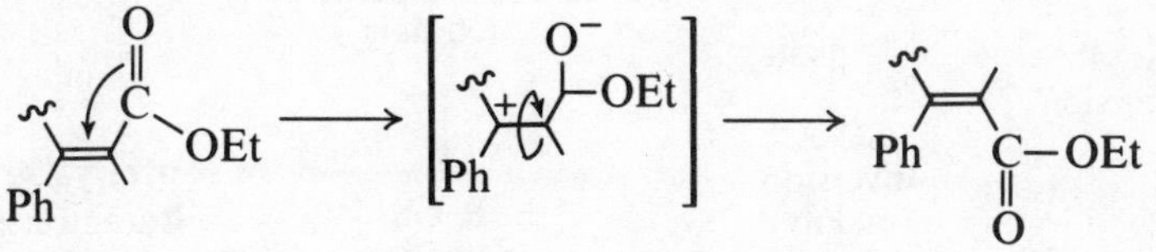

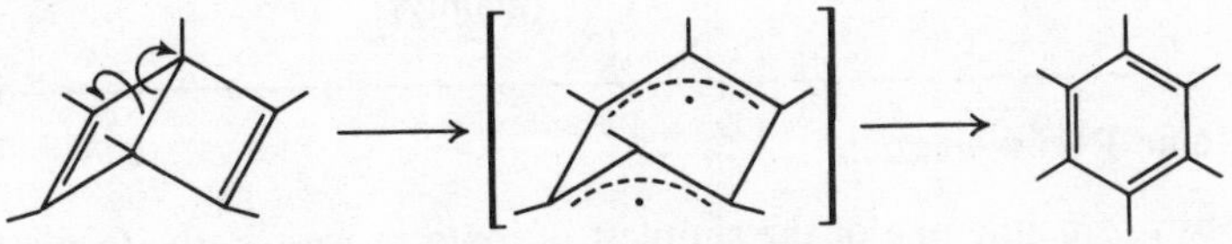

$\Delta V^{\ddagger} = -34\cdot6\,\text{cm}^3\,\text{mol}^{-1}$
$\Delta \bar{V} = -22\cdot5$

There is an overall volume contraction as the puckered bicyclohexadiene system becomes planar but it is somewhat surprising that the transition state is even more compact and the isomerization is very strongly accelerated by pressure.[392] The volume of activation is independent of solvent so a dipolar transition state is unlikely. The explanation may be that the transition state has acquired the flat geometry of the benzene ring but not established the bulky π-system which ultimately increases its volume; however, there are indications that the data will have to be revised.*

4.2.6 Ionic Reactions

Under this heading are placed the multitude of processes during which a change in the charge on one or more atoms is an integral feature of the pathways by which they occur. Besides simple heterolyses and the reverse, ionic recombination, are included nucleophilic and electrophilic displacements, additions and eliminations and complex multistage reactions which incorporate these components. Interactions of both reagents and transition states with the solvent are important so it is frequently found that rates are highly solvent-dependent and activation volumes dominated by electrostriction considerations which in turn may vary widely with temperature. Great care is therefore necessary to disentangle significant mechanistic interpretations from any kinetic data but, as the following examples will show, activation volumes have contributed significantly to our understanding of these reactions (Table 4.9).

Table 4.9. Mechanistic scheme for solvolytic reactions[105]

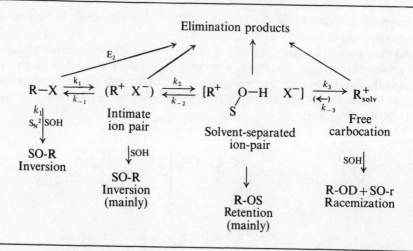

4.2.7 Ionic Recombination

While in principle one of the simplest of ionic processes, the formation of a covalent bond between anion and cation is also usually one of the fastest. An exception is the reaction of OH^- and other nucleophilic anions with various

---

*A recent re-examination by Brower estimates $\Delta V^{\ddagger} \sim -16 \, cm^3 \, mol^{-1}$: however, poor yields of product tend to reduce the precision attainable.

resonance-stabilized carbocations (such as the triphenylmethane dyes, crystal violet, and malachite green) to form the colourless carbinols.[98] From the tables of partial molar volumes, volumes of reaction between oppositely charged ions are positive since the electrostrictive effects are removed upon reaction: for example,

$$H_3O^+ + OH^- \longrightarrow 2H_2O, \quad \Delta\bar{V} = +21 \text{ cm}^3 \text{ mol}^{-1}$$

It is surprising therefore that for these reactions, Table 4.13, volumes of activation are zero for crystal violet and actually negative for other analogues,[99] although volumes of reaction are indeed positive.[100] Furthermore, entropies of activation are negative and roughly correlate with $\Delta V^{\ddagger}$, indicating similar origins of these values. The result is unexpected and points to some completely different process to that written above as being the slow step. This is quite likely to be solvent reorganization during ion-pair formation, probably solvent-separated and involving some quite structured arrangement.

$$\Delta V^{\ddagger} \quad 0 \text{ to } -20$$
$$— \Delta\bar{V} — \quad \sim +30$$

R = NMe$_2$ = Crystal violet
R = H   = Malachite green

Table 4.10

	$\Delta V^{\ddagger}$ (cm^3 mol^{-1})	Reference
Crystal violet + OH$^-$	0.0	99
Malachite green + OH$^-$	-12	
Bromophenol blue + OH$^-$	-14.9	
Phenolphthalein + OH$^-$	-19.7	
(reverse)	-10.6	

4.2.8 Acid–base Reactions

While volumes of reaction for many proton transfer processes are known from equilibrium measurements (section 3.1.2) few rate data are available since the reactions often approach the diffusion-control limit.[94] Perturbation methods such as temperature-jump and pressure-jump are required in which the system at equilibrium is very rapidly perturbed, for example, by an electrical discharge which raises the temperature by a few degrees, and the rate of re-attainment of equilibrium at the higher temperature can then be determined by fast spectrophotometry.[479] The neutralization of Alizarin Yellow, an indicator, has been studied in this way but the volume of activation obtained reveals a more complicated pathway than might at first be imagined[95]

Alizarin Yellow
(AH⁻)

$$\Delta V^{\ddagger} = -5\cdot1 \quad\Big\updownarrow\quad \Delta V^{\ddagger} = +10\cdot3$$

(A⁼) $+ H_2O$

Overall, the rate of the forward reaction is retarded by pressure (rates are of the order $10^{10}\ M^{-1}\ s^{-1}$) and the reverse reaction accelerated but a careful analysis of the kinetics reveals these values to be the resultant of several discrete steps each with an activation barrier:

$$[AH^-], [OH^-] \underset{\longleftarrow}{\longrightarrow} AH^-_{aq} + OH^-_{aq} \underset{\longleftarrow}{\longrightarrow} AH^-\cdots H_2O\cdots OH^-$$

$$\Delta V^{\ddagger} +13\cdot3 \qquad\qquad -3\cdot1 \qquad\qquad \Big\updownarrow \qquad +4\cdot6$$

$$A^{\overline{=}} + 2H_2O$$

Hydrogen-bonded reagents ($[AH^-], [OH^-]$) have first to be stripped of solvent at the donor and acceptor sites to give the aquated but otherwise open ions which then approach and exchange a proton through a water molecule before drifting apart. It appears that the first step contributes most to the overall volume of activation, namely the breaking of hydrogen bonds which is opposed by pressure. The inferred positive volume of activation for the proton transfer suggests a late transition state and is due to charge delocalization in the product, reducing electrostriction. Another fast reaction studied by a similar technique[96] is the proton transfer from (p-nitrophenyl)nitromethane to tetramethylguanidine.[96]

$$\Delta V^{\ddagger} = -13 \cdot 2 \text{(mesitylene)}, -17 \cdot 8 \text{(toluene)} - 16 \cdot 3 \text{(anisole)}$$
$$\Delta \bar{V} = -15 \cdot 9 \qquad\qquad -25 \cdot 5 \qquad\qquad -29 \cdot 3$$

This reaction is unusual in showing a very large deuterium isotope effect $(k_H/k_D = 45)$ and it is believed that quantum mechanical tunnelling is responsible for a significant part of the rate: that is, reaction may take place without the system having attained the classical activation energy due to the uncertainty in position of the proton, a particle of very low mass. This interpretation is supported by volumes of activation which are rather constant (ca. $-16 \, \text{cm}^3 \, \text{mol}^{-1}$) in several solvents, while the volumes of reaction vary widely (from -16 to -30). It is inferred that activation involves the approach of the reagents and transfer of the proton by tunnelling—necessarily involving a 'bare', unsolvated proton, which is followed by diffusion apart of products and solvation reorganization which depends on the medium and contributes to $\Delta \bar{V}$.

The volumes of activation for reactions between an anionic base and a neutral acid, for which no overall change in charge occurs, yield information concerning the position of the transition state on the 'reaction coordinate'. The association of base and acid leads to an initial reduction of volume which is offset by the reduction of electrostriction as the charge becomes delocalized. The volume rises again as the product is approached and may become positive, especially if charge on the product anion is more delocalized than on the reagent anion and which is associated with less solvent structuring, fig. 4.7.

On this basis, a strongly negative value of $\Delta V^{\ddagger}$ is indicative of a 'middle' position of the transition state but a value which is zero or positive is indicative of a late position. The hydrolysis of chloroform provides an example. The mechanism is believed to partake of a rate-determining decomposition of the conjugate base to give dichlorocarbene (3).[152] The volume of activation for ionization may be studied by the loss of an isotopic label (this prevents reversibility because the label is diluted in the water which is the solvent) giving $\Delta V_1^{\ddagger} = +9 \, \text{cm}^3 \, \text{mol}^{-1}$ characteristic of a very late transition state.[149] From the overall volume of activation, $+16 \, \text{cm}^3 \, \text{mol}^{-1}$ (section 4.1)[150] the difference,

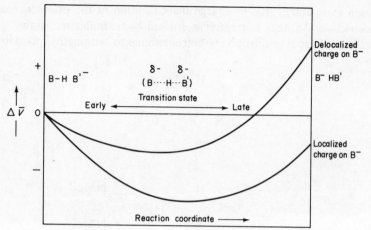

Fig. 4.7. Volume profile of a proton transfer reaction (schematic)

$$\text{CHCl}_3 + \text{OH}^- \underset{}{\overset{k_1}{\rightleftarrows}} \text{CCl}_3^- + \text{H}_2\text{O} \xrightarrow{\text{slow}} :\text{CCl}_2 + \text{Cl}^- \longrightarrow \text{products}$$

(D)   **(3)**

$+7\,\text{cm}^3\,\text{mol}^{-1}$, must be attributed to the slow step, and is quite consistent with fission of the trichloromethyl anion. Furthermore, the positive value of the deprotonation is indicative of charge delocalization onto d-orbitals in chlorine, a factor long held to account for the acidity of chloroform.[151] By contrast, proton removal from fluoroform has $\Delta V^{\ddagger} = +3\,\text{cm}^3\,\text{mol}^{-1}$ since it cannot use d-orbitals for charge delocalization in the anion.

Several enolization rates have been measured under pressure[149] such as that of acetophenone (OH^- in water), $\Delta V^{\ddagger} = -1\,\text{cm}^3\,\text{mol}^{-1}$. This small value indicates a late transition state and, since it is the same with ethanol/ethoxide as the base, it is inferred that no great change in electrostriction has occurred between reagent and transition state. That is, the charge and hence the proton is largely transferred in the transition state. By contrast, the value for indene (**4**) ($-4\,\text{cm}^3\,\text{mol}^{-1}$) may argue a more central position.

The use of a neutral base causes charge separation and hence the electrostrictive term will dominate the measured activation volume. A slow proton transfer is believed to be the rate-determining step in reactions of carboxylic acids with diphenyldiazomethane.[102] In dibutyl ether the rate is increased by pressure,

$\Delta V^{\ddagger} = -13\,\text{cm}^3\,\text{mol}^{-1}$ consistent with electrostriction of solvent around the incipient charges possibly also offset somewhat by stretching of the C–N bond as it loses multiple bond character.[103] The relatively small value points to an early transition state and little charge development ($\Delta \bar{V}$ in this solvent should be < -30). Proton transfers from carbon acids are, as a rule, far slower than from oxygen acids and invariably give rise to delocalized ions. Typical examples include enolizations,

$$\underset{CD_3}{\overset{O}{Ph-C}} + \bar{O}H \longrightarrow \underset{CD_2}{\overset{O}{Ph-C^{\ominus}}} + HDO \qquad -1$$

$$\Delta V^{\ddagger}(\text{cm}^3\,\text{mol}^{-1})$$

$$\underset{Me}{\overset{Me\quad NO_2}{C}}\; H \cdots B \longrightarrow \underset{Me}{\overset{Me}{C}}{=}N{\overset{O}{\underset{O}{}}}\; BH^+$$

$B = \bar{O}Ac \qquad -1$

$B = $ 2,6-dimethylpyridine $\qquad -31$

formation of cyclopentadienide and related ions, stabilized by aromaticity

indene + $\bar{O}Me$ (4) $\longrightarrow$ indenide anion + MeOH $\qquad -4$

and the formation of ylides, carbanions stabilized by adjacent d-orbitals.

$$\underset{H_3C\quad CH_3}{\overset{O}{S}} + \bar{O}H \longrightarrow \underset{H_3C\quad CH_2}{\overset{O}{S^{\ominus}}} + H_2O \qquad +2$$

$$CHCl_3 + OH^- \longrightarrow \underset{Cl}{\overset{Cl}{C^{\ominus}}}{=}Cl + H_2O \qquad +9$$

The ionization of 2-nitropropane by OAc^- has $\Delta V^{\ddagger} = -1$ but with 2,6-dimethylpyridine as base (in the non-polar solvent dibutyl ether) it is $-31\,\text{cm}^3\,\text{mol}^{-1}$.[153] This value may be anomalous since the reaction is suspected of being assisted by quantum mechanical tunnelling.[154] The symptom is that $\Delta V^{\ddagger}$ differs for the isotopic species Me_2CHNO_2 and Me_2CDNO_2, $\Delta V^{\ddagger}_H/\Delta V^{\ddagger}_D = 0.78$, and the primary kinetic isotope effect falls with pressure.[155]

The transfer of hydride ion, H^-, is formally analogous but there are fundamental differences between this and proton transfer. Hydride transfer, which constitutes an example of oxidation/reduction, must take place directly between donor and acceptor and a protic solvent cannot aid the transfer as it can that of a proton. Rates are frequently slow, the reason being as follows. The molecular orbitals involved in transfer of hydrogen between two first-row atoms are constructed from H($1s$) and two $2p$ (or hybrid) atomic orbitals in the combinations,

Proton transfer involves only one electron pair which is associated with the bonding orbital, ψ_1 but hydride transfer requires a further electron pair to be accommodated in the non-bonding ψ_2 which contributes no further stability but does contribute repulsion.

A central transition state and longer partial bonds than in proton transfer are predicted, so that volumes of activation should be moderately negative.[155] This is the case for borohydride reduction of a ketone[149] for which the slow step is:[150]

$$\Delta V^{\ddagger} = -11 \text{ cm}^3 \text{ mol}^{-1}$$

The Cannizzaro reaction is more complicated and involves a pre-equilibrium addition of base to an aldehyde followed by hydride transfer to a second aldehyde molecule. The overall reaction is strongly pressure-accelerated due partly to the

$$\Delta \bar{V} \approx -12 \qquad \Delta V^{\ddagger} \approx -15$$

$$\text{overall } \Delta V^{\ddagger} = -27 \text{ cm}^3 \text{ mol}^{-1}$$

effect on the equilibrium but also on that of the hydride transfer for which substantial bonding has occurred at the transition state.

A further example, the hydride transfer from the leuco form of the dye, crystal violet (**5**), to chloranil (**6**), is again anomalous. The transfer of hydride is assisted by quantum-mechanical tunnelling and the isotope effect is pressure-dependent (k_H/k_D (in MeCN) = 11.2 at 1 bar, 8.0 at 2 kbar) so that $\Delta V_D^{\ddagger}$ is a more correct measure of the volume change as deuterium is less subject to tunnelling than the light isotope. This is largely influenced by the development of charge but bond formation must also contribute significantly to such a large value[156] (Table 4.11)).

$$\Delta V_H^{\ddagger} = -25 \text{ cm}^3 \text{ mol}^{-1}$$
$$\Delta V_D^{\ddagger} = -35 \text{ cm}^3 \text{ mol}^{-1}$$

4.2.9 Ionization and Solvolysis

The reverse of ion recombination, ionization, is so inextricably bound up with solvolytic reactions that under this heading will be considered all nucleophilic displacements at a saturated carbon in which the incoming nucleophile is derived from the protic solvent SOH, by loss of a proton:

$$R\text{–}X \xrightarrow[\text{SOH}]{\text{Solvent}} R\text{–}OS + HX$$

The structural part of the substrate, R, may be primary, secondary, or tertiary alkyl, aralkyl, or cycloalkyl; the leaving group, X, is halide, sulphonate, or an 'onium ion in the commonest examples while the solvent, which must be hydroxylic, may be water, an alcohol, or a carboxylic acid. Reactions of this type are among the most deeply explored and thoroughly documented of any, and despite the seeming simplicity of the process, have in fact revealed great complexity. The early studies of Ingold[104] and coworkers led to a systematization of a great deal of data in terms of two basic mechanisms; the bimolecular (S_N2) route involves rearface attack from solvent or other nucleophile resulting in bonding to carbon while simultaneously the leaving group departs, probably assisted by solvent. Alternatively, the unimolecular (S_N1) pathway may prevail

Table 4.11. Activation volumes for proton- and hydride transfers

Acid	Base	Solvent	T (°C)	$\Delta V^{\ddagger}$ (cm³ mol⁻¹)	Reference
Alizarin yellow					
(p-Nitrophenyl)nitromethane	OH⁻	H₂O		−5.1	95
	Tetramethyl-guanidine	Mesitylene		−13.2	96
		Toluene		−17.8	
		Anisole		−16.3	
Benzoic acid	Diphenyl-diazomethane	Bu₂O	25	−13	103
PhCOCD₃	OH⁻	H₂O	27.4	−1	149
PhCOCD₃	OEt⁻	EtOH	19.5	−1	
PhCOCD₃	OC₆H₄Cl⁻	MeOH/H₂O	100.5	−5	
PhCD₂CN	OAc⁻	MeOH	48.8	−4	
PhCD₂COOMe	OMe⁻	MeOH	27.7	−3	
CD₃SOCD₃	OH⁻	H₂O	71	+2	
Me₂CHNO₂	OAc⁻	H₂O	21	−1	
Me₂CDNO₂	OAc⁻	H₂O	65	−1	
Me₂CHNO₂	2,4,6-Trimethyl-pyridine	H₂O; tBuOH	29	−31	155
Me₂CDNO₂	2,4,6-Trimethyl-pyridine	H₂O; tBuOH	29	−40	149
Indene-1,1,3-d_3	OMe⁻	MeOH	29.5	−4	149
PhC≡CD	OH⁻	MeOH/H₂O	27	−1	
Ph₂CD₂	OMe⁻	MeOH/DMSO	109	−3	
H⁻-Donor	H⁻-Acceptor				
(Me₂N—C₆H₄—)₃CH	Chloranil	Bu₂O		−25	156
(Me₂N—C₆H₄—)CD	Chloranil	Bu₂O		−35	
PhCHO—OH⁻	PhCHO	MeOH/H₂O	100	−27	149
BH₄⁻	PhCOCH₃	isoPrOH	22	−11	

when ionization between carbon and the leaving group constitutes the rate-determining step and in a subsequent, fast process the solvent coordinates to the intermediate carbocation:

Many criteria were formulated in order to assist in distinguishing between these types, but inconsistencies were found: some substrates behaved as predicted for the S_N2 mechanism by one criterion and as for the S_N1 by another. Subsequent work by Winstein[105] and emphasized by Sneen[106] suggests the operation of the more complex scheme, Table 4.9, in which products are seen to arise either directly by the classical S_N2 route or by attack of solvent on an ion-pair which results from the high dielectric environment of the substrate together with nucleophilic assistance from the solvent at carbon and electrophilic assistance at the leaving group. Ion pairs are species comprising a cation and anion still within mutual electrostatic interaction and may be 'intimate', i.e. juxtaposed within a solvent cage, or 'solvent-separated.[107,130] In the extreme case, they may drift apart before the cation becomes stabilized by coordination of solvent, giving rise to the classical S_N1 pathway. The particular mechanism favoured by a given system will depend upon a complex interplay of cation stability (favouring ionization), solvation, solvent basicity, and steric effects, all of which may be sensitive to both temperature and pressure. Hundreds of activation volume measurements have been made on solvolytic reactions[221,223] and it is clear that their interpretation is fraught with difficulty. However, added to other data, a coherent picture may emerge in certain cases.

Selected values of $\Delta V^{\ddagger}$ for solvolytic reactions are given in Table 4.12 below, from which it is apparent that all are accelerated by pressure with $\Delta V^{\ddagger}$ in the range -10 to $-20 \, \text{cm}^3 \, \text{mol}^{-1}$. Of the individual processes depicted in scheme 4.9, ionization to the intimate ion-pair, k_1, would certainly be expected to be pressure-accelerated on account of the electrostriction of solvent at the charge centres. So also would the S_N2 process, k_I, for the transition state contains a

transition state k_1 transition state k_I (s_N2)

solvent molecule in the process of becoming bonded and there are partial charges on the nucleophilic centres. It is likely that the subsequent separation of ions, k_2, k_3, would be accompanied by further solvation and electrostriction: as mentioned above (4.2.4) for triphenylmethane dyes, processes believed to be ion-pair formation from dissociated ions, and thus akin to k_{-3}, k_{-2} seem to be associated with small or negative values of ΔV^+. It follows that in solvolysis, k_2, k_3 might contribute a small or positive value if either of these processes were rate-determining or prior to the rate-determining step. It is usually assumed that product forming steps k_{II}, k_{III}, and k_{IV} are fast. To look first at extreme cases, methyl and primary halides should be prime candidates for the concerted S_N2 route (Table 4.15 below) while tert. alkyl and especially aralkyl compounds must certainly adopt an ionization mechanism. Tert. butyl halides were adopted by Ingold[104] as prime examples of his S_N1 route and by Winstein and Grunwald[146] also in developing their solvent-sensitivity parameter, m_s. Cumyl chloride was chosen by Brown as a framework for measuring conjugative substituent effects, (σ^+), the central assumption being the favourability of the ionization process in solvolysis.[116] No clear distinction between these supposed mechanistic differences can be made on the basis of ΔV^+ and this comment must apply generally to this type of reaction at least at the present state of knowledge. A further complication which may vitiate comparison between values is their sensitivity to solvent. In the few examples where pure solvents have been used, ΔV^+ appears to become less negative the higher the dielectric constant (e.g. MeOH, HCOOH[108,109]). Much of the published data have been derived from rates in mixed aqueous solvents which defy comparison and it appears that the structuring of the solvent is of paramount importance. With rather similar cosolvents such as water and glycol the volume of activation of benzyl chloride is insensitive to composition.[110] In aqueous acetone, ethanol, or dimethyl sulphoxide,[111-113] the same compound hydrolyses with activation volumes which vary in an unpredictable way with water concentration, fig. 4.8, extrema being observed. Hyne has calculated partial molar volumes of reagents and transition states for different solvent compositions and showed that the behaviour of $\Delta \bar{V}^+$ and $\Delta \bar{V}_g$ were similar in the case of p-chlorobenzyl chloride (for which S_N2

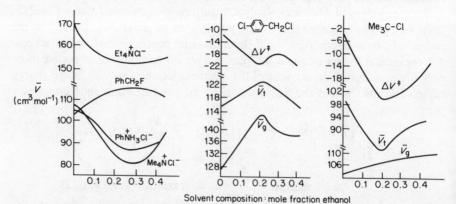

Fig. 4.8. Effect of solvent composition (EtOH–H_2O) on (a) partial molar volumes of some solutes (b) and (c) partial molar volumes of substrate ($\bar{V}_g$), and transition state ($\bar{V}_t$) and activation volume, ΔV^+ ($= \bar{V}_g - \bar{V}_t$) for some solvolyses. (After Hyne et al.[111])

behaviour might be expected) but very different for t-butyl chloride (ionization) and in which $\Delta V^{\ddagger}$ showed a solvent profile resembling that of a salt, anilinium chloride. The implication here is that the volume of activation depends upon specific solvation of both substrate and transition state and, in mixed aqueous solvents the solvent structure varies erratically with composition, as do partial molar volumes. Although one could not with confidence assign different mechanisms to these compounds on the basis of $\Delta V^{\ddagger}$ alone, a comparison between volumes of covalent halide and transition state can be made which is potentially more revealing.[111] A similar analysis of benzyl chloride hydrolysis in aqueous acetone and DMSO reveals that here also activation volumes show extrema and there is no great significance to be attached to an individual value of $\Delta V^{\ddagger}$—the difference between two large quantities, each of which vary considerably with solvent.[113, 114] Even if no mechanistic change occurs in the broad sense, a series of related compounds may exhibit differences in the pressure-dependence of their hydrolysis rates, for example, differing in the position of the transition state on the reaction coordinate according to the degree of bond-making or bond-breaking and also the extent of nucleophilic participation by solvent. The hydrolyses of substituted cumyl chlorides, undoubtedly *via* ion-pairs, have volumes of activation which roughly correspond with σ^{+}-values of the substituents.[115]

Since, by definition[116] for this reaction,

$$\log k_x - \log k_{\mathrm{H}} = -4.54\sigma_x^{+} \tag{4.41}$$

where $\rho^{+} = -4.54$, there is an expectation of a variation of the reaction constant, ρ^{+} with pressure:

$$\frac{\Delta V_x^{\ddagger}}{\Delta V_{\mathrm{H}}^{\ddagger}} = -\Delta\Delta V^{\ddagger} = \frac{2.303}{RT}\sigma_x^{+}\left(\frac{\partial\rho^{+}}{\partial P}\right)_T \tag{4.42}$$

The value of the reaction constant does, in fact, change in the predicted manner becoming less negative with pressure. It would appear that the driving force for these hydrolyses, as pressure is increased, derives less from electron release than from a more favourable entropy of activation. A similar conclusion may be drawn from hydrolyses of substituted benzyl chlorides,[117] less prone to ionization than cumyl chlorides (though still probably reacting via ion-pairs or a mixture of mechanisms); volumes of activation are more negative the stronger the electron-withdrawing group present. Differences in temperature can also render comparisons invalid. In general, and in pure solvents, the activation volume appears to become less negative at lower temperatures—indeed, the hydrolysis of benzyl chloride at 0° has $\Delta V^{\ddagger} = +5$.[117] This latter reaction shows unusual features which may, indeed, be present more generally if sought. The slope of the Arrhenius plot at 1 bar is strongly curved and becomes positive below 4 °C. This effect, which is much reduced at higher pressures, fig. 4.9, is ascribed to quantum mechanical tunnelling which is normally only associated with proton transfer.

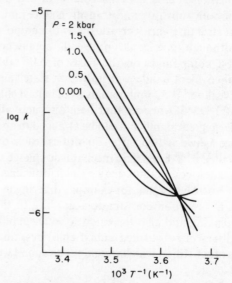

Fig. 4.9. Effect of pressure on the Arrhenius plot for the hydrolysis of benzyl chloride. (After Hyne et al.[117])

The low mass of the proton results in indeterminacy of its position and the ability to 'penetrate' the activation barrier when insufficiently activated in the classical sense.[101] If this interpretation is indeed correct, it implies the electrophilic involvement of solvent in the transition state, e.g.

The raising of the activation volume at low temperatures is no doubt to be ascribed to changes in solvent structuring around the substrate or transition state, and in mixed solvents can result in extrema being observed.

The solvolysis of 2-adamantyl compounds has been proposed as a model for unassisted ('limiting') ionization since the cage structure at the rear side prevents the approach of solvent.[119] It is likely that the slow step is the formation of the intimate ion-pair for which (formolysis of the tosylate), $\Delta V^{\ddagger} = -7 \, cm^3 \, mol^{-1}$.[108] This value is considerably less negative than those for other secondary substrates (e.g. isopropyl, -15; 2-pentyl, -11; 3-methyl-2-butyl, -12[109]) which is consistent with the absence of nucleophilic participation. However, the range of values for 1-substituted adamantyl compounds, which likewise can receive no rearface nucleophilic assistance, is considerable (-9 to -20) so that such generalizations may be unjustified.[121]

The involvement of ion-pairs on the reaction coordinate leading to a solvolytic reaction appears well justified by numerous criteria. It is believed that return to the reagent state can only occur by ion recombination within an intimate or 'tight' ion-pair, while solvolysis can occur with the solvent-separated or 'loose' species. If the leaving group or the cation are bidentate, return may be detected as a rearrangement, for example:

$$PhCH-OPNB \underset{return}{\overset{}{\rightleftharpoons}} \left(\begin{array}{l} Ph \\ \overset{}{\diagup} \\ \text{ } + \quad OPNB^- \\ Me \quad \text{tight i.p.} \end{array} \right) \overset{SOH}{\longrightarrow} \left(\begin{array}{l} Ph \\ \overset{}{\diagup} \\ \text{ } + \cdots SOH \cdots OPN\bar{B} \\ Me \quad \text{loose i.p.} \end{array} \right)$$

rearrangement
$$\Delta V^{\ddagger} - 13 \text{ cm}^3 \text{ mol}^{-1}$$

Solvolysis
$$\Delta V^{\ddagger} = -14 \text{ cm}^3 \text{ mol}^{-1}$$

$$Ph_2CH-S-C\equiv N \rightleftharpoons (Ph_2\overset{+}{CH} \quad S\bar{C}N) \longrightarrow (Ph_2\overset{+}{CH}\cdots SOH \cdots S\bar{C}N)$$

$$Ph_2CH-N=C=S \qquad Ph_2CHOS$$
$$\Delta V^{\ddagger} = -12 \text{ cm}^3 \text{ mol}^{-1} \qquad \Delta V^{\ddagger} = -16 \text{ cm}^3 \text{ mol}^{-1}$$

The difference in activation volume between solvolysis and rearrangement is small and indicates that electrostrictive interactions with the solvent are essentially complete at the transition state for rearrangement[491] but further slight volume reduction occurs *en route* to solvolysis. The loose ion-pairs are known to have a smaller partial molar volume than intimate pairs (section 3.2.3), and free ions are further favoured by pressure which affects the site of alkylation of sodium fluorenone oxime,[493] another bidentate nucleophile.

The effects of steric hindrance around the reaction centre have been examined by placing alkyl groups on the ortho-positions of benzyl and β-phenylethyl chlorides.[122] There appears to be a general trend towards a more negative volume of activation with the larger alkyl groups which perhaps provide steric hindrance to coplanarity of the carbocation and reduce charge delocalization. This is to be expected from the greater electrostriction associated with a localized charge, compared to a delocalized one.

Neighbouring group participation in solvolysis provides a further route to products. It is necessary that there be a donor group so placed that internal S_N2

displacement can occur as the slow step.[123] Three-, five-, or six-membered ring formation is most favourable and the cyclic intermediate usually passes to acyclic products in a fast step. Assisted and the normal unassisted routes may occur together competitively. The participating group, Y, provides an electron-pair

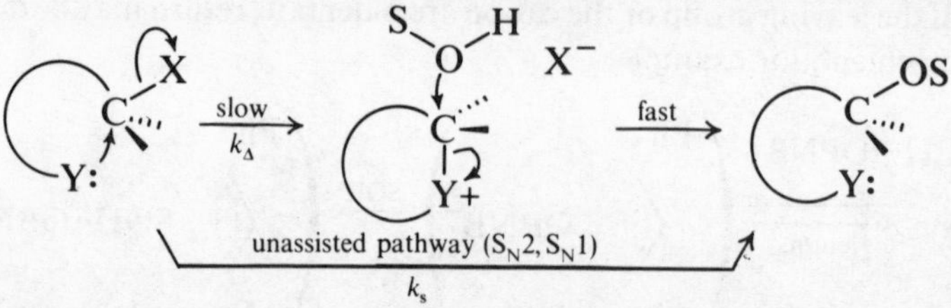

which may be an unshared pair, or from a π-system, frequently an aryl group, or even σ-electrons. Among the characteristics of this route are: enhanced rates, retention of configuration, and the possibility of rearrangement (i.e. transfer of Y to the reaction centre).

The solvolysis of exo-norbornyl compounds is a prime example of σ-participation, and rates are 10^3 times those of the endo isomers which are not so placed for assistance to ionization. The assisted solvolysis is accompanied by a

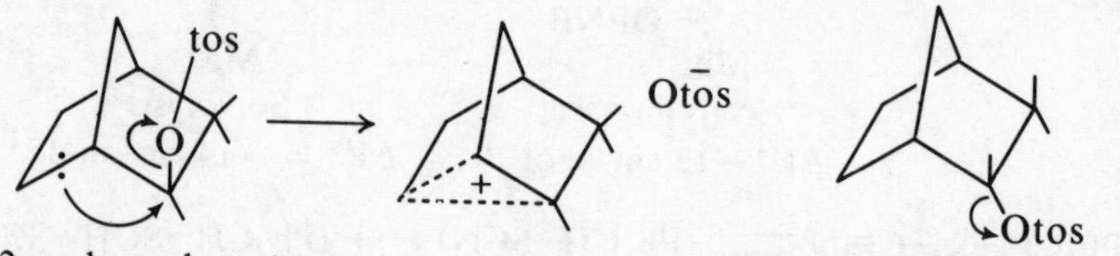

exo-2-norbornyl tosylate
$\Delta V^{\ddagger} = -14 \cdot 3 \text{ cm}^3 \text{ mol}^{-1}$

endo-2-norbornyl tosylate
$-17 \cdot 8 \text{ cm}^3 \text{ mol}^{-1}$

less negative activation volume because of the delocalized charge and the same is found for cyclopropyl participation in [3,1,0]-bicyclohexanes.[124, 125, 129] Where the participation involves attack by an anion, charge dispersal rather than charge separation results and even less negative values accrue since electrostriction is now diminished, as shown by the examples,[128]

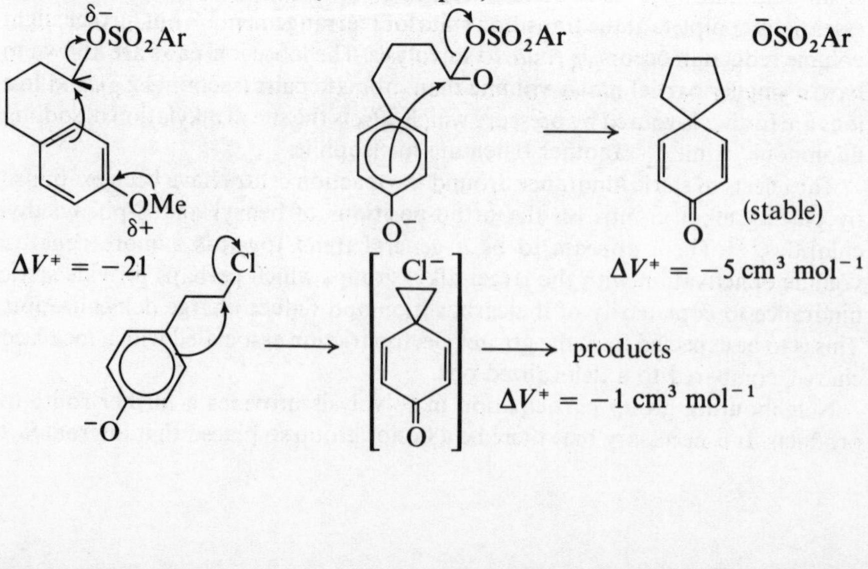

$\Delta V^{\ddagger} = -21$

(stable)

$\Delta V^{\ddagger} = -5 \text{ cm}^3 \text{ mol}^{-1}$

products

$\Delta V^{\ddagger} = -1 \text{ cm}^3 \text{ mol}^{-1}$

However, the neutral 4-methoxy analogue, which is known to receive π-assistance to ionization by the rate-enhancement criterion, has $\Delta V^+ = -21\,\text{cm}^3\,\text{mol}^{-1}$ due to charge separation in the transition state. The effect of neighbouring group participation on ΔV^+ remains somewhat uncertain therefore with a tendency towards a less negative value being likely. Indeed, rates of solvolysis of some 1-aryl-2-propyl sulphonates have been partitioned into assisted, k_Δ, and unassisted, k_s, components by the method of Schleyer[126] (deviations from a linear free energy plot) and components of the activation volume obtained for each process:

$$k_{\text{obs}} = k_s + k_\Delta$$
$$\Delta V^+_{\text{obs}} = \Delta V^+_s + \Delta V^+_\Delta$$

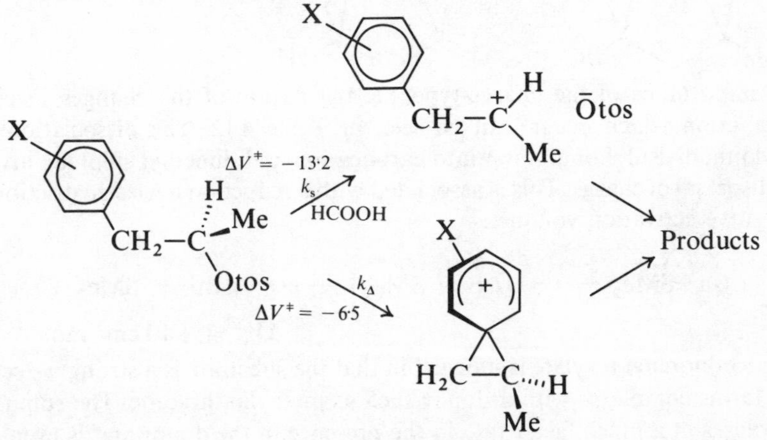

The value of ΔV^+_s, -13.2, fits well with values for other unassisted solvolyses but ΔV^+_Δ, the volume change for the assisted pathway, is only $-6.5\,\text{cm}^2\,\text{mol}^{-1}$.[127] It seems likely that this value reflects an earlier transition state with less charge development than is found with external nucleophilic displacements, since ring-formation itself should make a negative contribution.

Solvolysis of γ-halogenoamines and related compounds leads to fragmentation, F, thus:

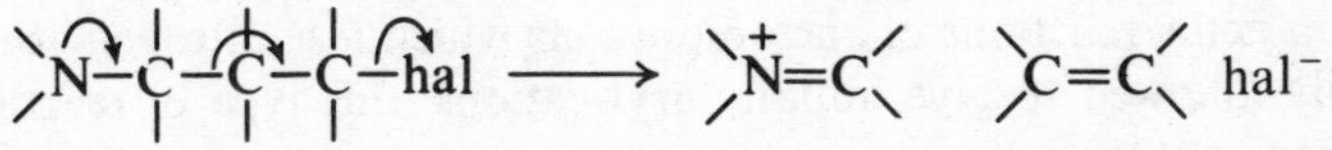

rates are much higher than those of corresponding carbon analogues despite unfavourable inductive influences of the nitrogen. That C–C bond fission is synchronous with C–hal fission, rather than subsequent is also suggested by the activation volumes which compared to the corresponding carbon homomorphs are 5–$10\,\text{cm}^3\,\text{mol}^{-1}$ less negative. This must be ascribed to positive contributions from C–C and possibly N–H bond stretching,[21] thus:

$$\Delta V^+ = -16.1 \text{ cm}^3 \text{ mol}^{-1}$$

$$\Delta V^+ = -20.3 \text{ cm}^3 \text{ mol}^{-1}$$

The importance of the charge-type, i.e. the nature of the changes in charge distribution which occur, can be seen in Table 4.12. The dissociation of *t*-butyldimethylsulphonium ion into carbocation and dimethyl sulphide involves the dispersal of charge. This is associated with a reduction in electrostriction and a positive activation volume:

$$t.\text{Bu}\overset{\frown}{-}\overset{+}{\text{S}}\text{Me}_2 \longrightarrow [\overset{\delta+}{t.\text{Bu}}\cdots\overset{\delta+}{\text{S}}\text{Me}_2] \longrightarrow t.\overset{+}{\text{Bu}} + \text{SMe}_2$$

$$\Delta V^+ = +10 \text{ cm}^3 \text{ mol}^{-1}$$

Trinitrofluorenyl tosylate is unusual in that the substrate is a strong π-acceptor and forms complexes with π-donor such as methylanthracene. The sulphonate solvolyses at a much faster rate in the presence of the donor and is even more strongly accelerated by pressure.[143] It must be admitted that the mechanism is not at present known and seems unlikely to involve ionization on account of the highly disfavoured 9-fluorenyl cation which would result and the further destabilizing nitro groups.

The variability of volumes of activation for solvolytic reactions, together with the solvent dependence and correlations noted with entropies of activation, support the conclusion reached by Laidler that the most important contributory factors are those related to the release or binding of solvent.[490]

Solvolysis of aromatic compounds does not in general occur, the best-known exception being reactions of diazonium ions which lose nitrogen at rates which are easily followed to give initially aryl cations; this type of reaction may be designated $S_N1(Ar)$:

Table 4.12. Activation volumes of some solvolytic reactions

	Solvent	T (°C)	$\Delta V^{\neq}$ (cm³ mol⁻¹)	Reference
1 MeBr	H₂O	30	−14.5	132
		60	−17.0	133
2 IsoPrBr	H₂O	25	−8.8 ($\Delta\bar{V} = -7.0$)	482
		40	−15.2	133
		50	−13.1	
		60	−10.0	
3 Tert.BuCl	D₂O	40	−9.07	135
	H₂O	0	−2.0	134
	aq. Me₂CO	25	−16.5	132
	60%H₂O–EtOH	0	−21.5	134
	MeOH	25	−25.5	142
4 CH₂ = CHCH₂Cl	H₂O	50	−11.4	136
	H₂O	25	−9.8 ($\Delta\bar{V} = -9.6$)	482
5 CH₂ = CHMeCH₂Cl	H₂O	50	−10.1	
6 MeCH = CHCH₂Cl	H₂O	12	−14.8	
7 PhCH₂Cl	100%H₂O	50	−9.9	112
	100%H₂O	0	+5.0	
	90%H₂O–MeOH	50	−12.1	137
	80%H₂O–MeOH		−15.6	
	70		−17.9	
	60		−19.8	
	25		−7.3 ($\Delta\bar{V} = -8.8$)	482
8	X = Me 50%H₂O–Me₂CO	50	−20.0	482
	H		−21.4	138
	Cl		−21.8	
	NO₂		−23.3	

X—⟨benzene ring⟩—CH₂Cl

Table 4.12 (cont.)

	Solvent	T (°C)	$\Delta V^{\ddagger}$ (cm^3 mol^{-1})	Reference
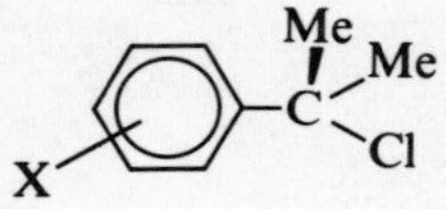 9 $X = H$	11%H$_2$O–Me$_2$CO	25	−12.0	115
3Me			−11.2	
3OMe			−10.9	
3-SMe			−18.2	
3-F			−19.7	
3-I			−20.4	
4-F			−10.9	
3-COOEt			−17.9	
3-Cl			−22.5	
4-Cl			−17.1	
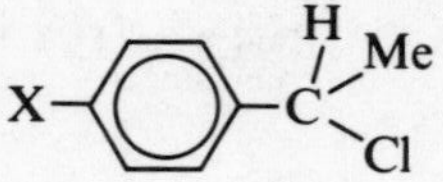 10 $X = H$	HCOOH	25	−7.8	127
Me			−7.1	
OMe			−7.3	
Cl			−9.1	
NO$_2$			−13.1	
11 Cyclopropylmethyl chloride	H$_2$O	16.6	−9.0	139
12 Cyclobutyl chloride	H$_2$O	30.5	−8.2	
13 Cyclopentyl chloride	H$_2$O	50	−14.7	
14 Cyclobutyl tosylate	30%H$_2$O–Me$_2$CO	40	−15.8	140
15 Cyclopentyl tosylate			−17.0	
16 Cyclohexyl tosylate			−17.5	
17 Cycloheptyl tosylate			−17.7	
18 Cyclooctyl tosylate			−17.5	
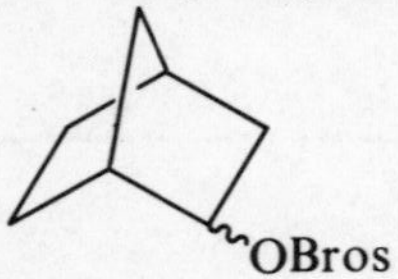 19 exo	6%H$_2$O–Me$_2$CO	25	−14.3	141
endo		40	−17.8	

254

20	anti	30% H$_2$O–Me$_2$CO	40	−14.0	140
	syn			−17.2	
21			25	−19.7	
22			40	−13.9	
23			25	−17.3	
24		2% H$_2$O–PrOH	40	−21.0	128
25			25	−5.4	
26		HCOOH	25	−6.9	108

Table 4.12 (cont.)

	Solvent	T (°C)	$\Delta V^{\ddagger}$ (cm^3 mol^{-1})	Reference
27	MeOH 15%H$_2$O–Me$_2$CO	25 25	−11.8 −9.0	
28	20%H$_2$O–EtOH	21.3	−10.4	131
29		12.4	−7.4	
30		49.6	−14.4	

31

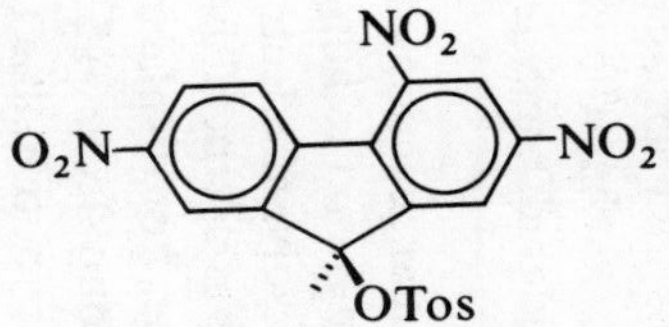

NO$_2$ / NO$_2$ / O$_2$N — fluorene — CH$_3$, OTos

AcOH (+ methyl anthracene) 55 −27.2
−33

143

32 tert.Bu—$\overset{+}{S}$Me$_2$

33 X — CH$_2$Cl (with X substituents)

X = H
Me
isoPr
tert. Bu

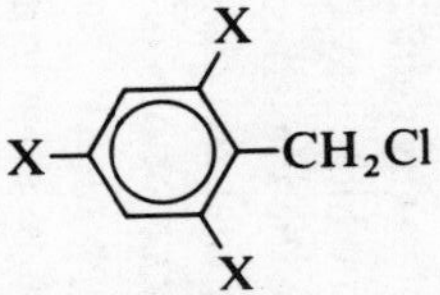

H$_2$O
90% H$_2$O—EtOH
20% H$_2$O—EtOH

60
25

+9.9
+13.1
−12.6
−17.3
−18.4
-15.7

144

The decompositions usually obey first-order kinetics cleanly, although product formation may be less simple than expressed above, polymers and tars being formed by the extremely reactive aryl cations. Bond fission and transfer of charge from nitrogen to carbon should combine to produce a transition state of considerably larger volume than the diazonium ion and this is found to be the case (Table 4.16). Volumes of activation for a series of p-substituted phenyldiazonium ions are all close to $+10\,cm^3\,mol^{-1}$ (temperatures from 30° to 70°).[301] An even larger volume increase is found for the decomposition of the ortho carboxylate (diazotized anthranilate) which is known to lose both nitrogen and CO_2 and form benzyne, a transient species which can be trapped by addition of a diene. This confirms the loss of both molecules synchronously rather than in two discrete steps.[302]

$$\Delta V^{\ddagger} = +15\,cm^3\,mol^{-1}$$

4.2.10 Non-solvolytic Nucleophilic Aliphatic Displacements

Under this heading are placed all reactions in which an external nucleophile, not simultaneously acting as solvent, displaces a suitable leaving group—halide, sulphonate, or 'onium ion for example—from a saturated carbon atom. Under the appropriate conditions, almost all anions may act as nucleophiles as may neutral species such as amines or sulphides so that the scope of this type of reaction is very wide. In all cases to be considered, the S_N2 mechanism may be assumed to be operating.

Volumes of reaction of many such bimolecular reactions have been measured. The principal distinction which may be drawn is that relating the charge type, as set out by Ingold[104] for the classification of solvent effects on rates.

Brower[192] has shown that the charge type is a crucial factor in determining volumes of activation for reactions of this type. For displacements upon neutral substrates by anions, type a, $\Delta V^{\ddagger}$ is commonly from -5 to $-10\,cm^3\,mol^{-1}$ which may be considered to arise from a contribution due to the association of the two molecules ($\Delta V^{\ddagger} \sim -15$) and the reduction in electrostriction due to dispersal of charge ($+5$ to $+10$). The latter effect should be more pronounced the more 'central' the transition state. The amount of electrostriction will, of course, depend upon the solvent. Examples are shown in Table 4.13 below. Other effects

harge type	Example	$\Delta V^{\ddagger}$

) Charge dispersal

$$I^- \curvearrowright CH_3 \overset{\frown}{-} Cl \longrightarrow \begin{bmatrix} \overset{\delta-}{I} \cdots \overset{|}{\underset{/\backslash}{C}} \cdots \overset{\delta-}{Cl} \end{bmatrix} \quad \text{small, negative}$$

) Charge separation

$$Et_3N: \curvearrowright CH_3 \overset{\frown}{-} Cl \longrightarrow \begin{bmatrix} \overset{\delta+}{Et_3N} \cdots \overset{|}{\underset{/\backslash}{C}} \cdots \overset{\delta-}{Cl} \end{bmatrix} \quad \text{large, negative}$$

:) Charge neutralization

$$I^- \curvearrowright CH_3 \overset{\frown}{\overset{+}{-}} NMe_3 \longrightarrow \begin{bmatrix} \overset{\delta-}{I} \cdots \overset{|}{\underset{/\backslash}{C}} \cdots \overset{\delta+}{NMe_3} \end{bmatrix} \quad \text{large, positive ?}$$

may be superimposed; if the anionic reagent has a highly delocalized charge while the displaced one is more localized, the reaction will be accompanied by greater electrostriction and $\Delta V^{\ddagger}$ will be the more negative, e.g.

$$\Delta V^{\ddagger} = -15 \text{ cm}^3 \text{ mol}^{-1}$$

Lithium halides as nucleophiles are associated with more negative volumes of activation than potassium halides:[168]

$$nPrCl + Li^+I^- \longrightarrow nPrI + Li^+Cl^-, \quad \Delta V^{\ddagger} = -22 \text{ cm}^3 \text{ mol}^{-1}$$
$$nPrCl + K^+I^- \longrightarrow nPrI + K^+Cl^- \qquad = -6$$

In acetone, both KI and LiI are strongly ion-paired ($K_a = 145$ and 180, respectively[455]) so differences in free iodide concentration or the need to produce it by dissociation cannot explain this. Further information concerning partial molar volumes of LiI may clarify this difference. Symmetrical substitutions, i.e. those in which nucleophile and leaving anion are the same, albeit differing in isotopic labelling, are found to have less negative activation volumes than unsymmetrical exchange reactions.[169] In the symmetrical process the charge is necessarily equally shared in the transition state but will otherwise be unequally shared and cause greater electrostriction. Indeed, $\Delta V^{\ddagger}$ may even become positive depending on the solvent, showing that de-solvation of the anion prior to attack is a necessary part of the reaction and contributes a positive volume change, the greater the more strongly solvated is the anion.[170,171]

Electrophilic catalysis by metal ions has been examined;[168] silver and mercury(II) will coordinate to the departing halide and facilitate bond breaking. The transition state is presumed to have carbocation character,

$$\Delta V^{\ddagger} = -16 \text{ cm}^3 \text{ mol}^{-1}$$

and activation volumes are again negative. It is possible that the dissociation of ion-pairs either prior to or during electrophilic attack gives rise to increased electrostriction and contributes to this volume. Similar conclusions may be drawn regarding the corresponding cationic charge-dispersal reactions, as far as information is available (entries 14–17).

The other charge type for which much data are published is the Menshutkin reaction, type b, typically between an amine and an alkyl halide for which ΔV^{+} is very solvent dependent and lies between -20 and $-50 \, cm^3 \, mol^{-1}$. This has been a favourite subject for study as it provides a probe into the effects of charge generation which may be extended into a wide range of solvents.[157–160] Values presumably contain contributions from bonding and from increased electrostriction, the latter being the greater and accounting for the solvent dependence. The Drude–Nernst equation predicts

$$V_e \propto \left(\frac{1}{\varepsilon^2} \cdot \frac{\partial \varepsilon}{\partial P} \right)$$

and data are available for testing this postulate (fig. 4.10).[172] A reasonable correlation is found although one solvent, acetone, appears exceptional as it is

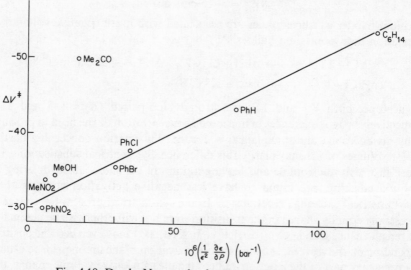

Fig. 4.10. Drude–Nernst plot for $Et_3N + EtI \rightarrow Et_4N^+ \, I^-$

also in the plot with the parameter q_p (fig. 4.2). Such discrepancies are likely if specific interactions are important, since the model is based upon the assumption of the solvent acting as a continuous dielectric medium. Ketones are known to interact strongly with nucleophiles and it is likely that the departing halide ion becomes covalently solvated, bringing about a further volume reduction. This example emphasizes the usefulness of pressure measurements as a probe of solvation effects, a notoriously difficult field in which to obtain definite information.

$$\underset{H}{\overset{Me}{\underset{\displaystyle H}{\overset{\displaystyle |}{\underset{\delta+}{Et_3N}\cdots\underset{}{\overset{}{C}}}}}\cdots\overset{\delta-}{I}\cdots\underset{\delta-O}{\overset{Me\ \ Me}{\overset{\diagdown\diagup}{C}}}$$

Hartmann, Heydtmann, and coworkers have examined the relationship between volumes of activation of several Menschutkin reactions in a series of solvents and the Kirkwood function $q_p (q_p = \partial(\varepsilon - 1)/(2\varepsilon + 1)/\partial P$, section 4.2.2).[166,167] A rough correlation exists when solvents of diverse types are included which is much improved if similar media are selected, fig. 4.2 above. The slope of this plot leads to an estimate of the dipole moment of the transition state after making assumptions of its radius and those of the reagents. A value of 7.8 D is obtained for the transition state from ω-bromoacetophenone and 2-methyl-pyridine, considerably larger than that of either component (1.92 and 3.14 D respectively) supporting the additional charge separation.

The dependence of $\Delta V^{\pm}$ on solvent compressibility has been examined (section 4.1.6)[167,173] and the relationship $\Delta V^{\pm} \propto \beta_T \cdot \bar{V}_{solv}$ found to hold. The intercept of such a plot is held to be a measure of $\Delta V_1^{\pm}$, the volume change of the reaction excluding solvent electrostriction and tends towards -20 to $-25\,cm^3\,mol^{-1}$.

It also appears that activation volumes correlate moderately well with rates, with entropies, and also free-energies of activation (fig. 4.11) for a series of solvents. It seems that as one proceeds towards 'faster' solvents (more polar, as judged by the empirical solvation scales $\mathscr{S}$[164,165] or X,[174] both of which have as their basis the rates of ionogenic reactions) volumes and entropies of activation become less negative and free energies lower, fig. 4.11. This could point to earlier transition states with less charge development; it also raises the possibility that, at least for highly ionogenic processes, the pressure effect is more a result of operating upon solvent structuring than the intrinsic volume change.[175] Furthermore, the value of $\Delta V^{\pm}$ expresses this better than does the entropy change, fig. 4.11.

The effects of steric hindrance in the Menschutkin reaction have been examined[176] using alkyl iodides and 2,6-dialkylpyridines. As the size of either reagent increases, so the activation volume becomes more negative. If orbitals defining the van der Waals radii of groups which come into contact during reaction can be made to 'interpenetrate', i.e. share the same space under the action of pressure, a larger diminution of volume will result, which is the view of Gonikberg[177] (although not without objections[178,179]). However, the finding that the *difference* in $\Delta V^{\pm}$ along a series of progressively more hindered systems is not solvent dependent adds support to the postulate. Another way of looking at this result is to regard the hindering alkyl substituents as non-polar medium-which contributes a more negative volume electrostriction than does the solvent. Other changes, however, occur as bulky groups are placed near the reaction centre. Both rates and enthalpies of reaction diminish, the larger the groups; which by the Hammond postulate implies a successively later transition state, i.e. one becoming more 'product-like'. This is reflected in volumes of activation which become progressively closer to the volumes of reaction:

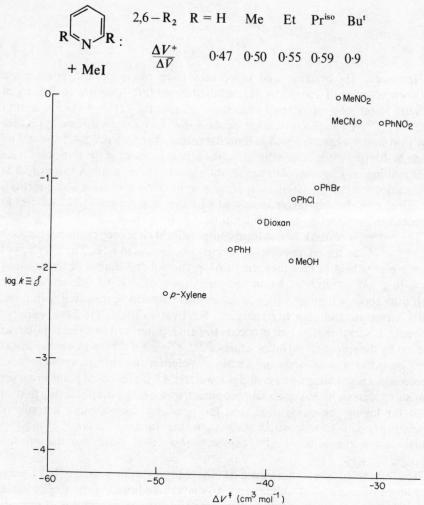

Fig. 4.11. Relation between rate and volume of activation for the reaction

in different solvents

Reverse Menschutkin reactions are also known, consisting of a displacement of quaternary nitrogen by a halide nucleophile:

$$\Delta V^{+}(PhNO_2, 65°) = +32 \text{ cm}^3 \text{ mol}^{-1}$$

As expected, the loss of dipolar character is accompanied by release of electrostricted solvent and a volume increase.[191]. Other estimates of $\Delta V_1^{\ddagger}$ are considerably smaller; on the basis of geometrical calculations, assuming a transition state bondlength for the newly forming bond, a value of $-6.5\,\text{cm}^3\,\text{mol}^{-1}$ was proposed[191] which has been modified to about -12[187] using a revised method of calculating van der Waals volumes[193] while a value of -11 is suggested by Kondo[173] from the solvent effects. If a reasonable mean of -10 to -15 is assumed for $\Delta V_1^{\ddagger}$, then $\Delta V_2^{\ddagger}$ is seen to lie in the range -10 to $-35\,\text{cm}^3\,\text{mol}^{-1}$ and to be very solvent dependent.

Internal displacements to form cyclic products (entries 16, 17) appear to be similar to acyclic analogues. The small value of $\Delta V^{\ddagger}$, -5.8, for cyclization of 4-chlorobutanol does not support a slow step analogous to a Menschutkin reaction (unless very little bonding has been accomplished):[190]

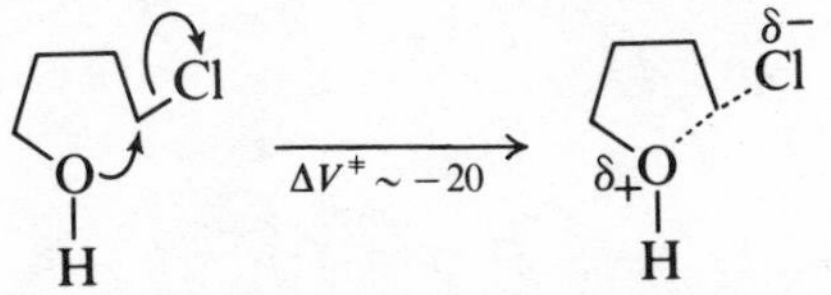

but rather that the removal of the proton precedes cyclization:

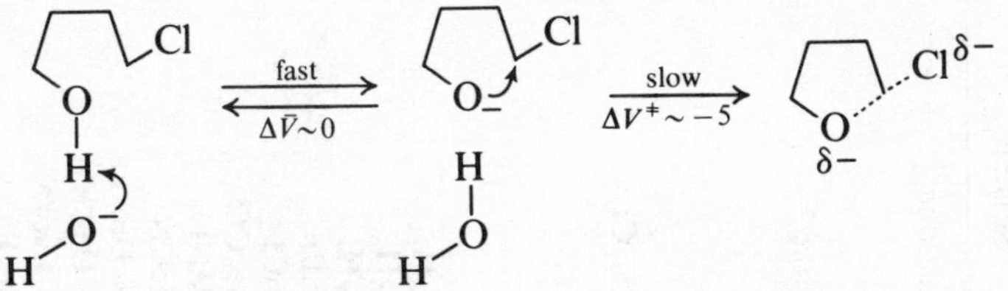

A few examples of reactions of charge type c have been studied under pressure; displacements at a methyl carbon in an 'onium ion are the most clearcut cases, since the presence of a β-hydrogen often leads to extensive elimination, Table 4.13,

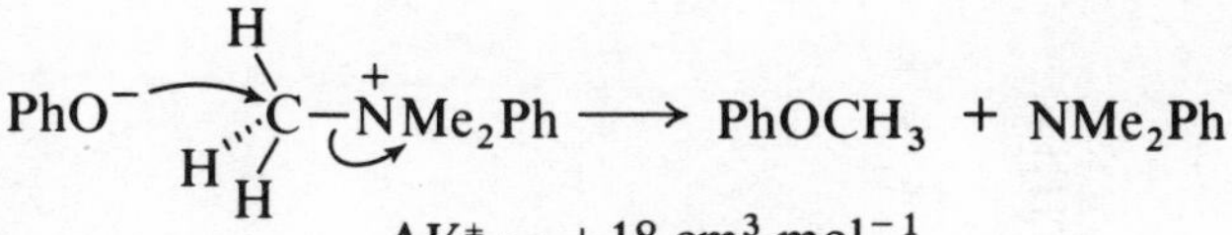

$$\Delta V^{\ddagger} = +18\,\text{cm}^3\,\text{mol}^{-1}$$
$$\text{(EtOH)}$$
$$+7\,\text{cm}^3\,\text{mol}^{-1}$$
$$\text{(H}_2\text{O)}$$

All such reactions proceed more slowly the more polar the solvent since the initial state is stabilized relative to the transition state and, as predicted, pressure strongly retards the reaction as a result of the loss of electrostriction. The overall activation volume is presumably the sum of a negative $\Delta V_1^{\ddagger}$ for association

Table 4.13. Activation volumes for non-solvolytic S_N reactions[a]

Charge type	Substrate	Nucleophile	Solvent	Temperature (°C)	$\Delta V^{\ddagger}$ (cm³ mol⁻¹)	Reference
1a	nPrCl	KI	Acetone	30	−6	168
2	nPrBr			20	−7	
3	PhCH₂Br			25	−9	
4	CH₂=CHCH₂Cl			—	−9	
5	isoPrBr			30	−11	
6	PrBr	LiCl		25	−25	
7	PrI				−22	
8	isoPrI				−27	
9	Et₃O⁺	(tetrahydrofuran)	CH₂Cl₂	0	−5	180
10	(cyclic RO⁺)				−10.5	
11b	EtI	Et₃N	C₆H₁₄	50	−58	181
			C₆H₆	50	−50	
			C₆H₆	25	−43	182
			PhBr	25	−40.5	
			PhCl	25	−37.8	
			Me₂CO	50	−53.8	183
			MeNO₂	25	−33.8	182
			MeCN	25	−32.1	
			1,4-Dioxan	45	−40.5	
			H₂O (1%)-Dioxan	25	−12.1	184
			MeOH	30	−22	185
				40	−23	
				50	−24	
				60	−25	
				70	−27	

No.	Substrate	Reagent	Solvent		ΔS‡	Ref.
12	pyridine	MeI	C_6H_6	30	−32.0	186
		EtI		50	−39.8	183
		$PhCH_2Br$		30	−35.3	175
13 b	2,6-disubstituted pyridine, R = H	MeI	Me_2CO	25	−21.9	176
	Me				−24.4	
	Et				−27.3	
	isoPr				−30.2	
	t. Bu				< −50	
14 b	3-isopropyl-5-methyl-thiazoline-2-thione (intramolecular)	MeI	MeCN	56	−25	187
15 a	nBuBr	OEt^-	EtOH	30	−2.2	188
	isoBuBr			45	−1.7	189
	EtBr			40	−2.7	190
16 a	2-chloro-tetrahydropyran (with OH)		H_2O	32	−5.8	
17 a	catechol dianion + bromobutyl (intramolecular)		MeOH		0	
18 c	$CH_3\text{–}\overset{+}{S}Me_2$	PhO^-	EtOH	76	+12	192
			H_2O	119	+7	
19 c	$CH_3CH_2\text{–}\overset{+}{S}Et_2$	OH^-	H_2O	109	+8	194
		OMe^-	MeOH	51	+15	
20 c	$CH_3\text{–}\overset{+}{N}Me_2Ph$	PhO^-	EtOH	83.6	+18	192
21 c	$CH_3\text{–}\overset{+}{N}Me_2Ph$	PhO^-	H_2O	128	+7	191
c	$CH_3\text{–}\overset{+}{N}PhMeEt$	I^-	$PhNO_2$	65	+32	

Table 4.13 (cont.)

Charge type	Substrate	Nucleophile	Solvent	Temperature (°C)	$\Delta V^{\ddagger}$ (cm³ mol⁻¹)	Reference
22 a	PhCH₂Cl	PhO⁻	PhOH	52	O-alkylation; O ortho-C-alkylation–11 para-C-alkylation–16	308
23	PhCH₂Cl	(2-oxidopyridine, N–O⁻ ring)	EtOH–H₂O		–9	308
24	CH₃–$\overset{+}{N}$(Ph)(CH₂Ph) (CH₂–CH = CH₂) Br⁻	Br⁻	CHCl₃	35	+14	191

ªComplete compilations of activation volumes for these reactions, see references 221 and 223.

(approximately $-10\,cm^2\,mol^{-1}$) and a strongly positive $\Delta V_2^{\ddagger}$, perhaps $+20$ to $+30$ which is solvent dependent and larger in the less polar solvents.

Nucleophiles with more than one reactive centre ('ambident anions') may show interesting pressure effects. The phenolate ion, for example, can act as a nucleophile at oxygen or at the ortho- or para carbons, the products of reaction with an alkyl halide being an ether or alkylphenol.[307] In phenol as solvent, C-alkylation becomes much more favourable at high pressure, the volumes of

RX = PhCH$_2$Cl: 13   52   35%

transition states for ortho- and para substitution by benzyl chloride being respectively 11 and 16 cm^3 smaller than that for O-alkylation[308] in alcohol and similar though smaller effects are found for allylation[309] in water. The explanation seems to lie in solvation of the possible transition states. During C-alkylation, the oxygen is becoming ketonic and is solvated by hydrogen bonding, (9); this is not possible during O-alkylation, (10), nor is the ring solvated and consequently its transition state occupies a larger volume. No such differences in

transition states for: C-alkylation   O-alkylation

solvation occur in alkylation of 2-pyridoxide, (11), which may occur on N or O. Whichever site is the centre of attack, the other is available for solvation so no such differences exist here as with phenolate.

4.2.11 Electrophilic Aliphatic Displacements

The displacement of one electrophile by another at saturated carbon is well known and usually involves metal derivatives as leaving groups and either metal or halogen as electrophile.[291] Both unimolecular and bimolecular pathways are recognized, thus:

S_E1

$$\underset{EtO_2C}{\overset{Ph}{>}}CH-HgBr \xrightarrow[slow]{} \underset{EtO_2C}{\overset{Ph}{>}}CH^- \underset{+HgBr}{} \xrightarrow[fast]{Hg^*Br_2} \underset{EtO_2C}{\overset{Ph}{>}}CH-Hg^*Br$$

$(Hg^* = radioactive)$

S_E2 (inversion)

$$Br-Br \quad \underset{R_2 \; R_3}{\overset{R_1}{C}}-SnR_3 \longrightarrow Br-\underset{R_2}{\overset{R_1}{C}} + \underset{Br}{\overset{SnR_3}{|}}$$

$$\underset{R_2 \; R_3}{\overset{R_1}{C}}-HgBr \quad Br-Br \longrightarrow \underset{R_2 \; R_3}{\overset{R_1}{C}}-Br + HgBr_2$$

Unlike S_N2 displacements for which inversion is the rule, S_E2 reactions may occur by either inversion or retention. The volume of activation for iododestannylation of tetramethylin, $\Delta V^{\ddagger} = -50 \, cm^3 \, mol^{-1}$, is very large and negative while the

$$I_2 + Me_4Sn \longrightarrow \left[\underset{H \; H}{\overset{H}{\underset{|}{\overset{|}{I \cdots I \cdots C \cdots SnMe_3}}}}^{\delta- \quad \delta+} \right] \longrightarrow I-CH_3 + ISnMe_3$$

volume of reaction is almost zero $(-2.8 \, cm^3 \, mol^{-1})$.[292] The transition state must be highly dipolar, similar to that for a Menschutkin reaction and it would be expected that $\Delta V^{\ddagger}$ would be solvent dependent since the value observed must be largely due to solvent electrostriction. This is taken as indicative of the inversion pathway since the retention transition state would have a smaller dipole moment and the ionic centres possibly would be less accessible to solvation. This latter route may be taken by reactions of tetramethyltin with mercury(II) chloride,

$$CH_3-SnMe_3 + HgCl_2 \longrightarrow \left[\underset{H \; H}{\overset{H \quad HgCl_2}{\underset{|}{\overset{\backslash}{C}}}} \underset{SnMe_3}{} \right] \longrightarrow CH_3HgCl + ClSnMe_3$$

since $\Delta V^{\ddagger}$ is considerably less negative, $-36 \, cm^3 \, mol^{-1}$, although in the absence of any more data for these reactions this conclusion is not firmly based, and may simply reflect a smaller degree of charge separation (Table 4.17).

Decarboxylation of carboxylate ions is a form of S_E reaction and is supposed to proceed by a unimolecular loss of CO_2 to form a carbanion which then abstracts a proton from solvent, HS:[304]

$$H_2C \underset{CO_2^-}{\overset{CO \cdot CH_3}{<}} \longrightarrow H_2C^- \overset{O}{\overset{\|}{C}}-CH_3 + CO_2 \xrightarrow{HS} CH_3-\overset{O}{\overset{\|}{C}}-CH_3 + S^-$$

The ease of loss of CO_2 increases with stability of the carbanion and the reaction is particularly facile for β-keto acids, malonic acids, α-nitrocarboxylates which form enolates, and trichloro- and tribromoacetates in which d-orbital stabilization is available. One might expect activation volumes to be similar to those for diazonium ion decompositions ($\Delta V^{\ddagger} \sim 10\, cm^3\, mol^{-1}$) (Table 4.14) in which also

Table 4.14. Volumes of activation for electrophilic substitutions at saturated carbon

Reaction	Solvent	T (°C)	$\Delta V^{\ddagger}$ (cm^3 mol^{-1})	Reference
$Me_4Sn + I_2$	nBu_2O	29	-50	292
$Me_4Sn + HgCl_2$	nBu_2O	30	-32	
$Bu_4Sn + HgCl_2$	MeOH	40	-36	
Decarboxylations				
$CH_2(COOH)_2$	H_2O	115	$+5$	303
	DMSO	100	-10	
	Dioxan	111	-7	
	THF	100	-10	
$Me_2C(COOH)_2$	H_2O	125	$+5$	
CH_3COCH_2COOH	H_2O	40	$+4$	
$PhCOCH_2COO^-$	H_2O	54	$+5$	
$CH_3COCMe_2COO^-$	H_2O	66	$+5$	
	MeOH	45	-3	
$PhC{\equiv}C{\cdot}COO^-$	H_2O	102	10	
	MeOH	102	10	
$NO_2C_6H_4CH_2COO^-$	H_2O	129	11	
	MeOH	114	10	
CF_3COO^-	H_2O	163	8	
	EtOH	163	0	
CCl_3COO^-	H_2O	80	10	
	EtOH	60	4	
CBr_3COO^-	H_2O	66	8	
	EtOH	30	6	

a small molecule is expelled and charge passed to the adjacent atom. Values found, however, are far more variable and are often solvent dependent (Table 4.15).[303] The loss of N_2 from a diazonium ion produces a phenyl cation in which charge is localized on carbon since the vacant sp^2 hybrid orbital is orthogonal to the aromatic ring. All easily decarboxylated acids give rise to anions in which the charge is delocalized so that changes in solvation during activation will depend greatly on the actual structure. Those carboxylates which give rise to rather poorly delocalized ions such as phenylpropiolate, (12), or p-nitrophenylacetate, (13), do indeed show $\Delta V^{\ddagger} \sim 10\, cm^3\, mol^{-1}$ and solvent insensitive. Those giving rise to enolate ions generally have $\Delta V^{\ddagger}$ smaller, and less positive in ethanol than in water indicating increased electrostriction, from which it is concluded that solvation in the transition state is tighter than in the carboxylate. No more than very general conclusions can be drawn from the existing data which is drawn from work at widely different temperatures and from both free acids and sodium

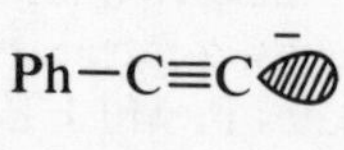

$$Ph-C\equiv C \ominus \quad\quad \text{(13 structure)}$$

(12) 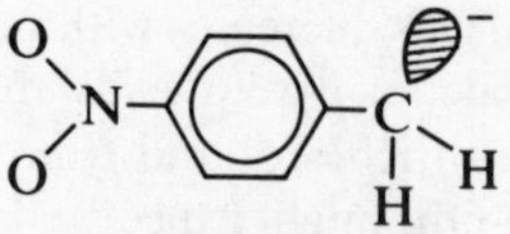   (13)

salts. If it is presumed that the former react as their anions, a prior dissociation step $(\Delta\bar{V} \sim -12\,cm^3\,mol^{-1})$ would be superimposed upon $\Delta V^{\ddagger}_{obs}$ making $\Delta V^{\ddagger}$ for the decomposition, *ca.* $+16\,cm^3\,mol^{-1}$. This is clearly impossibly large and it is likely that a cyclic transitions state is involved, the β-carbonyl group accepting the acidic proton from the undissociated acid,[305]

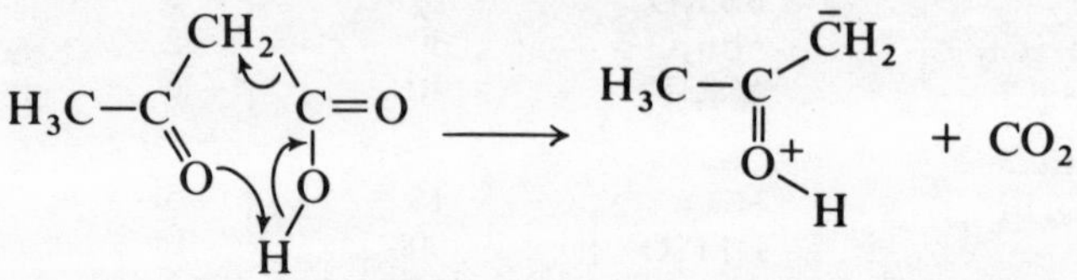

The solvent effect on $\Delta V^{\ddagger}$ supports a dipolar transition state which must be necessary for a dissociative process such as decarboxylation to show a negative activation volume in non-polar solvents. While decarboxylations occurring by one-bond fission have $\Delta V^{\ddagger} \sim +10\,cm^{3\,mol-1}$, the value for the fragmentation of β-bromoangelate, (14), $+18\,cm^3\,mol^{-1}$, clearly falls outside that range and points to a concerted two-bond process[387]

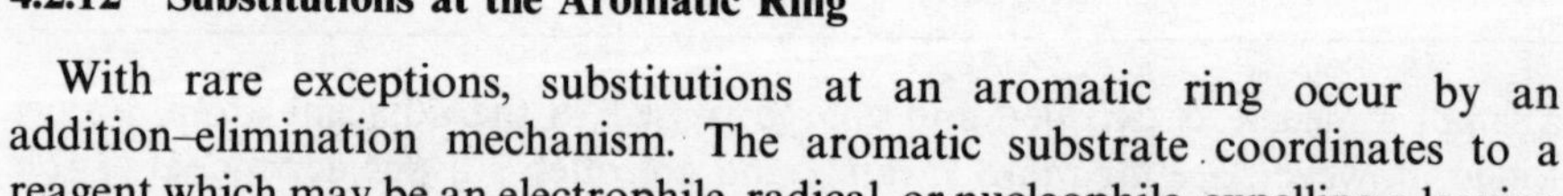

$$\begin{array}{c}H \\ Br\end{array}C=C\begin{array}{c}CO_2^- \\ H\end{array} \longrightarrow \left[\begin{array}{c}H \\ Br^{\delta-}\end{array}C\equiv C\begin{array}{c}CO_2^{\delta-} \\ H\end{array}\right]^{\ddagger} \longrightarrow Br^- + HC\equiv CH + CO_2$$

4.2.12 Substitutions at the Aromatic Ring

With rare exceptions, substitutions at an aromatic ring occur by an addition–elimination mechanism. The aromatic substrate coordinates to a reagent which may be an electrophile, radical, or nucleophile, expelling a leaving group of the same category in a subsequent step. Electrophilic substitution, in general, occurs as follows:

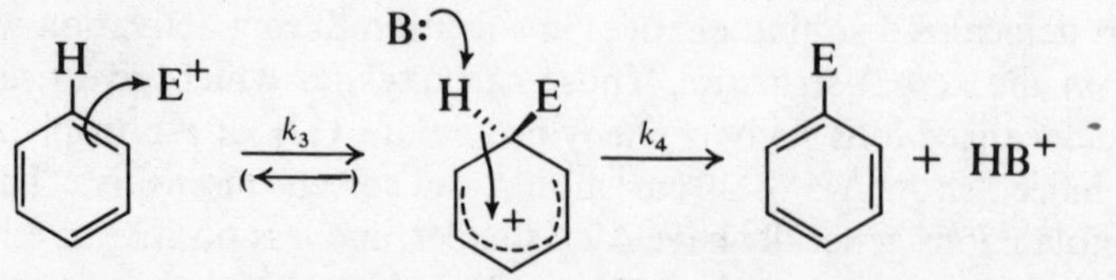

Addition of an electrophile causes the formation of a benzenium ion a species which may be long-lived in super-acid solution[207] and detectable by spectroscopy. The benzenium ion is a very strong acid and loses a proton to a base restoring the aromatic system. As a rule, k_4 is fast and k_3 is rate-determining:

$$\text{rate} = k[\text{ArH}][\text{E}^+]$$

Thus, second-order kinetics are often observed although more complex situations are sometimes found as, for example, in nitration. The nitronium ion, NO_2^+, which is the active electrophile in nitric acid systems, is enormously reactive and is formed by the series of equilibria,

$$2HNO_3 \underset{k_{-1}}{\overset{k_1}{\rightleftharpoons}} H_2NO_3^+ + NO_3^-$$

$$H_2NO_3^+ \underset{k_{-2}}{\overset{k_2}{\rightleftharpoons}} NO_2^+ + H_2O$$

then

$$\text{ArH} + NO_2^+ \xrightarrow{k_3} \text{Ar}\underset{H}{\overset{NO_2^+}{\diagdown}} \xrightarrow{k_4 \text{ fast}} \text{ArNO}_2 + H^+$$

For substrates bearing electron-withdrawing, deactivating substituent groups (halogen, $-COR$, $-CN$, $-NO_2$), k_3 is rate determining and the rate law,

$$\text{rate} = k[\text{ArH}][HNO_3]$$

is obtained or, when $[HNO_3]$ is in excess and constant, rate $= k[\text{ArH}]$ and first-order conditions are observed. Reactive substrates bearing alkyl, alkoxy, amino, etc. substituents scavenge NO_2^+ as soon as it is formed so the rate-determining step becomes NO_2^+ formation, i.e. k_2, since k_1, k_{-1} must both be fast being proton transfers. Now we find

$$\text{rate} = k_{\text{obs}}[HNO_3]$$

or, in excess nitric acid,

$$\text{rate} = k_{\text{obs}}$$

and zero-order conditions are observed. The kinetic analysis of activation volumes is as follows.

First-order conditions Assuming NO_2^+ to be formed under equilibrium conditions,

$$\text{rate} = k_3[\text{ArH}][NO_2^+]$$

$$[NO_2^+] = k_2/k_{-2}[H_2NO_3^+] = k_1/k_{-1} \cdot k_2/k_{-2} \cdot \frac{[HNO_3]^2}{[NO_3^-]} \qquad (4.43)$$

$[HNO_3]$ is constant and, in the early stages of the reaction at least, NO_3^- is also hence

$$\text{rate} = k_3 \cdot \frac{k_1}{k_{-1}} \cdot \frac{k_2}{k_{-2}} [\text{ArH}] = k_3 K_1 K_2 [\text{ArH}] \qquad (4.44)$$

hence

$$\Delta V_{obs}^{\ddagger} = \Delta V_3^{\ddagger} + \Delta \bar{V}_1 + \Delta \bar{V}_2$$

where $\Delta \bar{V}_1$, $\Delta \bar{V}_2$ are volumes of reaction for the two equilibria.

Zero-order conditions With k_2 rate-determining,

$$\text{rate} = k_2[H_2NO_3^+]$$

$$= k_2 \cdot \frac{k_1}{k_{-1}} \frac{[HNO_3]^2}{[NO_3^-]}$$

and

$$\Delta V_{obs}^{\ddagger} = \Delta V_2^{\ddagger} + \Delta \bar{V}_1$$

This should be dependent upon reaction conditions, solvent, temperature, etc. but not upon the nature of the substrate.

The pressure dependence of these nitration reactions should depend upon the kinetic order observed,

$$\underset{\text{(first order)}}{\Delta V_{obs}^{\ddagger}} \quad - \quad \underset{\text{(zero order)}}{\Delta V_{obs}^{\ddagger}} \quad = (\Delta V_3^{\ddagger} - \Delta V_{-2}^{\ddagger})$$

It is found that activation volumes for nitration of toluene by nitric acid in acetic acid (zero order) and of chlorobenzene and benzene under the same conditions (first order) are -10 and $-23.5\,\text{cm}^3\,\text{mol}^{-1}$ respectively[208] so that

$$(\Delta V_3^{\ddagger} - \Delta V_{-2}^{\ddagger}) \sim -12\,\text{cm}^3\,\text{mol}^{-1}.$$

This is the difference in activation volume for coordination of NO_2^+ to the aromatic molecule or to water. A possible explanation for this considerable value is that $\Delta V_{-2}^{\ddagger}$ be rather small since the water molecule involved is removed from its hydrogen-bonded solvation sphere, a process contributing a positive volume change. If $\Delta V_{-2}^{\ddagger}$ were about $-2\,\text{cm}^3\,\text{mol}^{-1}$ then $\Delta V_3^{\ddagger}$ would be about -10, the expected range for bond-formation with charge dispersion. In support of this,[209] $\Delta V_{obs}^{\ddagger}$ for the Friedel–Crafts benzoylation of benzene is $-11.4\,\text{cm}^3\,\text{mol}^{-1}$. This proceeds by coordination of the acyl cation, formed rapidly by the action of the Lewis acid on benzoyl chloride, to benzene and thus k_{obs} may be identified with k_3.

Aromatic reactivity is usually obtained from competitive reactions and expressed as a partial rate factor, f:

$$f = \frac{\text{Rate or amount of product for attack at one substrate position}}{\text{Rate or amount of product from attack at one position in benzene}}$$

Thus, toluene, nitration in 10% aq. acetic acid by HNO_3, 45°;

$$f\,(\text{ortho}) = 42, \quad f\,(\text{meta}) = 2.5, \quad f\,(\text{para}) = 58$$

expresses the observation that under specified conditions, each single site in toluene reacts at a rate compared to a single site in benzene by factors 42, 2.5, and 58, respectively.[210] The effects of pressure on f have been examined for several reactions which lead to an estimate of the difference in activation volume for attack at each position in a monosubstituted benzene:

$$\delta \Delta V^{+} = -RT\left(\frac{\partial \ln f}{\partial P}\right)_{T} \tag{4.45}$$

The effects are expected to be small since it is predicted that for a reaction in which no overall change in charge occurs, the reaction constant ρ should be independent of pressure.[211] That the regiospecificity of aromatic substitution is somewhat pressure dependent is seen from Table 4.15.[212-215]

Table 4.15. Differences in volumes of activation for $\delta \Delta V^{+}$ for nitration in acetic acid of monosubstituted benzenes[212]

Substituent	$\delta \Delta V^{+}$ o	$(cm^3\,mol^{-1})$ m	p
H	0	0	0
Me	2.1	1.2	3.2
t. Bu	0.6	1.6	3.5
F	0.1	-3.1	0.4
Cl	-1.2	-5.0	-0.5
Br	-2.0	-5.2	-1.1
Ph	6.2		6.6

The clear distinction between the positive values of $\delta \Delta V^{+}$ for activated positions (i.e. those for which $f > 1$) and the negative values for deactivated positions ($f < 1$) is striking. It points to a dependence of the reaction constant, ρ, on pressure since

$$\log(s.f) = \rho \sigma^{+}$$

(σ^{+} is the enhanced substituent constant appropriate to aromatic substitution[116] and s is a statistical factor). It has been argued that values of ρ should be independent of pressure for reactions between an ion and a neutral molecule.[456] Possibly aromatic substitution does not fall into this category since the nitronium ion needs to be generated from neutral precursors. Other evidence clearly indicates[45] that partial rate factors and reaction constants do change with pressure. Since values of $\delta\Delta V^{\ddagger}$ presumably originate in changes in solvation of the transition state, concentration of charges at ortho- and para-positions may result in the differences in values according to position.

Nucleophilic displacement from an aromatic ring is not, in general, a facile reaction and demands the presence of strongly electron-withdrawing groups such as $-NO_2$ or very good leaving groups such as $-N_2^+$.

(15)

$$\Delta \bar{V} - \text{ve} \qquad \Delta V^{\ddagger} + \text{ve ?}$$

An addition–elimination mechanism occurs,[219] the first step being rate-determining. Addition of methoxide to trinitroanisole giving the adduct, (15), (k_1) has been followed by stopped-flow[131] and has $\Delta V^{\ddagger} = -7.2\,\text{cm}^3\,\text{mol}^{-1}$ consistent with a bond-forming process. The overall volume of activation for the bromide displacement is very much more negative than this, -15 to $-30\,\text{cm}^3\,\text{mol}^{-1}$ (in methanol) which would be due to the large volume change for the addition (assuming an equilibrium is set up) since the expulsion of Br^- would by analogy be expected to show a small positive $\Delta V^{\ddagger}$. This result would be explained if the transition state of the slow step were close to the intermediate rather than the product. Displacement by a neutral nucleophile, such as a secondary amine, results in net charge separation so that observed activation volumes are even more strongly negative, and highly solvent- and temperature-dependent (Table 4.16).

The $S_N 1$ mechanism for aromatic substitution occurs in one well-documented case, the decompositions of aryldiazonium ions. The resulting aryl cation rapidly coordinates a nucleophile to form products.

$$Ar-N_2^+ \xrightarrow{\text{slow}} Ar^+ + N_2 \xrightarrow[\text{fast}]{BF_4^-} Ar-F$$

Bond fission with no change in charge will be characterized by a positive volume of activation which is borne out experimentally, Table 4.16. Rather similar values, $+9$ to $+11\,cm^3\,mol^{-1}$, are found for a series of substituted benzenes and there is

Table 4.16. Activation volumes of aromatic substitution reactions

(i) Electrophilic		Solvent	T (°C)	$\Delta V^{\ddagger}$ ($cm^3\,mol^{-1}$)	Reference
X—⟨benzene⟩—H $+HNO_3$	X = Me H Cl Me	AcOH MeNO₂	0 20	−10 −22.0 −23.5 −10	208
Cl⟨benzene⟩Cl				−20.5	
PhH + PhCOCl/AlCl₃		PhCOCl	29.5	−11.4	209

(ii) Nucleophilic		Solvent	T (°C)	$\Delta V^{\ddagger}$ ($cm^3\,mol^{-1}$)	Reference
Cl, NO₂, NO₂ (benzene)	nBuNH₂ t-BuNH₂	EtOH	35	−25 −35	217
Br (naphthalene)	Piperidine OMe⁻	Piperidine 60%H₂O, Piperidine MeOH	188 188 174	−57 −39 −23	194 218
Br (naphthalene)	Piperidine MeO⁻	Piperidine 60% aq. piperidine MeOH	189 186 174	−68 −44 −30	194 218
(quinoline) N Br	Piperidine MeO⁻	Piperidine 60%H₂O Piperidine MeOH	27 0 22 43 0 50	−25 −23 −19 −20 −10 −15	194 218
Br (quinoline) N	Piperidine	Piperidine 60%H₂O Piperidine	184 184	−69 −45	194 218
(quinoline) N Cl	Piperidine MeO⁻	Piperidine 60%H₂O piperidine Cyclohexane MeOH	51 51 105 60	−33 −23 −41 −14	

Table 4.16. (*cont.*)

(ii) Nucleophilic		Solvent	T (°C)	$\Delta V^{\ddagger}$ (cm^3 mol^{-1})	Reference
X—⟨benzene⟩—$\overset{+}{N}\equiv N$ BF$_4^-$	X = H	H$_2$O	29	+10	194, 302
	p–Me		49	+9.0	
	p–Cl		61	+10.8	
	m–Cl		45	+11.4	
	p-SO$_3^-$		50	+10.7	
	p–NO$_2$		60	+9.4	
	m–NO$_2$		69	+9.2	
⟨benzene⟩ N$_2^+$ BF$_4^-$ COOH		DMSO	25	+15	

no apparent relationship between $\Delta V^{\ddagger}$ and the substituent.[194] The decomposition of diazotized anthranilic acid leads to benzyne, (15), which is shown to be formed transiently by interception with anthracene to form triptycene.[219] The volume of activation is +15 which suggests that both bonds are breaking

synchronously. Aromatic radical substitution is known, e.g. phenylation,[220] but appears not to have been studied under pressure.

4.2.13 Acid-catalysed Reactions

There are many examples, mainly of displacement reactions, in which the presence of acid is essential or at least increases the rate of reaction. Invariably, this implies that a leaving group on the substrate becomes protonated thereby facilitating its departure in a nucleophilic displacement. The great majority of studies of acid-catalysed reactions have been carried out in aqueous solvents in the presence of strong acid (H_3O^+) or else a weak acid which furnishes both H_3O^+ and the undissociated HA as proton donors. Those reactions which respond only to H_3O^+ are termed 'specific acid catalysed' while those which respond to all acids, in proportion to their concentration and acid strength, are termed 'general acid catalysed'.[240] More germane to the present discussion is the pathway for the displacement which follows the protonation step which may be taken as an equilibrium, since proton transfers to heteratom bases in water are very fast, and will usually lie far to the left. However, the protonated substrate may show either of two types of behaviour. The leaving group may depart in a unimolecular step or it may be displaced by attack of solvent, normally water. These two mechanisms are known as (A1) and (A2) and are analogous to the classical S_N1 and S_N2 processes discussed previously (section 4.2.8):

$$R\text{--}L + H_3O^+ \underset{}{\overset{K_1}{\rightleftharpoons}} R\text{--}LH^+ + H_2O$$

then, either

$$R\text{--}LH^+ \xrightarrow[k_2]{\text{slow}} R^+ + LH \longrightarrow \text{products} \tag{A1}$$

or

$$R\text{--}LH^+ + H_2O \xrightarrow[k_2]{\text{slow}} R\text{--}OH + LH + (H^+) \tag{A2}$$

The involvement of a solvent molecule in the transition state of the slow step cannot be determined by the kinetics and indirect evidence must be used to determine which pathway is followed by a given reaction. In a general way, both the entropy and the volume of activation are indicative of mechanism and give somewhat similar information; the latter is more easily related to a physical model. The volume of activation in either case is a composite quantity,

$$k_{\text{obs}} = K_1 k_2 [R\text{--}L]$$

and

$$\Delta V_{\text{obs}}^{\ddagger} = \Delta \bar{V}_1 + \Delta V_2^{\ddagger}$$

where $\Delta \bar{V}_1$ is the volume change for the pre-equilibrium proton transfer. For both A1 and A2 mechanisms, $\Delta \bar{V}_1$ will be the same but in the A1 slow step, a bond is broken and charge is usually dispersed making $\Delta V_2^{\ddagger}$ positive. The A2 slow step involves coordination of a water molecule so that $\Delta V_2^{\ddagger}$ should be negative. For similar reasons, $\Delta S^{\ddagger}$ tends to be positive for A1 reactions, associated with a relaxation of solvation and an increase in freedom of the system whereas the involvement of solvent in the A2 transition state results in a loss of translational freedom and $\Delta S^{\ddagger}$ is negative. Typical ranges of values are:

	$\Delta V^{\ddagger}$ (cm^3 mol^{-1})	$\Delta S^{\ddagger}$ (J K^{-1} mol^{-1})
A1	-2 to $+4$	$+20$ to $+80$
A2	-15 to -8	-35 to 0

Some individual cases will now be considered. It is assumed that $\Delta \bar{V}_1$ is quite small, perhaps ± 0–$2\,cm^3\,mol^{-1}$ since it is simply a proton partition equilibrium between two oxygen bases as a rule. These equilibria cannot be studied directly and more definite information on this point would be very desirable.

Ester and amide hydrolysis There is no doubt that this occurs by addition–elimination ($A_{Ac}2$ mechanism), analogous to that for basic hydrolysis (section 4.2.14) but with two additional protons in the transition state. Thus the reaction is of the A2 type. The precise location of protons on the intermediates,

$$\Delta V^{\ddagger} = -7\cdot4\ cm^3\,mol^{-1}$$
$$\Delta S^{\ddagger} \sim -100\ JK^{-1}\,mol^{-1}$$

(16) and (17), is not certain but it is known that the carbonyl oxygen exchanges with water during reaction and the alcohol formed contains the intact ester alkoxy group, facts which are readily explained by this scheme.[241] Simple ester hydrolyses catalysed by strong acid in water have $\Delta V^{\ddagger} = -5$ to $-7\,cm^3\,mol^{-1}$ although much more negative values are observed in aqueous-organic media, Table 4.17.[242] This suggests that there is a considerable contribution from electrostriction and the transition state of the slow step may be written as (18) leading to (17).

(18)

The hydrolysis of butyrolactone appears similar to the acyclic esters but propiolactone has $\Delta V^{\ddagger} = +2.5\,cm^3\,mol^{-1}$. This indicates a change of mechanism and argues plausibly in favour of an A1 pathway which becomes favourable when, in addition to weakening of the carbonyl-oxygen bond by protonation, ring-strain can be relieved by unimolecular ring-opening.[243] The acid-catalysed

$$\Delta V^{\ddagger} = +2{\cdot}5\ cm^3\ mol^{-1}$$

hydrolysis of amides normally proceeds by an analogous $A_{Ac}2$ (A2) process, amine rather than alcohol the leaving group. Volumes of activation are around $-10\,cm^3\,mol^{-1}$ depending upon solvent composition and temperature.

Acid-catalysed esterification is essentially the reverse of hydrolysis, alcohol displacing water from protonated acid and so the volume of activation will be similar to that for hydrolysis. In the absence of strong acid, esterification can be auto-catalysed. A similar $A_{Ac}2$ mechanism operates, but now the protonating

acid is neutral (autoprotolysis) and the pre-equilibrium is highly favoured by pressure since charge is created. Activation volumes for self-catalysed esterification are therefore highly negative, -20 to $-30\,cm^3\,mol^{-1}$.[277-279] These reactions resemble uncatalysed ester hydrolyses, Table 4.20 ($\Delta V^{\ddagger} = -22.3\,cm^{-1}\,mol^{-1}$ for hydrolysis of ethyl pivalate by H_2O[279]). Since an equilibrium is set up, a large excess of one component, coupled with high pressure, would be needed for rapid product formation for preparative purposes.

Acid-catalysed hydrolysis of anhydrides has been examined[244] in dioxan–water mixtures. At very low water concentration the positive values of $\Delta V^{\ddagger}$ are interpreted as indicating an A1 mechanism but this goes over to A2 as the water content is increased and $\Delta V^{\ddagger}$ becomes very strongly negative.

Table 4.17 Activation volumes for acid–catalysed hydrolyses

Carbonyl compounds	Solvent	Temperature (°C)	$\Delta V^{\ddagger}$ (cm³ mol⁻¹)	Reference
CH₃COOEt	H₂O	40	−7.4	242
	H₂O 80%, dioxan		−9	
	60%		−11	
	40%		−14	
	20%		−15.9	
CH₃COOtBu	H₂O	35	−9.3	258, 261
	H₂O	60	0.0	261
CH₃COOMe	H₂O	40	0.4	262
(β-lactone)	H₂O	0	−7.9	
(γ-lactone)	H₂O	35	−9.1	243, 263
		25	+2.5	
(diester, X)	H₂O	35	−8.4	264, 261
X = H	H₂O, 0.017M, dioxan	60	+6.7	244
	0.107		+5.0	
	1.07		−5.6	
	10.0		−17.8	
	27.6		−25.4	
X = p–OMe	1.0		+4.8	
X = p–Cl	1.0		−9.5	
X = p–NO₂	1.0		−9.0	

Reaction	Conditions		Temp	Value	Ref
CH₃CONH₂	H₂O		55	−9.4	261
PhCONH₂	1M HClO₄		80	−12.1	
	4M HClO₄			−16.0	
CH₅CONH-tBu	H₂O, 0.2MHCl		80	−1.9	265
	1.0			−9.2	
Self-catalysed esterifications					
CH₃COOH+EtOH	None		50	−21.5	278
			60	−22.5	
+nPrOH			75	−23.4	279
+nBuOH			80	−32.6	278
+isoPrOH			75	−20.0	
+Me₂CH·CH₂OH			75	−18.9	
+2-butanol			75	−20.5	
			75	−14.2	
Me₃C–COOH+EtOH			75	−22.0	279
			80	−26.2	
Additions					
CH₂ = CHCOOH	H₂O		80	−14.0	280
			90	−15.8	
CH₃CH = CHCOOH			85	−18	282
CH₂ = CH₂	H₂O		180	−15.5	283
CH₃CH = CH₂			100	−9.6	284
			180	−21.9	286
(CH₃)₂C = CH₂	H₃O		35	−11.5	287
Me₂C = CHCOCH₃	MeOH/H⁺		30	−14.5	
			30	−23	
Eliminations					
CH₂–CH₂COOH (−H₂O)	H₂O		80	−9.6	280
—OH			90	−11.1	

Table 4. 17 (cont.)

Solvent	Temperature (°C)	$\Delta V^{\ddagger}$ (cm^3 mol^{-1})	Reference
CH$_3$CH–CH$_2$COOH (–H$_2$O) $\;$ OH	H$_2$O	85	−15	
CH$_3$CH–CH$_2$CHO (–H$_2$O) $\;$ OH	H$_2$O	30	−5.8	285
Me$_2$C–CH$_2$COCH$_3$ (–MeOH) $\;$ OMe	MeOH/H$^+$	30	−13	287
Ethers and acetals				
Et$_2$O	H$_2$O	200	−10.0	246
		161	−8.5	266
H$_2$C—CH$_2$ O	H$_2$O	0	−5.9	245
		15	−7.4	
		25	−7.9	
		40	−8.9	
MeCH—CH$_2$ O	H$_2$O	0	−8.4	
Me$_2$C—CH$_2$ O	H$_2$O	0	−9.2	
ClCH$_2$CH—CH$_2$ O	H$_2$O	25	−8.5	267

Compound	Reagent	Temp (°C)	Value	Ref
MeCH—CH₂ [a] (O)	MeOH/H⁺	25	−9.4	268
HOCH₂CH—CH₂ [a] (O)			−14.7	
ClCH₂CH—CH₂ [a] (O)			−9.1	
BrCH₂CH—CH₂ [a] (O)			−10.7	
H₂C—CH₂ (O)	HNO₃	25	−17.3	270
ClCH₂CH—CH₂ (O)			−15.0 / −14.2	
CH₂—CH₂ / CH₂—O	H₂O	25	−5.5	267
CH₂—CHMe / CH₂—O		25 / 40	−11.5 / −9.9	271
CHMe—CH₂ / CH₂—O		25 / 40	−11.3 / −9.7	

Table 4. 17 (cont.)

	Solvent	Temperature (°C)	$\Delta V^{\ddagger}$ (cm³ mol⁻¹)	Reference
$\underset{NH_2^+}{H_2C\!-\!CH_2}$ (ring)	H_2O	21	+1.8	249
$\underset{NH_2^+}{CHEt\!-\!CH_2}$ (ring)			-2.5	
$\underset{NH_2^+}{CEt_2\!-\!CH_2}$ (ring)	H_2O		-4.4	
$CH_2(OMe)_2$	H_2O	25	-0.5	271
$CH_2(OEt)_2$			0.0	
$MeCH(OEt)_2$			+1.5	
$CH(OMe)_3$	H_2O	0	+1.8	
		15	+2.4	
1,3,5-trioxan		100	-1.8	272, 273
Paraldehyde		35	+3.0	273
Pyranosides				
Me–α–D–galacto	H_2O	59.5	+5.4	254
Me–β–D–galacto		50.5	+4.9	
Et–β–D–galacto		50.5	+5.0	
Ph–β–D–galacto		50.5	0	
Me–β–D–gluco		59.5	+6.2	
Me–α–D–gluco		100	+5.1	275
Ph–β–D–gluco		50.5	+2.9	254
Me–α–D–manno		50.5	+3.6	
Me–β–L–arabino		50.5	+6.1	

Furanosides

Me-α-D-galacto	35	−3.6	254
Me-β-D-galacto	38.5	−3.9	
Et-β-D-galacto	35	−4.4	
Ph-β-D-galacto	5	+1.3	
Me-β-L-arabino	35	−2.0	
Me-α-D-xylo	21.5	−4.6	
p-cresyl-β-D-gluco	45	+3.4	
p-anisyl-β-D-gluco	45	+3.5	
sucrose	25	+6.0	274
α-D-glucose-1-phosphate	25	+4.3	276
glucopyranose,			480
mutarotation α → β		−11.7	
β → α		−10.8	

[a] Mixed products, both normal and abnormal ring opening.[269]

$$Ph-C\overset{O}{\underset{O-COPh}{\diagdown}} \quad \underset{\longleftarrow}{\overset{H^+}{\rightleftharpoons}} \quad Ph-C\overset{\overset{O}{\|}}{\underset{\underset{H}{\overset{+}{O}-COPh}}{\diagdown}} \quad \overset{A1}{\longrightarrow} \quad Ph\overset{+}{C}O + HOCOPh$$

$$\Delta V^{\ddagger} = +7 \text{ cm}^3 \text{ mol}^{-1}$$

$$A2 \Big\downarrow H_2O$$

$$Ph-C\overset{O}{\underset{OH}{\diagdown}} + HOCOPh$$

$$\Delta V^{\ddagger} = -25 \text{ cm}^3 \text{ mol}^{-1}$$

Hydrolyses of ethers, acetals, orthoesters, and epoxides Although this heading includes compounds of very different character, their hydrolyses all have in common the protonation of oxygen attached to a saturated carbon, and the departure of an alcohol moiety as leaving group by either A1 or A2 routes:

$$\overset{|}{\underset{|}{C}}-OR \quad \underset{\longleftarrow}{\overset{H^+}{\rightleftharpoons}} \quad \overset{|}{\underset{|}{C}}-\overset{+}{O}\overset{R}{\underset{H}{\diagdown}} \quad \overset{H_2O}{\longrightarrow} \quad \overset{|}{\underset{|}{C}}-OH \ + \ ROH$$

Simple ethers are rather inert to hydrolysis but the conventional cleaving agent, aqueous HI, suggests that the combination of strong acid and strong nucleophile is together more potent than strong acid alone and that the A2 mechanism would be likely. This is supported by the volume of activation; for diethyl ether in perchloric acid, $\Delta V^{\ddagger} = -8.5 \text{ cm}^3 \text{ mol}^{-1}$ at 160°,[245] -10 at 200°.[246] When HI is used rather than an acid with a non-nucleophilic anion, $\Delta V^{\ddagger}$ becomes positive, $+1.0 \text{ cm}^3 \text{ mol}^{-1}$, since the nucleophile is now charged and charge neutralization results. Epoxides are far more reactive than acyclic or larger-ring cyclic ethers since bond angle strain is relieved on ring opening.

$$H\overset{}{\underset{H}{\diagdown}}\overset{\delta+}{O}\cdots\overset{\overset{Me}{|}}{\underset{\underset{H}{|}}{C}}\cdots\overset{\delta+}{O}\overset{H}{\underset{Et}{\diagagup}} \qquad \overset{\delta-}{I}\cdots\overset{\overset{Me}{|}}{\underset{\underset{H}{|}}{C}}\cdots\overset{\delta+}{O}\overset{H}{\underset{Et}{\diagagup}}$$

A2 transition state S_N2 transition state

Ethylene oxide hydrolyses to ethanediol under both acidic and alkaline conditions. The acid-catalysed reaction is specific to H_3O^+ and the solvent isotope effect, $k_{D_2O}/k_{H_2O} \sim 2$, since D_3O^+ is a stronger acid than H_3O^+.[247] The entropy of activation $\Delta S^{\ddagger} = -30 \text{ J K}^{-1} \text{ mol}^{-1}$[248] suggesting an A2 mechanism which is confirmed by the acceleration of rate under pressure, $\Delta V^{\ddagger} = -6$ to $-9 \text{ cm}^3 \text{ mol}^{-1}$ (0–40°). Similar values for both $\Delta S^{\ddagger}$ and $\Delta V^{\ddagger}$ indicate an analogous mechanism for 1,2-epoxypropane and 2-methyl-1,2-epoxypropane, (19). This is somewhat surprising since the latter hydrolyses 600 times faster than epoxyethane and would be expected to be a strong candidate for an A1 mechanism as a tertiary carbocation could be formed. Oxetan, (20), is similarly inferred to use the A2 mechanism ($\Delta V^+ = -5.5$) when other criteria are ambiguous. The nitrogen analogues of epoxides are aziridines, secondary amines which form stable salts. The aziridinium ions undergo ring-opening in water, the rate being essentially that of the pure protonated species and activation volumes can be obtained which are independent of the protonation step.[249]

Dialkoxymethanes (acetals and ketals) and trialkoxymethanes (ortho-esters) hydrolyse extremely rapidly by a hydronium ion-catalysed pathway. Mechanisms have been extensively studied[250]

Both alkoxy groups depart intact as shown by the retention of chirality[251] and isotopic tracers.[252] The hydrolyses of substituted benzaldehyde acetals is moderately accelerated by electron-donating ($+M$) substituents,[253] $\rho = -2$ to -4, entropies of activation are positive and the secondary deuterium isotope effect (for $RC(OR')_2$) is $k_H/k_D = 1.15$, supporting the development of carbocation

$$\overset{|}{\underset{H(D)}{}}$$

character. Thus, the A1 mechanism is indicated and is favoured by the stabilization afforded the cation by the second alkoxy group (oxocarbonium ion).

$$\Delta V^+ = +13 \text{ cm}^3 \text{ mol}^{-1}$$

Subsequent steps in the conversion to aldehyde via the hemiacetal are all relatively fast.

$$CH_3\overset{+}{C}HOMe$$

$$\downarrow H_2O$$

$$
CH_3\underset{OMe}{\overset{OH}{CH}} \quad \underset{\longleftarrow}{\overset{H^+}{\rightleftharpoons}} \quad CH_3\underset{\underset{H}{\overset{+}{O}Me}}{\overset{OH}{CH}} \quad \longrightarrow \quad CH_3\overset{+}{C}HOH + MeOH
$$

$$\downarrow -H^+$$

$$CH_3CHO$$

In accordance with this scheme, volumes of activation tend to be positive although values in the main fall within the range -2 to $+6\,cm^3\,mol^{-1}$, Table 4.20 becoming more negative at higher temperature.

Glycosides are acetals and their hydrolysis has been the subject of especial interest on account of their biochemical relevance.[253] Previous work has shown that entropies of activation for alkyl pyranosides, (21), are in the range 40–$60\,J\,K^{-1}\,mol^{-1}$ while those for furanosides, (22), are from -10 to $-40\,J\,K^{-1}\,mol^{-1}$. A change of mechanism is indicated but there are further possibilities than simply A1 and A2 since glycosides are unsymmetrical acetals and may protonate on either the anomeric oxygen or the ring oxygen leading to ring opening or ring retention respectively in either case.

Volumes of activation support conclusions based on entropies and fall into two groups, $+3$ to $+6\,cm^3\,mol^{-1}$ for alkyl pyranosides and -2 to -4 for alkyl furanosides.[254, 256] The alkyl pyranosides are therefore assigned an A1 mechanism and, since rate-determining loss of methanol appears to be confirmed by an oxygen kinetic isotope effect,[25] ring retention evidently occurs. Alkyl furanosides are thought to react by an A2 route though at present it is not possible to distinguish between ring-opening and ring-retention pathways. The mechanistic change is probably due to ring size but information concerning non-carbohydrate analogues is lacking. S_N2 reactions in cyclopentane systems can be 100–200 times faster than those in corresponding cyclohexanes.[255] Furanosides are very much more reactive than the pyranosides and it seems that the bimolcular mechanism is here especially favourable. Aryl glycosides, both pyranosides and furanosides, appear to fit the A1 scheme (fig. 4.12) representing a mechanistic change in the furanoside series. If this is due to the presence of the aryl group it is likely to imply ring-opening and further delocalization of charge by the aromatic ring.

289

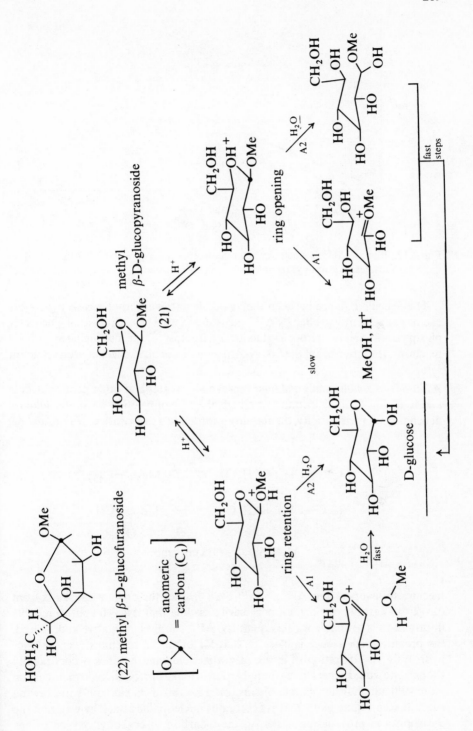

methyl β-D-glucopyranoside

(21)

(22) methyl β-D-glucofuranoside

$$\left[\begin{matrix} O \\ \diagdown \\ O \end{matrix} \diagup \bullet \ = \ anomeric \ = \ carbon \ (C_1) \right]$$

ring opening

ring retention

MeOH, H⁺

D-glucose

slow

fast steps

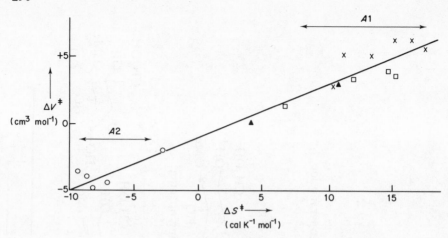

Fig. 4.12. Plot of $\Delta V^{\ddagger}$ against $\Delta S^{\ddagger}$ for hydrolysis of glycosides.[256] Key: ×, alkyl pyranosides; ▲, aryl pyranosides; ○, alkyl furanosides; □, aryl furanosides

The delicate balance between the available pathways in glycoside hydrolysis can be conveniently probed by $\Delta V^{\ddagger}$ measurements. The hydrolysis of glucose-1-phosphate[258] and of sucrose to glucose and fructose[259] are also included with A2 reactions, the latter being the first recorded instance of an effect of pressure upon rate.[260]

Additions to double bonds and their regression Water and other protic solvents add to olefins forming alcohols or ethers, the ease of reaction for simple alkenes depending on their basicity, the stability afforded to a carbocation. If a normal A2

$$Me_2C{=}CH_2 + H_3O^+ \underset{k_{-1}}{\overset{k_1}{\rightleftharpoons}} Me_2\overset{+}{C}{-}CH_3$$

$$Me_2\overset{+}{C}{-}CH_3 + H_2O \overset{k_2}{\longrightarrow} Me_2C{-}CH_3$$
$$\underset{\quad\quad\quad OH}{}$$

$$\Delta V^{\ddagger}(35) = -11{\cdot}5 \ cm^3 \ mol^{-1}$$

mechanism is operating, $\Delta V^{\ddagger}_{obs} = \Delta \bar{V}_1 + \Delta V^{\ddagger}_2$; since the proton transfer is from oxygen to carbon, one cannot necessarily assume that $\Delta \bar{V}_1 \sim 0$ and no reliable information is available for this quantity. $\Delta V^{\ddagger}_2$ should be negative and, indeed, the pressure dependence is close to that for other A2 reactions such as ester hydrolysis.[284] Hydration of α,β-unsaturated acids and ketones occurs readily though one would expect the carbonyl group to make protonation on the α-carbon more difficult than for simple olefins, on account of its electron-withdrawing effect. It seems quite likely that synchronous nucleophilic attack by water on the β-carbon and protonation either on the α-carbon or carbonyl oxygen brings about hydration in a single step, the transition states being

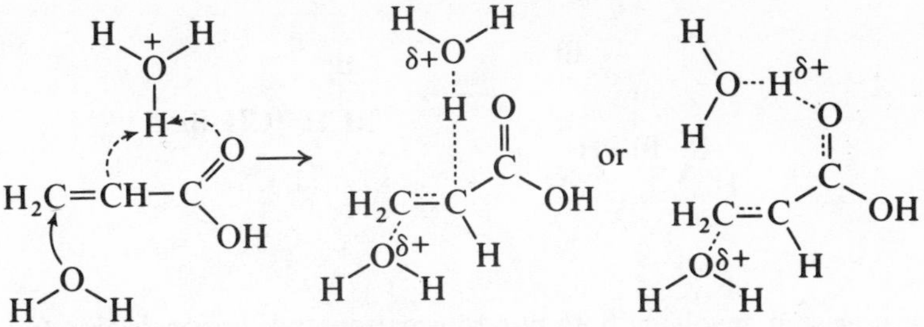

similar to that proposed for acetaldehyde hydration.[289]

Substantial negative volumes of activation are observed (-14 to -25) consistent with the binding of at least one water molecule[280] in the transition state. That more may be involved is suggested by Bunnett's ω-parameter, derived from the dependence of rate on water activity.[288] The hydration of mesityl oxide, $Me_2C{=}CHCOMe$, shows $\omega \sim 5$-8 which may be interpreted as showing the involvement of this number of water molecules.[287]

The reversibility of all steps in these hydrations leads to the possibility of acid-catalysed dehydration of alcohols to alkenes, especially facile if the product is conjugated. It is presumed that the same transition state is formed in both forward and reverse reactions (principle of microscopic reversibility). The volume

$$Me_2C{=}CHCOMe \; \rightleftharpoons \; Me_2\overset{\underset{\displaystyle OH}{|}}{C}{-}CHCOMe$$

$$\Delta V_f^{\ddagger} \rightarrow \; = \; -14.5\,cm^3\,mol^{-1}$$
$$\Delta V_f^{\ddagger} \leftarrow \; = \; -5.5$$
$$\Delta \bar{V} \rightarrow \; = \; -9.0$$

profile shows that elimination also has a negative volume of activation despite the overall increase of volume due to the release of a water molecule and, since this cannot readily be explained by electrostriction, as charge is being delocalized, it further suggests the coordination of more water in reaching the transition state.

Halogen addition is mechanistically similar to acid-catalysed hydration of alkenes. The reaction is generally believed to occur by initial displacement of halide ion by attack of the π-electron pair on halogen:

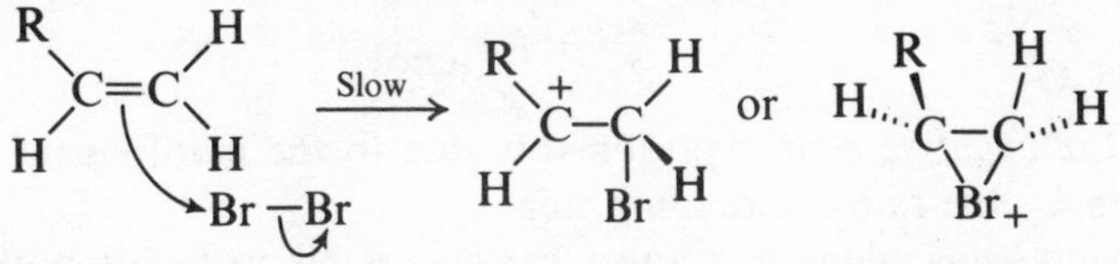

A carbocation (the more stable one if an unsymmetrical olefin is used) or else a cyclic bromonium ion is formed intermediately which is rapidly converted to dibromide.

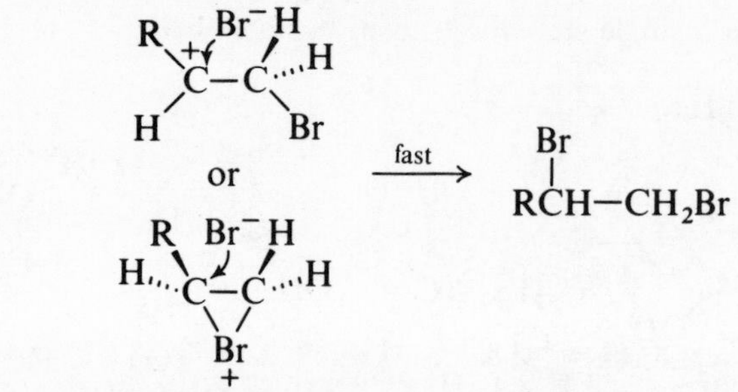

$$\text{(structures)} \xrightarrow{\text{fast}} \begin{array}{c} Br \\ | \\ RCH-CH_2Br \end{array}$$

The slow step, involving both bond-formation and -fission, is akin to an S_N2 reaction between neutral substrates (e.g. the Menschutkin reaction) and should be dominated by electrostriction caused by charge separation, the bonding effects cancelling. Volumes of activation are large and negative as predicted but exact values are not available since the reaction is difficult to conduct such that rates are those of the simple step shown above. The equilibrium, $Br_2 + Br^- \rightleftharpoons Br_3^-$, interferes and has a negative volume which has not been disentangled from $\Delta V_{obs}^{\ddagger}$.[306]

Ad$_E$2 reactions

			$\Delta V^{\ddagger}$ (cm^3 mol^{-1})	Reference
PhCH = CHPh + Br$_3^-$	MeOH	0°	< -20	306
HOCH$_2$CH = CH$_2$ + I$_3^-$	H$_2$O	0°	< -15	
	MeNO$_2$	0°	< -40	

Acid-catalysed rearrangements. One of the most important reactions under this heading is enolization, the initial step in acid-catalysed reactions α to a carbonyl group. The net reaction is the transfer of a proton from α-carbon to carbonyl oxygen and is reversible. For the enolization of acetone, $\Delta V^{\ddagger} = -2.1$ cm^3 mol^{-1}

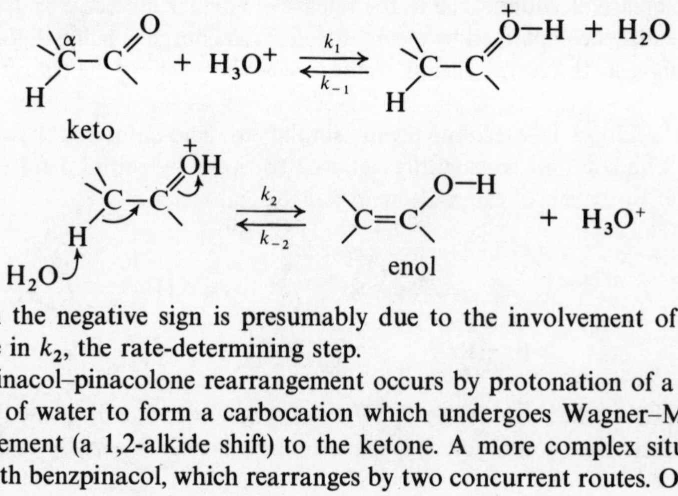

in which the negative sign is presumably due to the involvement of a water molecule in k_2, the rate-determining step.

The pinacol–pinacolone rearrangement occurs by protonation of a 1,2-diol, and loss of water to form a carbocation which undergoes Wagner–Meerwein rearrangement (a 1,2-alkide shift) to the ketone. A more complex situation is found with benzpinacol, which rearranges by two concurrent routes. One is the

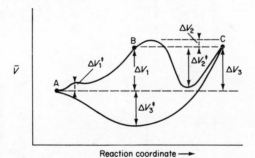

direct rearrangement as described for pinacol; the other proceeds *via* the epoxide, (23), which may be isolated and its conversion to benzpinacolone, separately studied.[290] The volumes of reaction and of activation for each step have been measured, giving the volume profile, fig. 4.13.

$$Ph_2C-CPh_2 \xrightarrow{k_3} PhC-CPh_3$$

Fig. 4.13. Schematic volume profile for the rearrangement of benzipinacol;

$$A = \begin{matrix} Ph_2C-CPh_2 \\ HO \quad OH \end{matrix} \qquad B = \begin{matrix} Ph_2C-CPh_2 \\ O \end{matrix} \qquad C = PhCOCPh_3$$

$\Delta V_1^{\ddagger} = +2.3 \qquad \Delta V_2^{\ddagger} = -16 \qquad \Delta V_3^{\ddagger} = -40 \, cm^3 \, mol^{-1}$

$\Delta \bar{V}_1 = +19 \qquad \Delta \bar{V}_2 = +1 \qquad \Delta \bar{V}_3 = +20$

(After Tamura and Moriyoshi[290])

Epoxide formation could arise by the internal displacement of protonated diol, in the trans conformer, (24), by a process which is an example of neighbouring group participation. Alternatively, the carbocation, (25), could be an intermediate:

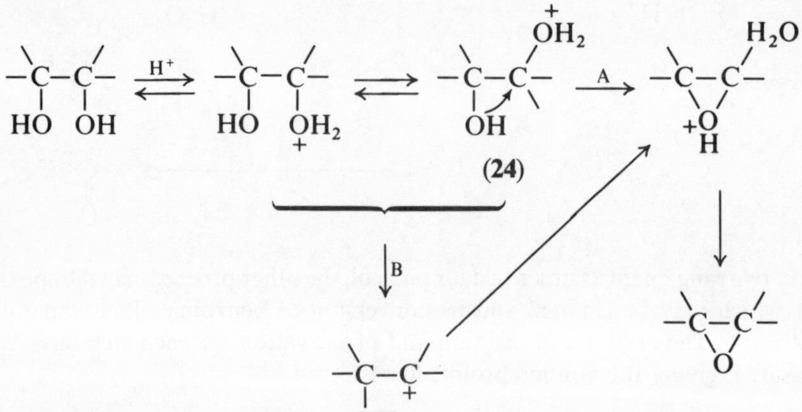

(24)

(25)

Internal displacements with charge dispersion usually show $\Delta V^{\ddagger} \sim 0$ to -5 whereas the experimental value for epoxide formation is $+2.3 \, \text{cm}^3 \, \text{mol}^{-1}$ which is accordingly assigned to the rate-determining loss of water, path B.

Rearrangement of epoxide to pinacolone *via* the protonated intermediate, (26), could occur in one or two steps:

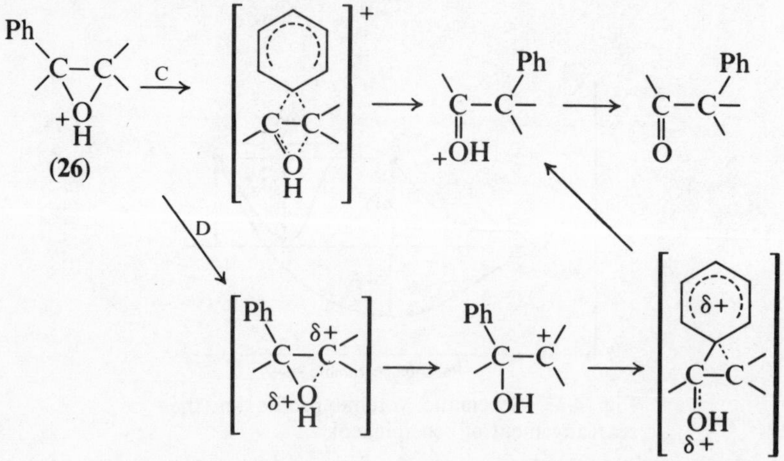

(26)

The volume of activation is large and negative, $-17 \, \text{cm}^3 \, \text{mol}^{-1}$, although the volume of reaction is almost zero. The necessity for a compact transition state rules out D and suggests that phenyl transfer is occurring in the slow step.

A second example of epoxide rearrangement shows a further point of interest. 1,1-Diphenyl-2-p-anisylepoxypropane rearranges thermally both to a ketone and an indene derivative. The latter is favoured by pressure. When conducted in a chiral solvent (diethyl (+)-tartrate), chirality is induced in the ketone product only when the reaction is carried out at high pressure.[492] Asymmetric induction of this type, induced by pressure, could well result from the stronger solvation brought about and deserves further investigation.

The degenerate Wagner–Meerwein rearrangement of the 1,2-dimethoxynorbornyl cation is slowed by pressure as judged from the narrowing of the NMR resonances. The reason must be the delocalization of charge in (26) which may be a transition state or a non-classical ion intermediate but which is associated with a lowering of electrostriction.[438]

(26)

The benzidine rearrangement has intrigued organic chemists for many years. A 1,2-diarylhydrazine (hydrazobenzene) rearranges under acidic conditions, the N-N′-linkage becoming a 4,4′-(benzidine) or, if para positions are blocked, 2,4′-(diphenyline) 2,N′-(ortho-semidine) or 4,N′-(para-semidine). The products obtained from different hydrazobenzenes and the history and current interpretation of the mechanism have been reviewed by Ingold.[104] The rearrangement is intramolecular and is normally catalysed by the successive addition of two protons to the molecule so that rate = k_3[hydrazobenzene] $[H^+]^2$ where $[H^+]$ represents $[H_3O^+]$ or some other appropriate acidity function. It appears that the doubly protonated molecule undergoes N,N-bond fission in a heterolytic sense but the two halves remain loosely bonded presumably by interaction of their π-systems; re-formation of covalent bonding then follows.

Benzidine may result directly from the transition state, (27), but the other products require rotation of the two rings before bond formation. Certain hydrazobenzenes, such as hydrazo-1-naphthalene, undergo rearrangement catalysed by only one proton, i.e. rate = k_2[hydrazobenzene] $[H^+]$, and in many cases the 1- and 2-proton routes may become of similar rate so that mixed orders are observed: in any mechanistic studies care must be taken to define the rate constants which are measured. Furthermore, other processes can intervene including disproportionation,

$$2\,ArNH-NHAr \xrightarrow{\ k_d\ } ArN=NAr+2\,ArNH_2$$

and oxidation by air, both involving hydrogen transfer:

$$ArNH-NHAr+O_2 \xrightarrow{\ k_0\ } ArN=NAr+H_2O$$

These possibilities may be summarized by the following kinetic scheme:

$$\text{Hb} + \text{H}^+ \underset{K_1}{\overset{K_1}{\rightleftarrows}} \text{HbH}^+ \underset{\text{H}^+}{\overset{K_2}{\rightleftarrows}} \text{HbH}_2^{++}$$

with vertical arrows $\uparrow k_d$ to **D** and $\uparrow k_o$ to **O** above, and $\downarrow k_1$ to $\text{B} + \text{H}^+$ and $\downarrow k_2$ to $\text{B} + 2\text{H}^+$ below.

where Hb, B represent hydrazobenzene and benzidine (including other rearrangement products), D and O are disproportionation and oxidation products, respectively. By appropriate techniques, the individual rates may be disentangled and measurements have been extended to high pressure.[329–333] Activation volumes for 1- and 2-proton pathways, for oxidation and disproportionation, are given in Table 4.18.

Table 4.18. Activation volumes for rearrangements of hydrazobenzenes

$X-\bigcirc-\text{NH}-\text{NH}-\bigcirc-Y$ Solvent	Temperature (°C)	1-proton pathway $\Delta V_{\text{I}}^{\ddagger}$ (cm^3 mol^{-1})	2-proton pathway $\Delta V_{\text{II}}^{\ddagger}$ (cm^3 mol^{-1})	Reference
Rearrangement				
X = 2-Me; Y = 2-Me	H$_2$O 4%, EtOH	25	−2.5	−7.5	329
X = 2-Br; Y = 2-Br			−10.7	−0.4	330
X = 2-OMe; Y = 2-OMe			−12		331
X = 2-Cl; Y = 2-Me			−6.8	−3.2	332
X = 2-OMe; Y = 2-Me			−8.5	—	333
Disproportion X = 2-OMe; Y = 2-OMe		$\Delta V_d^{\ddagger}$ +5			331
Oxidation X = 2-Me; Y = 2-Me		$\Delta V_0^{\ddagger}$ −40			329

The measured activation volumes for rearrangements are composite quantities and from the scheme it would be inferred that

$$\Delta V_{\text{I}}^{\ddagger} = \Delta \bar{V}_1 + \Delta V_1^{\ddagger}$$
$$\Delta V_{\text{II}}^{\ddagger} = \Delta \bar{V}_1 + \Delta \bar{V}_2 + \Delta V_2^{\ddagger}$$

No measurements of volume changes in protonation equilibria of hydrazines appear to have been made but one may assume $\Delta \bar{V}_1 \sim -5\,\text{cm}^3\,\text{mol}^{-1}$, typical of monoprotonation of a nitrogen base in water by H_3O^+: this value should be more negative in the aqueous-ethanolic solvent used.

Protonation of pyridines (Table 3.9) in water and methanol is accompanied by $\Delta \bar{V} \sim +5$ and $+10$ respectively. If the latter value is applicable here, $\Delta V_1^{\ddagger}$ would be, on the whole, positive. Furthermore, $\Delta \bar{V}_2$ would be expected to be larger than $\Delta V_1^{\ddagger}$ by comparison with the first and second protonation equilibria of, for example, 1,2-diaminoethane ($ca.$ $+7$ and $+10\,\text{cm}^3\,\text{mol}^{-1}$ respectively). Therefore values of $\Delta V_2^{\ddagger}$ must be even more positive, i.e. the two-proton transition state must occupy a greater volume than reagents by perhaps $+10\,\text{cm}^3\,\text{mol}^{-1}$. It would be expected that this value would result from the two rings in the transition state lying farther apart than solvation distance, and farther apart for the 2-proton route than the 1-proton. It is apparent that the value of ΔV^+ will be very sensitive to the interannular distance in the transition state; a simple calculation shows that the volume increases by about $25\,\text{cm}^3\,\text{mol}^{-1}$ for each $0.1\,\text{Å}$ in separation. The differences in activation volume between the various substrates might well be due to this factor and even, at least for the 2-proton values, shows a trend with substituent. The mechanism of disproportionation is less certain; it may be acid-catalysed (also base- or un-catalysed under appropriate conditions) and is first-order in hydrazo compound although two molecules are involved in product formation. The following sequence is suggested by Ingold.[328]

The redox process is the rapid hydride transfer from hydrazobenzene to the cation formed by heterolysis of the protonated hydrazobenzene an alternative pathway to the rearrangement. Although normally a minor pathway, it can be important; 4,4′ diiodohydrazobenzene gives 100% disproportionation.[334] Again, the activation volume $\Delta V^{\ddagger}_{obs} = \Delta \bar{V}_1 + \Delta V^{\ddagger}_d = +5\,cm^3\,mol^{-1}$ for dimethoxy-hyrazobenzene supports a bond-breaking process, $\Delta V^{\ddagger}_d \sim +10\,cm^3\,mol$, for the rate-determining step in accordance with this scheme.

The Orton rearrangement is the apparent migration of chlorine in N-chloroacetanilide, to the aromatic ring under the specific catalytic influence of HCl. There is no doubt that the rearrangement is intermolecular and chlorine is produced intermediately which may be isolated or allowed to chlorinate some other species:[328]

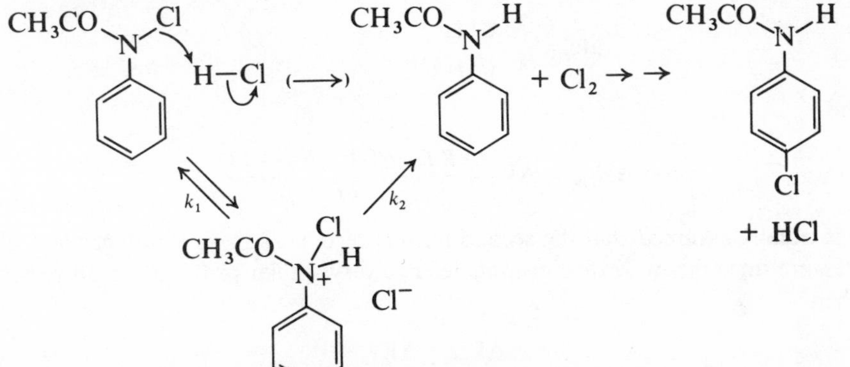

The first stage appears to be rate-determining, the second being a normal electrophilic aromatic substitution. The rearrangement is retarded by pressure, $\Delta V^{\ddagger} = +5\,cm^3\,mol^{-1}$ [337,355] which is not expected for a nucleophilic displacement on hydrogen as shown above. It seems likely that a pre-protonation by HCl, K_1, is followed by rate-determining heterolysis of the N-Cl bond, k_2, accompanied by charge neutralization. Then $\Delta V^{\ddagger}_{obs} = \Delta \bar{V}_1 + \Delta V^{\ddagger}_2$; $\Delta \bar{V}_1$ for protonation of an amine by H_3O^+ in an alcoholic solvent would be about $-10\,cm^3\,mol^{-1}$ and the charge-neutralization brought about by the displacement on chlorine (similar to an S_N2 reaction, charge type c, Table 4.11) would contribute about $+15$ in agreement with the observed value.

4.2.14 Base-catalysed Reactions

There is a certain ambiguity concerning the function of a base catalyst; on the one hand it may serve to remove a proton, i.e. as a Brønsted base, while on the other it may act as a general nucleophile adding on to a π-system and thereby increasing the reactivity of the substrate. Examples of both types of behaviour are discussed in this section.

Double bond migration in alkenes is promoted by rather strongly basic systems such as tertiary butoxide in dimethylsulphoxide. Allylic shifts are brought about by proton removal and, in an aprotic medium, the system probably passes through a hydrogen-bonded intermediate:

(I) $\xrightleftharpoons[k_{-1}]{k_1}$ (intermediate) $\xrightleftharpoons[(k_{-2})]{k_2}$ (II)

If the isomer (I) is transformed into the much more stable (II), i.e. we can ignore k_{-2}, then

$$\text{rate} = \frac{k_1[\text{I}]}{(k_{-1}/k_2)+1}$$

and

$$\Delta V^{\ddagger}_{\text{obs}} = \Delta V^{\ddagger}_1 + \frac{RT\partial \ln[(k_{-1}/k_2)+1]}{\partial P}$$

It is usually assumed that the second term is zero, i.e. k_{-1}/k_2 is independent of pressure since the two rate constants refer to very similar protonation processes when

$$\Delta V^{\ddagger}_{\text{obs}} \sim \Delta V^{\ddagger}_1$$

and we are observing essentially the activation volume for an acid-base reaction in this case involving a carbon acid and an oxygen base. Δ^3-Cyclohexenes undergo two consecutive isomerizations,

and activation volumes for several such reactions have been determined and found to lie in the range -20 to $-25\,\text{cm}^3\,\text{mol}^{-1}$ (Table 4.22 below).[335,336] The anionic intermediate is seen as possessing a considerably lower volume than the reagents. The interpretation of this result is not entirely clear since the reagent state of the potassium t-butoxide appears to be in the form of inactive trimers which means that a pre-equilibrium dissociation to the monomeric base—of unknown volume change—may need to be considered. It would seem likely that such a de-aggregation step would be accompanied by an increase in volume. The negative volume of activation observed (and also negative entropy of activation) point to a compact, perhaps heavily solvated transition state which may be typical of the dissociation of a carbon acid, though as yet there is little information on this point.

Enolization An analogous rearrangement of great importance is base-catalysed enolization which constitutes a pre-equilibrium step in α-substitutions of carbonyl- and nitro-compounds at high pH.

Products may arise from either enol or enolate; the initial reaction is the dissociation of a carbon acid and, surprisingly, little information concerning the volume change is available. Proton transfer from carbon to a nitrogen or oxygen base probably involves a negative volume change as shown by reactions of acetophenone in D-exchange with anionic bases, $\Delta V^{\ddagger} \sim -3\,\text{cm}^3\,\text{mol}^{-1}$ (Table 4.18) and also some dissociations of nitro and other relatively acidic compounds. The simple exhange between (p-nitrophenyl)nitromethane and tetramethyl-guanidine,

a fast reaction followed by the temperature-jump method,[338] is accelerated by pressure, $(\Delta V^{\ddagger} = -17.8, \Delta \bar{V} = -25\,\text{cm}^3\,\text{mol}^{-1})$ which is ascribable to elec-trostriction since the base is neutral, and the same applies to enolization of 2-nitropropane, e.g.

$$\Delta V^{\ddagger} = -28\,\text{cm}^3\,\text{mol}^{-1}$$

This reaction is unusual in that ionization by hindered bases seems to be aided by quantum-mechanical tunnelling which is pressure-dependent, fig. 4.14 below. The primary kinetic isotope effect for this reaction, k_H/k_D, diminishes from the abnormally high value of 16.3 at 1 bar to about 11 at 1.2 kbar.[339] Thus values of $\Delta V^{\ddagger}$ for 2-nitropropane and 2-nitropropane-1d differ (Table 4.22). the former being anomalous since in addition to accelerating the reaction, pressure produces a diminution in the amount of tunnelling. Enolization of carboxyl compounds is much less facile than that of ketones since the negative charge already present inhibits further dissociation. Nevertheless at high temperature, isotopic substitution of carboxylic acids in D_2O occurs presumably by this route,

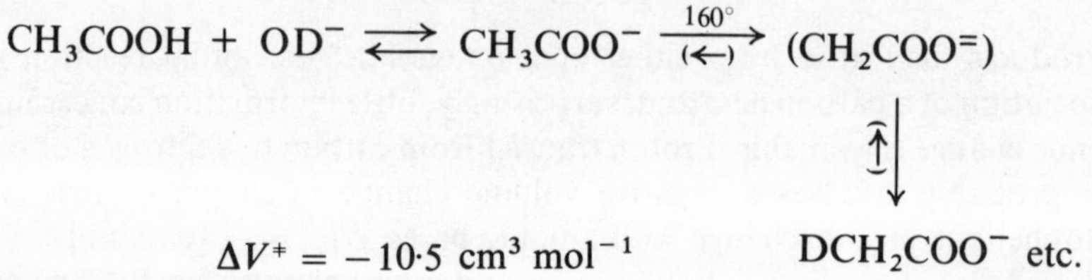

$$CH_3COOH + OD^- \rightleftharpoons CH_3COO^- \xrightarrow[\scriptscriptstyle(\leftarrow)]{160^\circ} (CH_2COO^=)$$

$$\Delta V^{\ddagger} = -10 \cdot 5 \text{ cm}^3 \text{ mol}^{-1} \qquad\qquad DCH_2COO^- \text{ etc.}$$

The reaction is base-catalysed and quite susceptible to pressure[340] since further charge creation occurs at each ionization step.*

Enolization is a prerequisite to the aldol condensation, the nucleophilic attack of enolate carbon on a carbonyl carbon with the eventual formation of a β-hydroxyaldehyde or -ketone:

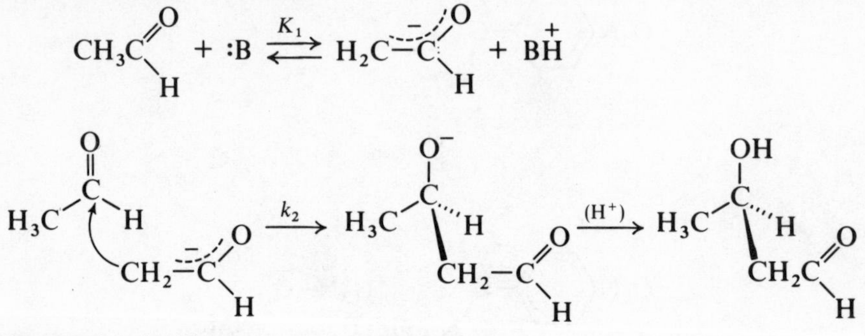

The reaction is effectively the dimerization of the carbonyl compound and so should show a negative volume of reaction. The volume of activation consists of two terms, $\Delta V^{\ddagger}_{obs} = \Delta \bar{V}_1 + \Delta V^{\ddagger}_2$: if :B is an anionic base (e.g. OH$^-$), there will be little volume change on enolization, perhaps $\Delta \bar{V} \sim -5$ cm^3 mol^{-1} while $\Delta V^{\ddagger}$ should resemble the value for basic hydrolysis of an ester, ~ -6 cm^3 mol^{-1}. Therefore, negative values of $\Delta V^{\ddagger}_{obs} \sim -10$ would seem reasonable and such

* This is in contrast to the similarly slow exchange of formic acid, which is not enolizable, and is postulated to occur by a molecular mechanism for which only a small volume change is expected:

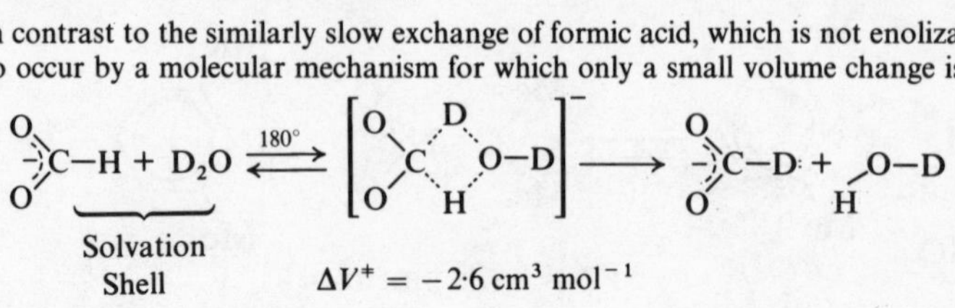

$$\Delta V^{\ddagger} = -2 \cdot 6 \text{ cm}^3 \text{ mol}^{-1}$$

values are found for the dimerization of isobutyraldehyde[341] and cyclo-hexanone.[342] It seems remarkable that values reported for n-butyraldehyde are positive[341] in contrast and suggests that further examination of this result should be undertaken to locate the source of the discrepancy.

The retro-aldol reaction of diacetone alcohol to two acetone molecules must proceed by the same pathway and carbon–carbon bond fission is rate-determining. This is associated with a positive activation volume of similar magnitude, $\Delta V^{\ne} = +6.9\,\text{cm}^3\,\text{mol}^{-1}$.[496]

β-Eliminations

A spectrum of mechanisms encompasses the loss of nucleophilic leaving group and a proton on the adjacent carbon which constitutes a β-elimination. Three principal types may be considered differing in the order in which the two bonds break,[293] namely:

E1: rate-determining loss of nucleophile (as in S_N1 reactions) followed by rapid proton loss to solvent, first-order kinetics being observed:

E2: rate-determining attack of base on β-hydrogen and concerted loss of leaving group. The stereochemistry is normally trans and second-order kinetics are observed:

E1cb: prior loss of β-proton to give a carbanion intermediate followed by loss of the nucleophile. The kinetics depend on whether the initial step is reversible and whether k_1 or k_2 is rate-determining:[294]

Detailed studies indicate that these broad mechanistic types present a greatly oversimplified picture and further ramifications of the scheme have been discussed by Bordwell[295] and by McLennan.[296] The question naturally arises as to whether elimination mechanisms may be distinguished by their activation volumes.

The classical E1 pathway partakes of the same rate-determining step as does the substitution (section 4.2.11). Substitution and elimination products arise by solvent attack on the reactive intermediate, presumably an ion-pair in most cases. This mode of elimination is not strictly base-promoted since the participation by the base occurs in the subsequent fast step.

The transition state will lie somewhere between reagent and ion–pair and its volume will contain a positive contribution due to bond-stretching and a negative one, usually larger, if charge separation occurs, due to electrostriction. Eliminations from tertiary 'onium ions' (charge dissipation) have $\Delta V^+ = +8$ to $+15\,\mathrm{cm^3\,mol^{-1}}$ [194] while those from neutral halides are between -10 and $-20\,\mathrm{cm^3\,mol^{-1}}$, all being solvent dependent. Different values are obtained for activation volumes measured by following disappearance of reagent and by following olefin formation. This is because there is a difference in activation volumes of the subsequent fast steps which therefore affects the product distribution though not the rate. The substitution product is favoured over elimination product as expected for an associative process, the difference in activation volumes being 7–$13\,\mathrm{cm^3\,mol^{-1}}$.[297] Volumes of activation are more negative than the volumes of reaction; the subsequent fast step involves an increase in volume of around $+9\,\mathrm{cm^3\,mol^{-1}}$ and similar volume data are obtained for eliminations in a variety of protic and aprotic solvents,[343] although in the latter the E2 mechanism may be favoured. No clear distinction appears possible here.

The concerted E2 process appears to take place in eliminations of primary halides with alkoxide ions in alcoholic or dipolar aprotic solvents.[295] Volumes of activation for reactions of this charge type are typically -10 to $-12\,\mathrm{cm^3\,mol^{-1}}$ (Table 4.22) and must be seen as the resultant of bond-making between base and β-hydrogen compensated by bond-stretching of the leaving group, together with charge delocalization. It is similar in these respects to an S_N2 reaction between anion and neutral substrate,[298] with additional space required for the bulky incipient double bond.

While olefin-forming $E1_{cb}$ eliminations have been extensively studied, the only examples for which activation volumes are available are retro-carbonyl additions,

$$O-C\overset{Ph}{\underset{CN}{\overset{\cdots H}{<}}} \quad \overset{OH^- \ k_1}{\underset{k_{-1}}{\rightleftharpoons}} \quad \bar{O}-C\overset{Ph}{\underset{CN}{\overset{\cdots H}{<}}} \quad \overset{slow}{\underset{k_2}{\longrightarrow}} \quad O=C\overset{Ph}{\underset{H}{<}} \ + \ CN^-$$

(28) H_2O $\Delta V^{\ddagger} = +12 \ cm^3 \ mol^{-1}$

$$O-C\overset{Me}{\underset{CH_2COCH_3}{\overset{\cdots Me}{<}}} \quad \overset{OH^-}{\rightleftharpoons} \quad \bar{O}-C\overset{Me}{\underset{CH_2COCH_3}{\overset{\cdots Me}{<}}} \quad \overset{slow}{\underset{k_2}{\longrightarrow}} \quad O=C\overset{Me}{\underset{Me}{<}}$$

(29) H_2O $\overset{-}{C}H_2COCH_3$

$$\Delta V^{\ddagger} = +6 \ cm^3 \ mol^{-1}$$

Base-catalysed eliminations of mandelonitrile, (28), and diacetonealcohol, (29), are positive.*[298,344] These values are the sum of the pre-equilibrium volume change, $\Delta \bar{V}_1$ and the volume of activation of the slow step, $\Delta V_2^{\ddagger}$. The former is probably small, so bond-breaking in the slow step dominates the observed value. It is considered that a continuum of transition states between E2 and E1cb may exist, the relative extent of fission of the C–H and C–L bonds, and amount of carbanion character, varying between the limiting mechanisms. There are a number of eliminations known which appear to occupy intermediate ('central') positions in this scheme by mechanistic criteria such as β–H isotope effects, effects of leaving group on rate, entropy of activation. For example, the reaction between 2-chloroethyl phenyl sulphone, (30), and acetate ion,

$$\underset{Ac\bar{O}}{\underset{H}{\overset{PhO_2S}{H-C-C-H}}} \overset{Cl}{\underset{H}{}} \longrightarrow \underset{\overset{\delta\delta-}{AcO}}{\underset{H}{\overset{PhO_2S}{H-C=C-H}}} \overset{\overset{\delta\delta-}{Cl}}{\underset{H}{}} \longrightarrow \underset{H}{\overset{PhO_2S}{C=C}}\overset{H}{\underset{H}{}} + Cl^-,$$

(30)   AcOH

shows a small sensitivity to substituents in the aryl group unlike a true E1cb type, a β–H isotope effect of $k_H/k_D = 2$, which could indicate either very little or very extensive proton transfer in the transition state; the Brønsted β-constant obtained from the effect of base strength on the reaction is 0.74 showing extensive proton transfer is occurring but a small sensitivity of rate to the leaving group suggests the C–Cl bond to be little affected. The volume of activation is $-1 \ cm^3 \ mol^{-1}$, intermediate between E2 and E1cb which characterizes the transition state as central, the C–H bond extensively broken, the C–Cl bond hardly stretched. Other examples are included in Table 4.22. The E1cb-like elimination of methanol from 1-methoxyacenaphthene

* The volume of activation for base-catalysed decomposition of diacetone alcohol is reported to be very solvent dependent as well as temperature dependent to an unusual degree; $\Delta V^{\ddagger}$ varies from -9 to $+8 \ cm^3 \ mol^{-1}$ in aqueous alcoholic solvents, the more negative values being associated with high water content and high temperature.[345-347]. This may reflect the importance of changes in solvation in the activation process.

fits the pattern for a central process, $\Delta V^{\ddagger} = +3 \, cm^3 \, mol^{-1}$. However, this value is greatly affected by the presence of a crown ether. These substances are complexing agents for Group I cations, in this case the potassium of the $K^+ OBu^{t-}$ used as base.[300] Crown ethers are known to affect rates and stereochemistry of concerted eliminations,[300] the inference being that the attacking base is the ion-pair rather than the free anion. This is evidently the case for the acenaphthene reaction and the value of $\Delta V^{\ddagger}$ in the absence of crown ether includes a negative contribution for electrostriction around the separating potassium ion. The complexed potassium ion no longer plays a part in the activation step and $\Delta V^{\ddagger}$ increases to $+15 \, cm^3 \, mol^{-1}$.[298] This is quite likely to be a general phenomenon in solvents of low polarity.

It appears, therefore, that, while volumes of activation may not be a reliable guide to carbocation character in elimination (i.e. the E1–E2 interface), there is evidence that the degree of carbanion character (between E2 and E1cb) may be assessed by this technique.[298] A comparison of values of $\Delta V^{\ddagger}$ for elimination of HBr from *cis*- and *trans*-β-bromostyrenes provides an application. The latter is believed to exhibit normal E2 behaviour while the *cis* compound has been claimed to show E1cb character[350] on account of a larger substituent effect on rates. Volumes of activation are, however, identical, $-5 \, cm^3 \, mol^{-1}$, suggesting similar mechanisms.[298] The less negative values than prototype E2 reactions normally exhibit may be the result of a 'central' mechanism, i.e. some carbanion character or may be due to the intrinsically large volume occupied by π-bonds of the products, here an alkyne rather than an alkene. The spontaneous fragmentation of β-bromoangelate is greatly retarded by pressure, $\Delta V^{\ddagger} = +18 \, cm^3 \, mol^{-1}$. This is consistent with a concerted E2 type reaction and quite distinct from the expectations for either possible two-step mechanism.

α-Eliminations. The basic hydrolysis of chloroform is known to lead to a species which acts as a source of dichlorocarbene, $:CCl_2$, and high pressure studies suggest that this is indeed produced free and not as a solvate since $\Delta V^{\ddagger} = +16\,cm^3\,mol^{-1}$, in accordance with the B1 process but too large a value for the B2[386]

$$CHCl_3 + OH \rightleftharpoons \bar{C}Cl_3 + H_2O$$

$$B1 \quad CCl_3^- \longrightarrow \left[\overset{\delta-}{Cl}\cdots\overset{\delta-}{\bar{C}Cl_2}\right]^{\ddagger} \longrightarrow :CCl_2 + Cl^- \quad \Delta V^{\ddagger} > +10$$

$$B2 \quad H_2O + CCl_3^- \not\longrightarrow (H_2\overset{+}{O}-\bar{C}Cl_2) + Cl^- \quad \Delta V^{\ddagger} \text{ small}$$

The formation of dimethylallylidine, (31), by an analogous process however, has $\Delta V^{\ddagger} \sim 0$ which must be ascribed to its dipolar nature which adds an electrostrictive component to the positive activation volume

$$Me_2\underset{\underset{Cl}{|}}{C}-C\equiv C-H + OH^- \rightleftharpoons Me_2\underset{\underset{Cl}{|}}{C}-C\equiv C^{\ominus} + H_2O$$

$$Me_2\underset{\underset{Cl}{|}}{C}-C\equiv C^- \longrightarrow \left[\begin{array}{c}\overset{\delta+}{Me_2C}-C\equiv C^-\\ \vdots\\ Cl^{\delta-}\end{array}\right] \longrightarrow Me_2\overset{+}{C}-C\equiv\bar{C}: \longleftrightarrow Me_2C=C=C:$$
$$Cl^-$$

$$(31)$$

In a similar way, the hydrolysis of difluoramine is plausibly by an α-elimination via fluoronitrene, (32). Direct formation of the dimeric product N_2F_2 from the intermediate anion would undoubtedly have a negative activation volume yet a positive one is observed due to N–F bond-breaking in the rate-determining step.

$$NHF_2 + OH^- \rightleftharpoons \bar{N}F_2 + H_2O \quad \Delta\bar{V}\sim 0$$

$$NF_2^- \xrightarrow{\text{slow}} :NF + F^- \quad \Delta V^{\ddagger} + ve$$
$$(32)$$

$$2NF_2^- \not\longrightarrow FN=NF + 2F^- \quad \Delta V - ve$$

Hydride-transfer and disproportionation. Several base-promoted redox reactions are known in which a hydride transfer is facilitated by initial coordination of base to a carbonyl compound. Examples are the Meerwein–Pondorff procedure and the Cannizzaro reaction:[348]

A non-enolizable aldehyde disproportionates to alcohol and acid by a process which is essentially similar to a $B_{Ac}2$ reaction, hydride ion being the leaving group. According to the scheme, above,

$$\Delta V_{obs}^{\ddagger} = \Delta \bar{V}_1 + \Delta V_2^{\ddagger} = -27 \, \text{cm}^3 \, \text{mol}^{-1} \,{}^{351}$$

Allowing $\Delta \bar{V}_1 = -12$ for the pre-equilibrium association, the hydride transfer has a strongly negative volume of activation, $\Delta V_2^{\ddagger} = -15 \, \text{cm}^3 \, \text{mol}^{-1}$ which is similar to that for reaction between a ketone and borohydride,

It appears that, in the absence of charge generation, hydride transfer proceeds with a considerable diminution of volume, unlike proton transfer for which small negative or positive values are found. The more compact transition state for hydride transfer has been explained[348] as a consequence of a smaller electron repulsion term than in the case of proton transfer. Hydride transfer between two electrophilic centres involves one electron pair while proton transfer between two nucleophilic centres involves two:

$$-\bar{N}: \overset{+}{H} : \bar{N}- \quad \text{proton transfer, } 4e^-$$
$$-\overset{+}{E} \, \bar{H}: \overset{+}{E}- \quad \text{hydride transfer, } 2e^-$$

A more complex example is the reduction of dibutyl ketone by butoxide ion[349] for which the stoichiometry is

$$Bu_2C{=}O + PrCH_2O^- \xrightarrow{(H+)} Bu_2CHOH + PrCHO$$

The reaction is second-order in both ketone and butoxide and is moderately accelerated by pressure, $\Delta V_{obs}^{\ddagger} = -13 \, \text{cm}^3 \, \text{mol}^{-1}$. A mechanism which agrees with the kinetics requires a rate-determining proton-exchange by two molecules of a ketone-alkoxide adduct:

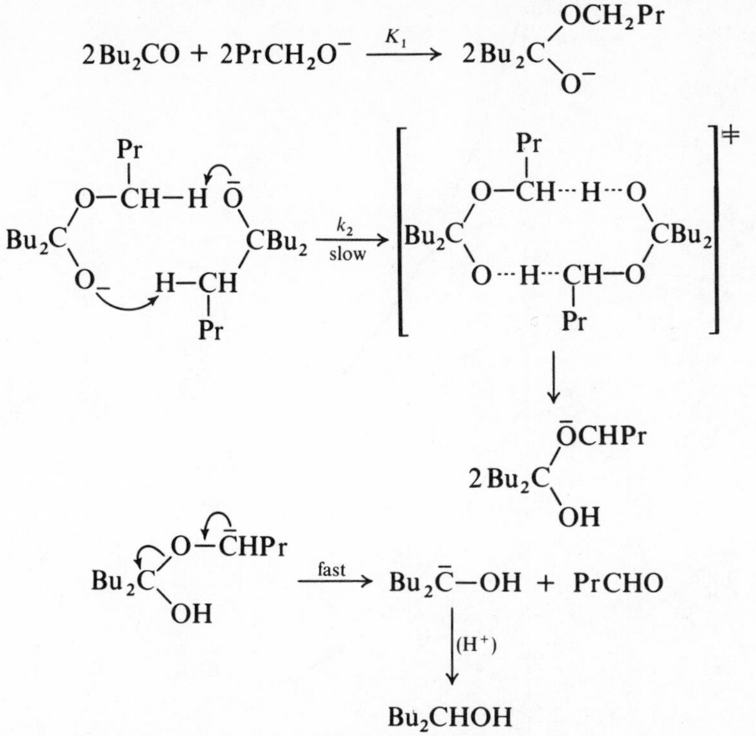

The redox step, which is fast, consists in the expulsion of electrophilic oxygen as aldehyde. The volume change for the preassociation, ~ -12, added to the volume of activation for proton transfer, -0 to -5, would seem to account for the observed value of ΔV^{+}.

Hydride transfer is the initial and rate-determining step in the reaction between leuco Crystal Violet and chloranil:

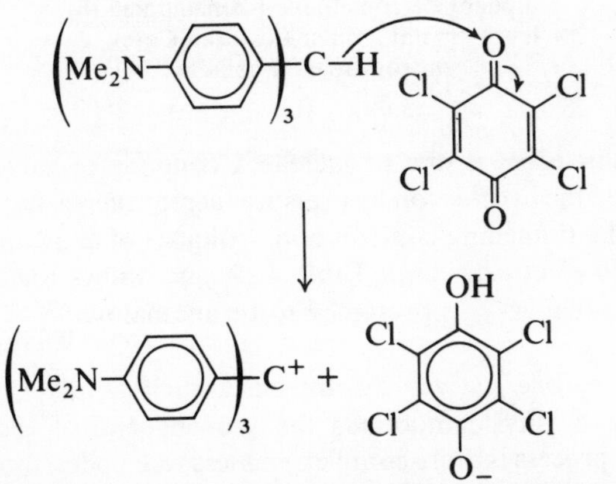

This reaction is believed to be assisted by tunnelling and, again, the abnormally large isotope effect ($k_H/k_D = 12$, 1 bar) is reduced by pressure, fig. 4.14. It is

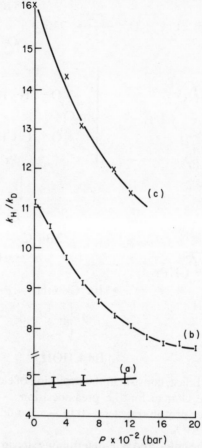

Fig. 4.14. Pressure-dependence of some primary kinetic isotope effects: (a) diphenyldiazomethane + benzoic acid, (b) leuco-crystal violet + chloranil, (c) 2-nitropropane + collidine

believed that this effect is due to increased coupling of solvent translational motion to that of the hydride ion by pressure thereby increasing its effective mass and reducing the tunnelling contribution. Volumes of activation are large and negative, due to electrostriction, Table 4.19, and values for the two isotopic species differ so that $\Delta V_H^{\ddagger}$ is presumed to be anomalous.[354]

Base-catalysed polymerization Strong bases such as alkyl lithiums initiate polymerization of vinyl compounds, the growing end of the chain being a carbanion. The process is more complex and less well understood than is radical polymerization (section 4.2.4). Strong acceleration by pressure has been found (e.g. $\Delta V^{\ddagger} = -36\,cm^3\,mol^{-1}$ for 2,3-dimethylbutadiene) but any interpretation

Table 4.19. Activation volumes of base-catalysed reactions

	Solvent	T (°C)	$\Delta V^{\neq}$ (cm³ mol⁻¹)	Reference
Double bond isomerizations:				
A→B	DMSO	80	−20.2	336
B→A			−19.2	
C→D			−22.5	
D→C			−22.5	
D→E			−24	
Ionization of carbon acids				
NO₂C₆H₄CH₂NO₂ + HN = C(NMe₂)₂	Toluene	30	−17.8	338
(ionizable H underlined)	Xylene		−14.6	
	Mesitylene		−13.2	
	Anisole		−16.3	
	Chlorobenzene		−13.0	

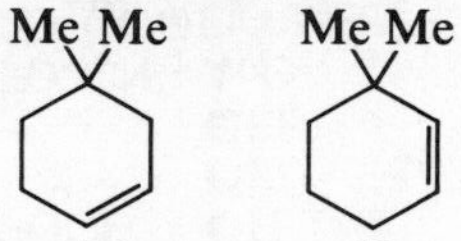

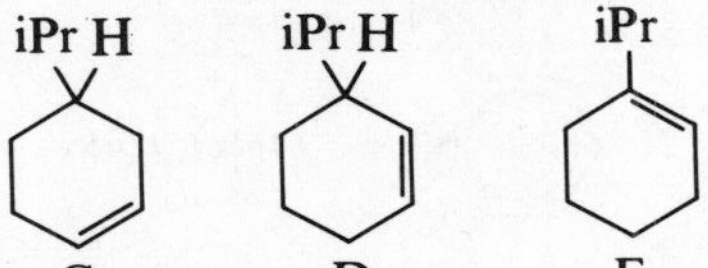

Me Me — A

Me Me — B

iPr H — C

iPr H — D

iPr — E

Table 4.19 (cont.)

	Solvent	T (°C)	$\Delta V^{\ddagger}$ (cm³ mol⁻¹)	Reference
Me₂CHNO₂ + (2,6-dimethylpyridine) + OAc⁻	tBuOH:H₂O, 60:40, 50:50	25	−31, −28	339
	H₂O,		−1	351, 149
Me₂CDNO₂ + (2,6-dimethylpyridine) + OAc⁻	60:40, 50:50		−40, −35	351, 149
	H₂O		−1	
CD₃SOCD₃ + OH⁻	H₂O		+2	
PhCOCD₃ + OH⁻	H₂O		−1	
OEt⁻	EtOH		−1	
OPh⁻	MeOH		−5	
PhCD₂CN + AcO⁻	MeOH		−4	351
PhCD₂COOEt + MeO⁻	MeOH		−3	
(indene, D, D D)	MeOH		−4	
Ph₃CD + OH⁻	MeOH-H₂O		−1	
Ph₂CD₂ + MeO⁻	MeOH-Me₂SO		−3	

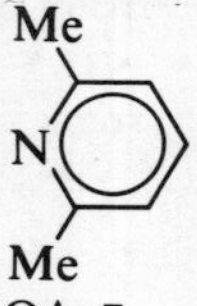

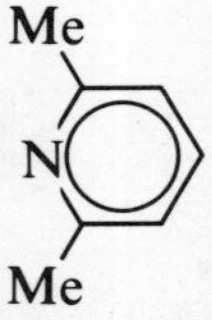

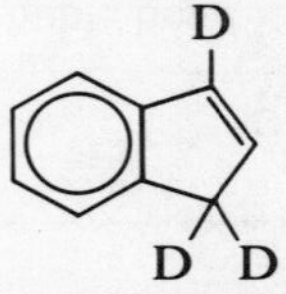

312

β-Eliminations

MeCH₂CHBrMe + EtO⁻	EtOH	48	−10	194
PhCH₂CH₂Cl + EtO⁻	EtOH	65	−12	298
cis-PhCH=CHBr+Na⁺ isoPrO⁻	isoPrOH	26	−6	
trans-PhCH = CHBr + Na⁺ isoPrO⁻		118	−5	
cis-BrCH = CHBr + Na⁺ MeO⁻	MeOH	37	−5	
ClCH₂CH₂SO₂Ph + OAc⁻	EtOH	50	−1	
1-Methoxyacenaphthene + *t*BuO⁻	*t*BuOH	80	+3	
1-Methoxyacenaphthene			+15	
+ crown ether				
Mandelonitrile + OH⁻	H₂O	25	+12	
Diacetonealcohol + OH⁻	H₂O	25	+6	

Hydride transfer

PhCOCH₃ + BH₄⁻	isoPrOH		−11	351
2PhCHO	OH⁻ , H₂O-		−27	
	MeOH		−13	349
Bu₂CO + BuO⁻	BuOH			

(Me₂NC₆H₄)₃CH + [tetrachloro-1,4-benzoquinone structure] MeCN 29 −25 354

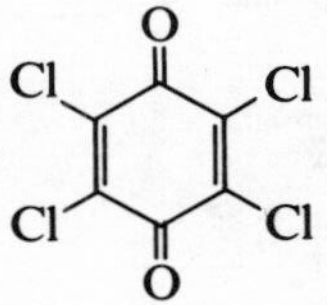

(Me₂NC₆H₄)₃CD + [tetrachloro-1,4-benzoquinone structure]   −35

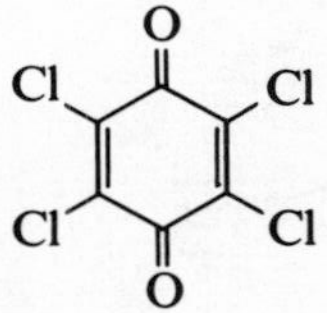

Anionic polymerization

CH₂ = CHMe-CHMe = CH₂ + BuLi	Heptane	−36	391
	Ether	−48	

$$\underset{R}{\overset{CH=CH_2}{|}} + \; BuLi \longrightarrow \underset{R}{\overset{\bar{C}H-CH_2Bu}{|}} \underset{Li^+}{} \xrightarrow{RCHCH_2} \quad \underset{BuCH_2}{\overset{R}{\diagdown}}\!\!\underset{}{\overset{}{CH}} \quad \underset{CH_2}{\overset{R}{|}}\!\!\overset{}{CH}{:}^{\ominus} \; \text{etc.}$$

needs to take into account de-aggregation of the base and ion-pair equilibria, besides changes in ionic character and it is very likely that ionogenesis accounts as much for the pressure effect as does the bond-formation of the propagation step.[391]

$$(BuLi)_n \; \rightleftharpoons \; nBuLi_{(solv)}$$
$$BuLi + M \; \longrightarrow \; Bu - M^- Li^+$$

4.2.15 Carbonyl- and Sulphonyl-substitutions

Hydrolyses and related nucleophilic substitutions of esters, amides, acid halides, and anhydrides are included in this section. As for nucleophilic displacements at saturated carbon, concerted (S_N2) and ionization (S_N1) routes may be considered but a further pathway is now available, namely addition–elimination *via* a tetrahedral intermediate, ($B_{Ac}2$). This is possible since the carbonyl and sulphonyl groups may increase the central co-ordination number, an option not available to a tetrahedral carbon.

The accepted mechanism, based on a wealth of evidence, for base-promoted hydrolysis of carboxylic esters such as ethyl acetate, is as follows:[195]

$$\Delta V^{\ddagger} \text{ (in water)} = -7 \cdot 6 \text{ cm}^3 \text{ mol}^{-1}$$

The existence of the tetrahedral intermediate, (33), was demonstrated by Bender's observation of [18]O exchange between carbonyl oxygen and water.[196] The full kinetic analysis can be complicated but, assuming the proton transfers are fast and not kinetically significant and also that k_{-2} is negligible, two cases may be considered for which the slow step is either k_1 or k_2. If k_1 is rate-determining, second-order kinetics are observed and

$$k_{obs} = k_1/(K_p + 1)$$

where $K_p = k_{-1}/k_2$; thus,

$$\Delta V_{obs} = \Delta V_1^{\ddagger} - \Delta \bar{V}_p$$

Now $\Delta V_1^{\ddagger}$, the activation volume for coordination of OH^- to the ester (k_1), should be negative since it is a bond-making process and, despite some loss of electrostriction from charge dispersion in the transition state one would expect a value around $-10\,cm^3\,mol^{-1}$. By contrast, the processes k_{-1} and k_2 are formally very similar (expulsion of an oxygen anion) so that $k_{-1} \sim k_2$ and $\Delta \bar{V}_p \sim 0$.

If k_2 were rate-determining, we have

$$k_{obs} = k_2\,K_e$$

where $K_e = k_1/k_{-1}$ and

$$\Delta V_{obs}^{\ddagger} = \Delta V_2^{\ddagger} + \Delta \bar{V}_2$$

Of these, $\Delta V_2^{\ddagger}$ would be positive since it is both dissociative and charge-dispersing while $\Delta \bar{V}_e$ would be correspondingly negative, pressure favouring association. The result should be a cancelling and a small, possibly positive value of $\Delta V_{obs}^{\ddagger}$. Table 4.23 sets out some values for ester hydrolyses, ranging from -5 to $-10\,cm^3\,mol^{-1}$, which suggest support for the first case and that k_1 is rate-determining. This is reasonable on chemical grounds since if $k_{-1}(\approx k_2)$ were slow, the tetrahedral intermediate would accumulate with time, which is not observed. Base-catalysed amide hydrolysis shows a similarly large negative value consistent with the $B_{AC}2$ process. Values contrast with those for the acid-catalysed routes which are usually positive.

In aqueous acetone, the volume of activation for ethyl acetate hydrolysis becomes more negative with increasing water concentration[198] for reasons which are not clear. However, anomalies exist in this type of medium and activation parameters are known to show extrema when plotted against composition (fig. 4.15); it is probable that these results carry information concerning changes in

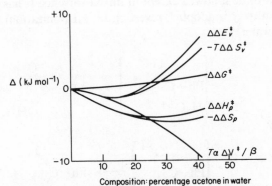

Fig. 4.15. A comparison of activation parameters at constant volume and at constant pressure for base-promoted hydrolysis of ethyl acetate in aqueous acetone. (After Tonnet and Whalley[198])

solvent structuring. Constant volume entropies and enthalpies of activation were obtained using eqs. (3.42)–(3.43) and are claimed to be easier of interpretation than the constant pressure analogues. The factor $T\alpha\Delta V^+/\beta_T$, which is the difference between the constant pressure and constant volume parameters, can be seen to fall quite sharply with increasing acetone concentration.

Hydrolysis of anhydrides by neutral aqueous solvent is also presumed to proceed by the $B_{Ac}2$ route, and so also is ethanolysis. Activation volumes are

strongly negative (Table 4.19), the more so in ethanol than water or the proportion of dioxan[200] in a mixed solvent. It is likely that the transition state is dipolar since the nucleophile is neutral.[203,244]

Hydrolyses of amides ($B_{Ac}2$) by alkali show a considerably more negative activation volumes than those of esters.[196] This is most likely due to a change in the value of $\Delta\bar{V}_p$ since partitioning of the tetrahedral intermediate now involves expulsion of either an oxygen base or a nitrogen base for which volume changes are not so likely to be the same.

Less assurance may be felt concerning the hydrolyses of acyl halides which are usually too fast to be followed except in mixed solvents. It has been suggested than an S_N2 pathway is adopted, reverting to S_N1 (ionization) with increasing proportion of water[199]

$$\Delta V^+ \text{ (H}_2\text{O-dioxan)} = -20 \text{ to } -30$$
$$\Delta\bar{V} \qquad\qquad\qquad = -13 \text{ cm}^3 \text{ mol}^{-1}$$

The volume of activation is far more negative than the volume of reaction, pointing to a highly polar transition state and becomes more negative still with increasing water content,[200] contrary to predictions from dielectric constant properties of the medium. Since $\Delta \bar{V}$ also is almost independent of the solvent composition it appears that the volume of activation is largely a reflection of solvent organization and specific interactions with the transition state, fig. 4.17.

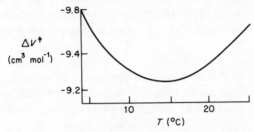

Fig. 4. 16. Pressure effect on the activation volume for hydrolysis of methanesulphonyl chloride in D_2O. (After Hine[102])

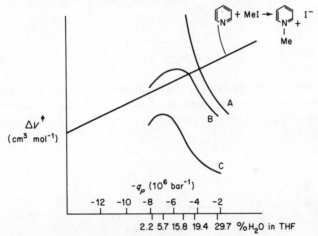

Fig. 4. 17. Effect of solvent composition on volumes of activation on some solvolytic reactions; (a) benzoyl chloride in aqueous acetone; (b) p-anisyl chloride; (c) benzoyl chloride in aqueous THF. Results for a typical Menshutkin reaction are included for comparison

The reaction constant, ρ, for hydrolyses of benzoyl chlorides which is about 2.1 at 1 bar increases to 2.8 at 1 kbar[199] and activation volumes of substituted benzoyl chlorides tends to be more negative when electron-withdrawing substituents are present. It seems probable that this reveals a change in the position of the transition state along the reaction coordinate both with substituent and with pressure change; a mechanistic change with solvent is also probable since a reversal of reactivity between benzoyl and anisoyl chlorides occurs, H > OMe at low water content to H < OMe at high, consistent with ionization becoming favourable. The hydrolysis of methanesulphonyl chloride although of similar magnitude becomes less negative with increasing water content. The solvent isotope effect, k_{H_2O}/k_{D_2O}, has been measured and found independent of pressure from which it is argued that in this case, the solvent hydrogen-bonded structure is not a crucial factor in determining $\Delta V^{\ddagger}$.[184] However, the activation volume varies with temperature featuring a minimum value at 15°, fig. 4.16,[102,202] which is similar to the minima in physical properties observed, and is held to be due to changes in the structure of water. At present, these lines of evidence appear to conflict.

The hydrolyses of acetylphosphates provide an example of another charge-type. The mono- or dianions may be generated by buffering suitably and the reactions are spontaneous in water.[206] Activation volumes are very small and appear to be inconsistent with the involvement of a water molecule. Probably a

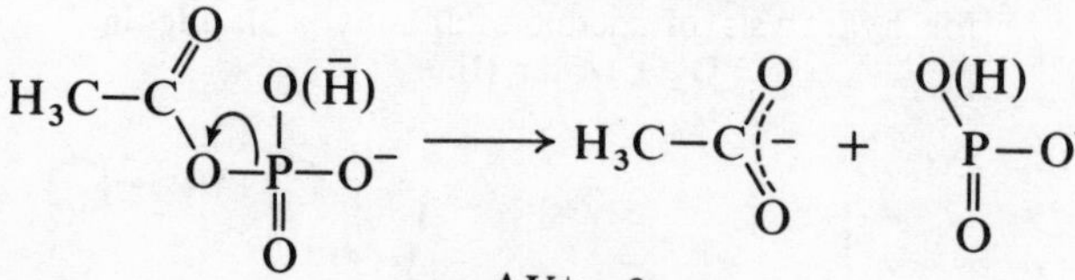

$$\Delta V^{\ddagger} \sim 0$$

unimolecular heterolysis occurs; and a cancelling of effects. The phenyl analogue, (34), however, hydrolyses with $\Delta V^{\ddagger} = -19\,\text{cm}^3\,\text{mol}^{-1}$, clearly showing a mechanistic change. This is probably ascribed to a $B_{Ac}2$ mechanism with water the nucleophile.

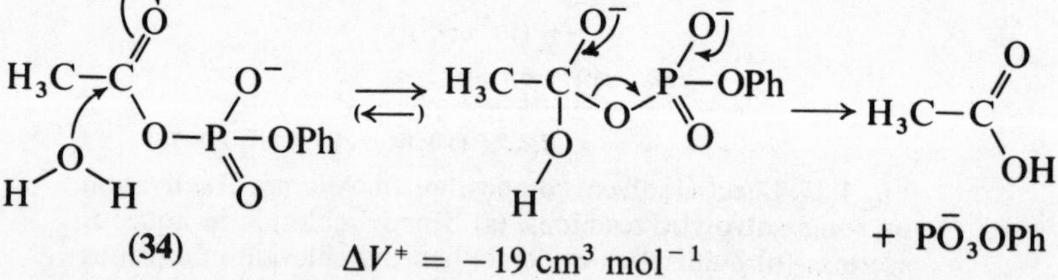

$$\Delta V^{\ddagger} = -19\,\text{cm}^3\,\text{mol}^{-1}$$

A more complex example illustrating the power of activation volume measurements is the hydrolysis of chloroacetyl hydrazide,

$$\text{ClCH}_2\text{CONHNH}_2 + \text{OH}^- \rightarrow \text{Cl}^- + \text{CH}_3\text{COOH} + \text{N}_2 + \text{N}_2\text{H}_4$$

From kinetic measurements a route was postulated in which the slow step was the fragmentation of the conjugate base. This clearly does not accord with the measured activation volume, $\Delta V^{\ddagger} = -5\,\text{cm}^2\,\text{mol}^{-1}$. Instead, an internal displacement is indicated[481] *via* an α-lactam intermediate.

4.2.16 Reactions of Transition-metal Coordination Compounds

Pressure studies have proved extremely valuable in elucidating mechanisms of substitution of ligands coordinated to a transition metal which are proving to be very unpredictable and subject to change with minor structural alteration. In solution, which usually is in water, the metal ion will exist with a covalently bound complement of ligands (which could be solvent) constituting the inner coordination sphere. These ligands are in turn hydrogen bonded to a second layer of solvent molecules, the outer coordination sphere beyond which may be considered a third layer still under the influence of the central charge with bulk water beyond. The attractive forces in the solvation layer effectively place the water molecules under high internal pressure which depends upon the charge which compresses them and also reduces their compressibility. A di- or tri-valent ion with its first solvation shell may be considered virtually incompressible and volume changes which accompany substitution are associated with the transfer of ligands between regions of high and low compression. This topic has been reviewed by Stranks,[358,460] by Kelm,[359] and by Swaddle.[486] The inner coordination shell usually contains six ligands arranged in an octahedral binding geometry and substitution by an external donor can take place by either an associative or a dissociative route.[360] The associative pathways may involve an intermediate of seven coordination (denoted A) or a synchronous exchange of ligands from outer to inner shell (denoted I_a). Similarly the dissociative route may

Table 4.20. Activation volumes of carbonyl and sulphonyl displacements

	Solvent	T (°C)	$\Delta V^{\neq}$ (cm³ mol⁻¹)	Reference
1 $MeCO_2Me + OH^-$	H_2O	25	−9.9	196
2 $MeCO_2Et + OH^-$		20	−8.8	197
3		20	−5.6	
4 $isoMeCO_2isoPr + OH^-$		30	−6.4	
5 $n\text{-}MeCO_2nBu + OH^-$		20	−6.6	
6 $MeCO_2isoBu + OH^-$		20	−5.6	
7 $MeCO_2nC_5H_{11} + OH^-$		20	−6.3	
8 $MeCO_2Et + OH^-$	100% H_2O	20	−5.8	198
	80% Me_2CO	10	−7.6	
	69	31	−8.9	
	57	43	−13.1	
			−16.8	
9	70% H_2O, EtOH	25	−15	204
10	57% H_2O, EtOH	25	−28	205
11 $MeCONH_2 + OH^-$	H_2O	25	−14.2	196
12 $EtCONH_2 + OH^-$			−16.9	
13 $MeCO \cdot OPO_2OH\ H_2O$	H_2O	25	−0.6	206
14 $MeCO \cdot OPO_3^=\ H_2O$	H_2O	25	−1	
15 $MeCO \cdot OPO_2OPh\ H_2O$	H_2O		−19	
16 $PhCH_2CO_2Et + MeO^-$	MeOH		−12	351

9: aromatic ring bearing CO_2Et and OMe substituents

10: bicyclic structure bearing $O \cdot COMe$

Substrate	Conditions	Temp.	Value	Ref.
17 ClCH$_2$CONHNH$_2$, OH$^-$				
PhCOCl H$_2$O	H$_2$O		-5	386
	H$_2$O 2.2%, tetrahydrofuran	20	-33.1	199
	29.7%, tetrahydrofuran		-41.0	
	H$_2$O, 4.3%, dioxan	25	-20.8	200
	29.3%, dioxan		-31.4	
X$-$◯$-$COCl H$_2$O				
X = OMe	H$_2$O 2.2%, tetrahydrofuran		-27.5	199
	29.7%, tetrahydrofuran		-31.2	
X = Br	2.2%, tetrahydrofuran		-33.8	
	6.1%, tetrahydrofuran		-28.3	
X = NO$_2$	2.2%, tetrahydrofuran		-43.0	
PhC(=O)$-$O$-$C(=O)Ph H$_2$O, 60	H$_2$O, 1M+dioxan		-13.3	244
	10M		-21.1	
	27M		-26.6	

pass through an intermediate of five-coordination (D) or a synchronous exchange (I_d), fig. 4.18.

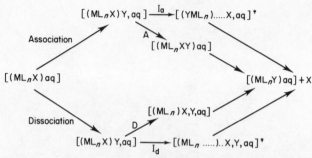

Fig. 4.18. Mechanisms of substitution for transition metal complexes; M = metal, L = non-exchanging ligand, X = exchanging ligand, Y = incoming ligand; parentheses ( ) enclose inner coordination sphere, brackets [ ] enclose outer solvation sphere

Furthermore it is assumed, and supported experimentally, that partial molar volumes of 5-, 6-, and 7-coordinate species are essentially the same[465] so that volume changes result from the release or coordination of ligand molecules and associated solvation. In many cases there is a parallel change in the volume and entropy during the activation step[99,466] but the former is the easier to interpret.

Th difference between I_a and I_d mechanisms lies in the prominence of bond-making over bond-breaking which has taken place in the transition state. The characteristics of each type are summed up in Table 4.21. A distinction between associative and dissociative processes may be made on the basis of their volumes

Table 4.21. Characteristics of inorganic substitution mechanisms in water[359]
$$ML_nX + Y \quad ML_nY + X$$

	Dissociative		Associative	
	D	I_d	A	I_a
In transition state:				
M–X bond	Broken	Stretched	Intact	Stretched
M–Y bond	Unformed	Partial	Formed	Partial
Rate dependent on nature of:	X	X and Y	Y	X and Y
$\Delta S^{\ast}$	Large +ve	Small	Large −ve	Small
$\Delta V^{\ast}$	Positive +10 to +30	Small positive	Negative	Small negative
$\Delta\beta^{\ast}$ ($cm^3\,mol^{-1}\,bar^{-1}$)	Positive 8×10^{-4}	Small $0\text{–}2 \times 10^{-4}$	Negative	Small

of activation which, at least for substitution of neutral ligands such as solvent, are negative for the former and positive for the latter. Changes in solvation of the exchanging ligands in these cases is usually small so that $\Delta V_{obs}^{+} \approx \Delta V_1^{+}$. The increase in volume associated with the D mechanism is a result of bond-breaking and forcing a coordinated molecule into the first solvation sphere and possibly simultaneously displacing a solvent molecule from this to an outer sphere. The reverse holds for associative mechanisms. The curvature of the plot of ln k against P gives the compressibility coefficient of activation, $\Delta\beta^{+}$, which in these reactions gives useful information. This is because partial molar volumes and compressibilities of water (or other ligands) are reduced in the progression from bulk solvent to inner coordination sphere (Table 4.22) as a result of the proximity of

Table 4.22. Estimated partial molar volumes, $\bar{V}$ and compressibilities β_T of water in the vicinity of M^{3+}

Environment	$\bar{V}$ $(cm^3\,mol^{-1})$	β_T $(10^{-2}\,kbar^{-1})$
Bulk liquid	18.0	4.58
Outer hydration sphere	15.6	1.4
Inner hydration sphere	15.0	0.8
Coordination sphere	14.5	ca. 0.7

the ionic charge permitting a prediction to be made of the value of $\Delta\beta^{+}$. The dissociation of a ligand molecule from direct coordination into the solvation layer will be accompanied by an increase in its volume and compressibility, hence $\Delta\beta^{+}$ will be positive while the binding of a free solvent molecule into a solvation or coordination sphere results in a negative value. Since reacting species are usually ionic, their activities will be sensitive to the ionic strength of the medium and, for the purposes of comparison, rate measurements should be carried out at constant ionic strength or corrected by the Debye–Hückel expression,

$$\ln k_1 = \ln k_0 + \frac{2z_A z_B \sqrt{I}}{1 + Ba\sqrt{I}} + 0.2[z_A z_B]\sqrt{I}$$

It may be predicted that, for the D mechanism, $\Delta\beta^{+}$ should take the value of the difference in compressibility of water in bulk (1 bar) and in the coordination sphere, ca. $8 \times 10^{-4}\,cm^3\,mol^{-1}\,bar^{-1}$ with a smaller value for the I_d process. The following examples will illustrate these principles. Water exchange is one of the most straightforward reactions to interpret:

$$(\overset{3+}{M}(NH_3)_5H_2{}^{18}O) + H_2O \rightarrow (\overset{3+}{M}(NH_3)_5H_2O) + H_2{}^{18}O$$

and the following metals have been investigated giving activation volumes thus: Ni $(+7.1)$,[477] Co $(+1.2)$, Rh (-4.1), Ir (-3.2), Cr $(-5.8\,cm^3\,mol^{-1})$. These values indicate associative pathways for the Cr, Ir, and Rh complexes but a dissociative mechanism for the Co compound, possibly on account of its smaller size and inability to become 7-coordinate.[361,362] Furthermore, $\Delta\beta^{+}$ is negative for the Cr,

Ir, and Rh compounds and positive for the Co analogue. The precision of measurement could not give accurate magnitudes for distinguishing I-types from A or D, though, except in the case of Cr for which an almost linear plot of $\ln k$ against P means that $\Delta\beta^{\pm}$ is very small and hence argues for the I_a mechanism.[363] Water exchange in $Co(en)_2(H_2O)_2$ brings about isomerization from *trans* to *cis* and for this reaction

$$\Delta V^{\pm}_{obs} = +14\,cm^3\,mol^{-1}, \quad \Delta\beta^{\pm} = 10 \times 10^{-4}\,cm^3\,mol^{-1}\,bar^{-1}\,[364]$$

(35)

Clearly, the D mechanism *via* the 5-coordinate intermediate, (35), best fits these data.

A few examples of non-aqueous neutral displacements have been recorded. Substitution of carbon monoxide by a phosphine from metal carbonyls in alkane solvents comes as near as possible to a neutral, non-solvated process.[377] Carbonyls of Mo and Cr react with first-order kinetics and $\Delta V^{\pm} = +10$ and $+15\,cm^3\,mol^{-1}$, respectively. Clearly, the D mechanism is implicated. $W(CO)_6$, on the other hand, reacts with second-order kinetics and has $\Delta V^{\pm} = -10\,cm^3\,mol^{-1}$ which must indicate that the A mechanism is employed. Exchange of solvent acetonitrile at Ni^{++} shows a moderate positive volume and entropy of activation ($+9.6\,cm^3\,mol^{-1}$ and $+37\,J\,K^{-1}\,mol^{-1}$) which is interpreted as due to the I_d mechanism.[145,476] Measurements were made on this fast reaction from relaxation times obtained by NMR, and analogous results were obtained for methanol as solvent.[478] Solvent exchange at Nb and Ta appears to make a distinction between oxygen and nitrogen donors (ethers, cyanides, and amines) which react by a D mechanism ($\Delta V^{\pm} + 15$ to $+30\,cm^3\,mol^{-1}$) and S, Se, and Te donors which have values consistent with A or I_a processes (-11 to -20).[474]

The association of iron (Fe^{3+}, $Fe(OH)^{2+}$) with Br^- is fast but can be followed by temperature-jump. There is very little pressure dependence of the rate and the rate process is interpreted as being ion-pair formation, i.e. association at a distance beyond the outer coordination sphere which would involve little change in solvation.[494]

Substitution of charged ligands by water involves additionally a charge separation so that now $\Delta V_{obs}^{\ddagger} = \Delta V_1^{\ddagger} + \Delta V_2^{\ddagger}$. The contribution from solvation, $\Delta V_2^{\ddagger}$, makes interpretation less easy. One method of disentangling these components is to compare volumes of activation with volumes of reaction. For a series of Co(III) complexes $\Delta \bar{V}$ is slightly more negative than $\Delta V^{\ddagger}$ while for Cr(III) substitutions the opposite is true. This is equivalent to saying that, in a plot of $\Delta V^{\ddagger}$ against $\Delta \bar{V}$, a linear relationship is observed with slope > 1 for Co

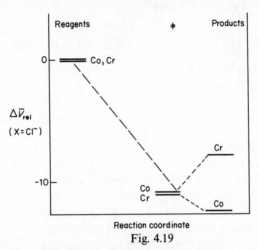

Fig. 4.19

and < 1 for Cr, fig. 4.19. For Cr, the transition state is more compact than products arguing for an associative mechanism whereas this trend is not observed for Co and may be considered consistent with the dissociative mechanism observed for water exchange although less unambiguously. Further support for this has been provided by Palmer and Kelm[365] for the reaction,

$$Co(NH_3)_5X^{3+n} \xrightarrow{\Delta V^{\ddagger}_{obs}} Co(NH_3)_5^{3+} + X^n$$

$$\downarrow \text{fast}$$

$$Co(NH_3)_5H_2O^{3+}$$

Assuming a limiting D mechanism,

$$\Delta V_{obs}^{\ddagger} = \bar{V}_I + V_{X_n} - \bar{V}_{RX}$$

and by measuring partial molar volumes of the substrate, $\bar{V}_{RX}$, and of the displaced anion, X^n, a value of the supposed volume of the penta-coordinate intermediate $\bar{V}_I$ is obtained. This value should be constant when X is varied and it is found to be so, confirming the model.

Changes in mechanism are sometimes apparent. Displacements of anionic ligands from Co(III) complexes are usually accompanied by large increases in volume which vary with the nature of the displaced ligand.[373] The simple D mechanism usually exhibited by Co(III) would have $\Delta V^{\ddagger}$ negative (bond breaking and increased electrostriction from the released anion) while the A mechanism should have values independent of the leaving group, in fact $\Delta V^+ \sim +22\,cm^3\,mol^{-1}$ for loss of electrostricted water around the hydroxide ion. However, compare the following:

$$[Co(NH_3)_5]^{3+}\,Br^-,\,OH^-$$

$$[Co(NH_3)_5Br^{2+}] + OH^- \xrightarrow{\text{D}} \times$$

$$\xrightarrow{\text{A}} \times$$

$$\big\uparrow\big\downarrow \Delta V_1 \qquad [Co(NH_3)_5Br,\,OH]^+$$

$$I_d cb\,[Co(NH_3)_4NH_2Br]^+ + H_2O \xrightarrow{\Delta V_2^+} [Co(NH_3)_5OH]^{2+} + Br^-$$

$$\Delta V_{obs}^+ = +8\cdot5\,cm^3\,mol^{-1}$$

A plausible mechanism is displacement from the conjugate base by water. Now,
$$\Delta V^+ = \Delta \bar{V}_1 + \Delta V_2^+;$$
evidence suggests $\Delta \bar{V}_1 \sim +18\,cm^3\,mol^{-1}$ due to water released during charge neutralization leaving $\Delta V_2^+ = -10$, compatible with an I_d mechanism. Similar arguments hold for displacement of sulphate rather than bromide for which $\Delta V^+ = +19.5$, $\Delta \beta^+ = +9\,cm^3\,mol^{-1}\,kbar^{-1}$—an even greater release of electrostricted water due to complete charge neutralization. On the other hand, the corresponding selenite, $Co(NH_3)_5, -OSeO_2^+$, shows an activation volume which is negative, $\Delta V^+ = -17\,cm^3\,mol^{-1}$. This must indicate a completely different mechanism and labelling shows the Se–O bond to be broken. A very large decrease in compressibility ($\Delta \beta^+ = -19\,cm^3\,mol^{-1}\,kbar^{-1}$) indicates considerable numbers of water molecules being taken up in solvating the leaving group.

Conjugate base formation may bring about a mechanistic change exemplifying the unpredictability of complex reactions. The first reaction is evidently of the I_d

$$Fe^{III}(H_2O)_6^{3+} + Cl^- \xrightarrow[(\leftarrow)]{} Fe^{III}(H_2O)_5Cl + H_2O;\ \Delta V^+ = +7.8$$

$$Fe^{III}(H_2O)_5OH^{2+} + Cl^- \xrightarrow[(\leftarrow)]{} Fe^{III}(H_2O)_4OHCl + H_2O;\ \Delta V^+ = -4.5$$

type while that of the conjugate base, I_a.[381] These reactions are fast and need to be followed by temperature-jump. The same is true of many water displacements from divalent cations, Ni^{2+}, Co^{2+}, etc., by nitrogen ligands. Volumes of activation of a great many such reactions are $+5$ to $+7\,cm^3\,mol^{-1}$ although volumes of reaction differ markedly both in magnitude and sign. This suggests a similar activation process for each, presumably outer-sphere association.[382]

$$Ni_{aq}^{2+} - NH_3 \xrightarrow[(\leftarrow)]{} NiNH_{3\,aq}^{2+}$$

Displacements by anionic nucleophiles would necessitate their de-solvation which would impart a substantial positive component to ΔV^+. This would depend in magnitude upon the electrostriction volume of the actual nucleophile. $Fe(phen)_3^{2+}$ and $Fe(bipy)_3^{2+}$ undergo ligand substitution by OH^- and by CN^- and in all cases $\Delta V^+ \sim +20$ cm^3 mol^{-1}. The two anions are known to have very different volumes of electrostriction (22 and 8 cm^3 mol^{-1}, respectively) although their intrinsic volumes must be similar. Desolvation therefore cannot play a rate-determining role and the I_d mechanism if inferred.[468]

Ligand-exchange in octahedral Nb and Ta complexes clearly proceeds by a pathway determined by the ligand. Exchange of oxygen and nitrogen donors (followed by NMR line-broading) is dissociative, $\Delta V^+ +15$ to $+30$ cm^3 mol^{-1} while for sulphur, selenium, and tellurium ligands it must occur by the I_d process, $\Delta V^+ -10$ to -20.[474]

A more complex case occurs in the hydrolyses of square planar platinum complexes whose rate law contains first- and second-order terms

$$(Pt\ dien\ X)^+ + Y^- \longrightarrow (Pt\ dien\ Y)^+ + X^-$$

$$\begin{array}{c} \overset{H_2}{N} \\ | \\ HN-Pt-Br \\ | \\ \underset{H_2}{N} \end{array}$$

$$Rate = k_1[\text{complex}] + k_2[\text{complex}][Y^-]$$

The individual rate constants can be separated and the separate volumes of activation determined, $\Delta V^+(1)$ and $\Delta V^+(2)$. It is found that $\Delta V^+(2) \sim -8$ cm^3 mol^{-1}, independent of pressure and the same volume of the transition state calculated for forward and reverse directions of the same reaction,

$$\bar{V}_+ = \bar{V}_{RX} + \bar{V}_{Y-} + \Delta V^+$$

confirming the extreme associative (A) nature of this reaction. For the first-order process, $\Delta V^+(1) \sim -18$ cm^3 mol^{-1} which is less easy of interpretation. While the kinetic form suggests a dissociative process, this would need to have the positive volume of activation ($+5$ cm^3 mol^{-1}) compensated by a large solvation term. At the time of writing, this and other aspects of inorganic mechanisms are in an active state of debate. Volumes of activation of these and other reactions are summarized in Table 4.23.

Redox reactions are familiar in complex chemistry and involve an exchange of, usually, one electron between two coordination compounds in suitable oxidation states, such as the process,

$$Fe(II)(H_2O)_6^{2+} + {}^*Fe(III)(H_2O)_6^{3+} \xrightarrow{1e^-} Fe(III)(H_2O)_6^{3+} + {}^*Fe(II)(H_2O)_6^{2+}$$

* Fe = isotopically labelled

Table 4.23. Activation volumes for reactions of transition metal complexes. All reactions are in water unless otherwise stated: exchangeable ligand underlined

	T (°C)	ΔV^* (cm^3 mol^{-1})	$\Delta\beta^*$ (cm^3 mol^{-1} kbar^{-1})	References
Co^{III}(NH$_3$)$_5$$\underline{X}$ + H$_2$O:				
X = NO$_3^-$	25	−6.3		
Br$^-$		−9.2	−2	367,
Cl$^-$		−10.6	−1	368,
SO$_4^=$		−18.5	−4	369,
NCS$^-$	88	−4.0		
N$_3^-$	75	+16.8		
H$_2$O		−5.8	0	370
Co^{III}(NH$_3$)$_5$$\underline{X}$ + OH$^-$:				
X = SO$_4^=$	15	+19.5		358,
SeO$_4^=$	25	−17.1		373
PO$_4^{3-}$	55	+28.9		
trans Coen$_2$XY + H$_2$O				
X = Cl, Y = Cl	25	+11		371
SeO$_3$H, H$_2$O* (retention)	55	+7.9		
SeO$_3$H, H$_2$O (inversion)	25	+8.0		358
H$_2$O, H$_2$O (inversion)	35	+7.6		
		+14		
Cr^{III}(NH$_3$)$_5$$\underline{X}$ + *H$_2$O				
X = *H$_2$O	25	−5.8	0	361
Cl$^-$		−10.8	−1.0	
Br$^-$		−10.2	−1.0	372
I$^-$		−9.4		

Reaction	T (°C)	$\Delta V^\ddagger$		Ref.
NCS^-	80	-8.6		363
urea		$+1.3$	0	374
$Cr^{III}(H_2O)_5X + H_2O$				
$X = H_2O$	45	-9.3		375
NO_3^-	25	-12.5		
I^-	25	-5.4		
$Cr^{III}(H_2O)_6 + ox \rightarrow Cr(H_2O)_4^+ ox + 2H_2O$	25	-2.2		
$Cr^{III}(H_2O)_4 ox + ox \rightarrow Cr(H_2O)_2 ox_2^- + 2H_2O$		-8.2		
$Cr^{III}(H_2O)_2 ox_2^- + ox \rightarrow Crox_3^{3-} + 2H_2O$		-10.0		
$Fe_{aq}^{3+} + Cl^- \rightarrow FeCl_{aq}^{2+}$	25	$+7.8$		381
$FeOH^{2+} + Cl^- \rightarrow FeOHCl^+$		-4.5		
$Ni_{aq}^{2+} + pada \rightarrow Ni(pada)^{2+}$	49	$+7.7$		383
$Ni_{aq}^{2+} + NH_3 \rightarrow Ni(NH_3)_{aq}$	30	$+6.0$		
$Ni(dien)_{aq}^{2+} + pada \rightarrow Ni(dien)_2 pada$	25	$+4.2$		385
$Co_{aq}^{2+} + gly \rightarrow Co(gly)^{2+}$	25	$+8.0$		
$Zn_{aq}^{2+} + gly \rightarrow Zn(gly)^{2+}$	10	$+7$		
$Cu_{aq}^{2+} + gly \rightarrow Cu(gly)^{2+}$ forward	25	$+12$		
reverse		-1.7		
$Zn_{aq} + pada \rightarrow Zn(pada)_{aq}$ (glycerol)	20	$+12.2$		384
Racemizations				
$Cr(ox)_3^{3-}$		-16.3		464
$Cr(ox)_2 phen^-$		-12.3		
$Cr(ox)_2 bipy^-$		-12.0		
$Cr(ox)phen_2^+$		-1.5		
$Cr(ox)bipy_2^+$		-1.0		470
$Fe(phen)_3$		$+15.6$		
$Ni(phen)_3$		-1.5		
Isomerizations (*trans → cis*)				
$Cr(ox)_2(H_2O)_2^-$		-16.6		471
$Cr(mal)_2(H_2O)_2^-$		$+8.9$		472
$Co(en)_2(H_2O)_2^{3+}$		$+14.3$		
Redox reactions				

Table 4.23 (cont.)

	T (°C)	$\Delta V^{\ddagger}$ (cm³ mol⁻¹)	$\Delta \beta^{\ddagger}$ (cm³ mol⁻¹ kbar⁻¹)	References
$Fe(II)_{aq} + Fe^*(III)_{aq}$		$-16\,(-12)$		358 (460)
$VO_2(nta)_2^{2-} + \tfrac{1}{2}H_2O_2$		-3.3		473
$Tl(I)_{aq} + Tl^*(III)_{aq}$		-14		366
$(NH_3)_5Co(III)Br + Fe(II)_{aq}$		$+8$		469
$Ir(NH_3)_5H_2O^{3+} + {}^*H_2O$	70	-3.2		362
$Rh(NH_3)_5\underline{H_2O}^{3+} + {}^*H_2O$	35	-4.1		361
$Fe(phen)_3^{2+} + 6H_2O \rightarrow Fe(H_2O)_6 + 3phen$	35	$+15.4$	0	376
$Ni(phen)_3^{2+}$		-1.2	0	470
$Mo(CO)_6 + Ph_3P$ (Isooctane)	103	$+10$		377
$Cr(CO)_6 + Ph_3P$ (Cyclohexane)	124	$+15$		
$W(CO)_6 + Bu_3P$	120	-10		
$Pt(dien)\underline{Br}^+ + X \rightarrow Pt(dien)X + Br^-$	25	$\Delta V_1^{\ddagger}$	$\Delta V_2^{\ddagger}$	
$\quad X = N_3^-$		-15	-8.5	378
$\quad OH^-$		-18		
$\quad NO_2^-$		-18	-6.4	
$\quad$ Pyridine		0		
$Pt(dien)\underline{Cl}^+ + N_3^-$		-17		
$\quad I = 0.2M\ NaClO_4$	25			
$\quad I = 2.0M$		-8.2		
$Pt(dien)N_3 + X^-$	25			
$\quad X = I^-$ (first-order path)		0		
$\quad$ (second-order path)		-12.2		
$\quad = NCS^-$ (first-order path)		0		
$\quad$ (second-order path)		-7.3		
$Pd(Etdien)I^+ + Br^-$	40			379
$\quad H_2O$		-10.2		
$\quad DMSO$		-9.2		
$\quad DMF$		-7.9		

MeOH

trans[Ir(CO)Cl(PPh$_3$)$_2$] + MeI

$\longrightarrow$ trans[Ir(CO)(PPh$_3$)$_2$ClIMe]

	−11.7	
	−15.2	380
DMF		25
Acetone	−20.5	
Chloroform	−19.2	
Chlorobenzene	−23.6	
Benzene	−29.8	
NbCl$_5$L + L* L = Me$_2$O		474
Me$_2$S		475
CoBr$_2$(PPh$_3$)$_2$ + PPh$_3$	−12.1	

A symmetrical reaction of this type has $\Delta G^0 = 0$ but the free energy of activation has been analysed by Stranks into four components[358] incorporating Coulombic work in bringing together the two ions, $\Delta G_{coul}^{\ddagger}$, ligand reorganization, $\Delta G_{int}^{\ddagger}$, solvent reorganization, $\Delta G_s^{\ddagger}$, and changes in activity due to the ionic strength of the medium, $\Delta G_{\gamma}^{\ddagger}$; hence

$$\Delta V^{\ddagger} = \Delta V_{coul}^{\ddagger} + \Delta V_{int}^{\ddagger} + \Delta V_s^{\ddagger} + \Delta V_{\gamma}^{\ddagger}$$

for $Fe^{2+} + Fe^{3+}$

$$-16 = -7.06 \quad -0.47 \quad -5.67 \quad -3.78; \quad \Delta V_{obs}^{\ddagger} = -12$$

Each component may be separately calculated by approximate procedures and the result is tolerably close to the observed activation volume. The model therefore seems reasonably accurate and describes the reaction as occurring between two complex ions which approach only until their outer solvation spheres are in contact at which electron exchange occurs without the breaking of any covalent bonds; the solvation shells rearrange to accommodate the new charges under whose influence they now appear. The redox process,

$$*Tl_{aq}^+ + Tl_{aq}^{3+} \longrightarrow *Tl_{aq}^{3+} + Tl_{aq}^+$$

could take place by a two-electron transfer ($\Delta V_{calc}^{\ddagger} \sim -25\ cm^3\ mol^{-1}$) or by two successive one-electron transfers ($\Delta V_{calc}^{\ddagger} \sim -14$, assuming the second is fast). The observed value of the activation volume is $-13\ cm^3\ mol^{-1}$ which therefore supports the second mechanism, a slow outer solvation sphere one-electron transfer.[366]

$$Tl_{aq}^+ + Tl_{aq}^{3+} \; \rightleftharpoons \; 2\,Tl_{aq}^{2+} \xrightarrow{\text{fast}} \text{products}$$

Other redox reactions take place by an inner-sphere complex which contains a bridging ligand. For reasons which are not entirely understood, volumes of activation are now positive. It is supposed that de-solvation of the more delocalized charge is responsible[469]

$$(NH_3)_5Co(III)\overset{++}{Br} + Fe(II)_{aq}^{++} \longrightarrow [NH_3)_5Co...Br...Fe_{aq}]^{4+}$$
$$\longrightarrow (NH_3)_5\overset{++}{Co}\;(II)H_2O + Fe(III)_{aq}^{++}\overset{++}{Br}$$
$$\Delta V^{\ddagger} = +8\ cm^3\ mol^{-1}$$

Several estimates of the temperature dependence of the volume of reaction have been made. Using the T-jump perturbation of the equilibrium binding of the murexide ion by various metals, as an example,[485]

$\Delta V^{\ddagger}$ (forward) $= +12$ (hence dissociative) and by measuring the enthalpy change at various temperatures and a series of pressures, by kinetic methods, $\partial \Delta \bar{V}/\partial T$ could be obtained from

$$\left(\frac{\partial \Delta H_T}{\partial P}\right)_T = \Delta \bar{V}_T - T\left(\frac{\partial \Delta \bar{V}_T}{\partial P}\right)_P$$

The value $-0.16\,cm^3\,mol^{-1}\,K^{-1}$ is remarkably constant for a number of anion-cation recombination reactions (-0.09 to -0.16) taken as due to a similar solvent reorganization process.

Racemization mechanisms A chiral coordination compound may be racemized during ligand exchange whether by A or D processes in which case polarimetry provides a convenient method for following the reaction. Racemizations of octahedral chelates offer several possibilities for intramolecular racemization.[460,461] The complete dissociation of one chelating ligand, almost certainly to be accompanied by ligand exchange and therefore an intermolecular process, would show a large volume increase as found for Fe(phen)$_3^{2+}$[462,463] ($\Delta V^{\ddagger} = +15\,cm^3\,mol^{-1}$). It should be noted that racemization of the analogous nickel compound, Ni(phen)$_3^{2+}$, has $\Delta V^{\ddagger} = -1.5\,cm^3\,mol^{-1}$.[83] This is interpreted not as due to a change in mechanism since on the grounds of similarities of enthalpy and entropy of activation the ligand dissociation process is still indicated. Instead, the difference is attributed to the low spin state adopted by nickel in the 4-coordinate intermediate and to its possessing a lower volume than the high spin state intermediate favoured by iron. If this is so, these results point to a possible pitfall in the interpretation of inorganic reaction mechanisms.

The fission of a single bond of a chelated ligand permits an intramolecular racemization to occur via a five-coordinate intermediate and an increase in volume. If the dissociated arm carries a charge, however, electrostriction may result in a negative volume of activation. An alternative 'twist' mechanism would involve no bond-breaking or changes in charge and $\Delta V^{\ddagger} \sim 0$.

Examples of both are known; some oxalate complexes of Cr racemize with a large negative volume of activation and no ligand exchange, pointing to the one-ended dissociation route and charge separation.[464] Others have a near-zero value interpreted as due to the twist mechanism. One factor influencing the availability of pathway is the nature of the ligand; a rigid molecule such as phenanthroline is sterically incapable of relinquishing only one of its coordinate bonds though this is possible for more flexible ligands such as dipyridyl. In many of these reactions, as with other displacements, acid-catalysis is often observed since protonation of a ligand increases its capability as a leaving group. Similar mechanisms must be considered for isomerizations of *cis* to *trans* isomers of an octahedral complex. Again mechanistic change may be experienced after an apparently minor structural change; the oxalato compound, $n = 0$, reacts by an associative route, $\Delta V^{\ddagger} = -16$ while the corresponding malonato compound, $n = 1$, uses a dissociative route, $\Delta V^{\ddagger} = +8.9\,\mathrm{cm^3\,mol^{-1}}$.[471]

4.3 PREPARATIVE CHEMISTRY AT HIGH PRESSURE

While the application of pressure may increase the rate of a reaction which has a negative volume of activation, in many cases it is far easier to increase the temperature to achieve the same end. There are numerous examples, however, of reactions which can only be successfully carried out under high pressure conditions which will be discussed in this section.[216,423] The advantages of high pressure synthesis, the uniqueness of products and the relatively small amount of energy needed to compress a liquid and maintain it in compression are offset by disadvantages inherent in the cost and engineering problems associated with scaling up apparatus to work in excess of 10 kbar and which can hold a suitably large volume of reagents. High pressure synthesis, however, is taking a more significant place in the chemist's repertoire as the following examples testify. A simple piston and cylinder apparatus (fig. 1.45a) or a capsule apparatus (fig. 1.45b) suffices for most of these reactions.

4.3.1 Cycloadditions

Furans are capable of acting as dienes in the Diels–Alder reaction but only give good yields with the most reactive dienophiles which add at low temperature, since the adducts are thermolabile. Pressure renders possible the synthesis of many otherwise inaccessible oxabicycloheptanes by this route.[402]

42%

The following (4 + 2) cycloadditions, which will not occur at atmospheric pressure on account of the lack of reactivity associated with one component, greatly extend the synthetic capabilities of the Diels–Alder reaction: aldehydes as dienophiles[431,432]

addition of α,β-unsaturated aldehydes to unactivated alkenes[433]

dipolar cycloadditions with nitronic esters[434]

γ-Pyrones normally lose CO_2 during cycloaddition. At high pressure this may be prevented and the bicyclic lactones successfully synthesized[435]

The 'ene' reaction is known to be a concerted 6π-electron cycloaddition[436] so should be assisted by pressure. Although no activation volumes have been

measured since temperatures normally $> 150°$ are required, products have been obtained in high yield under pressure[437]

β-Pinene

Under the very high pressure conditions of the anvil press, certain polycyclic aromatics undergo self-addition in the solid state with consequent changes in their semiconductor properties. This capability seems to depend on there being a sufficiently close spacing between HOMO and LUMO of the π-system. Thus, anthracene undergoes no reaction while pentacene forms a polymer, possibly 35, above 200 kbar which shows none of the long wavelength absorption (500–700 nm) of the monomer, clear evidence for breaking of the conjugated system.[425] Other polycyclic hydrocarbons only react in the presence of acceptor molecules such as iodine, presumably by way of the charge-transfer complexes which are known as crystalline species. Perylene at 200 bar in the presence of iodine forms a dimer with a cage structure,[426] and pyrene forms an even more complex tetramer. Even benzene derivatives can be made to take part in cycloadditions: with hexachlorocyclopentadiene unique 2:1 adducts[403] are

35

obtained. Some dimer of the diene is also formed. Cyclooctatetraene will dimerize at 12 kbar in two formally 'forbidden modes'.[404]

4.3.2 Steric Hindrance

Pressure may be able to overcome steric factors preventing reaction at normal conditions. The quaternization of 2,6-diakylpyridines is very greatly accelerated by pressure and will take place readily at 10 kbar.[176] Other potentially useful

heterocyclic syntheses include the formation of oxadiazoles,[446]

and dihydrotriazines[447]

$$RCHO + 3R'CN + NH_3 \xrightarrow{\geq 2\,kbar}$$

Self-catalysed esterification and hydrolysis of esters can be induced by pressure making this an attractive proposition for hindered or acid-sensitive compounds[405]

$$Me_3C-COOH + EtOH \xrightarrow[kbar]{8-20} Me_3C-COOEt + H_2O$$

4.3.3 Additions

Alcohols add to cyanides to give iminoethers, the equilibrium greatly favouring product at high pressure. In the presence of an amine, these in turn are converted to amidines or can trimerize to sym-triazines.[406]

Pressure will also promote Michael additions of alcohols to α,β-unsaturated acids.[407]

$$Me_2CHOH + MeCH{=}CHCOOR \rightarrow MeCH{-}CH_2COOR$$
$$\overset{|}{\underset{OCHMe_2}{}}$$

4.3.4 Heterogeneous Catalysis

Adsorption of a reagent on a solid surface should be assisted by pressure although as yet little is known of this effect in purely liquid systems. The isomerization of m-xylene to the *para* isomer over a zeolite catalyst is accelerated some 5-fold at 5 kbar,[408] with a pseudo-volume of activation of $-40\,\text{cm}^3\,\text{mol}^{-1}$. This reaction probably proceeds by coordination of a cationic site on the catalyst to the aromatic ring:

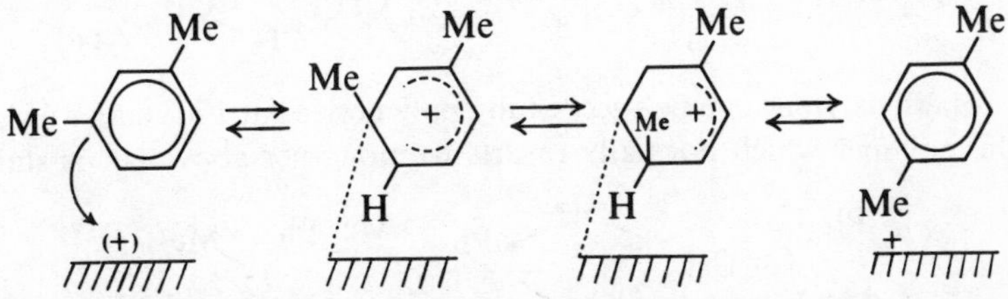

Alkane isomerization over an aluminium halide catalyst is also promoted by pressure,[409] e.g. n-pentane $\rightarrow$ isopentane. There is an unusual rate-pressure curve, a maximum appearing at 3–4 kbar. Evidently at least two processes are affected by pressure and in opposite directions though a satisfactory explanation is lacking.

4.3.5 Polymerization

Polymerizations are normally accelerated by pressure and many unusual products may be obtained.[424] Copolymers are sometimes formed when at atmospheric pressure only homopolymers are produced, e.g. between styrene and phenyl isocyanate.[410] Cyanides polymerize under somewhat extreme conditions

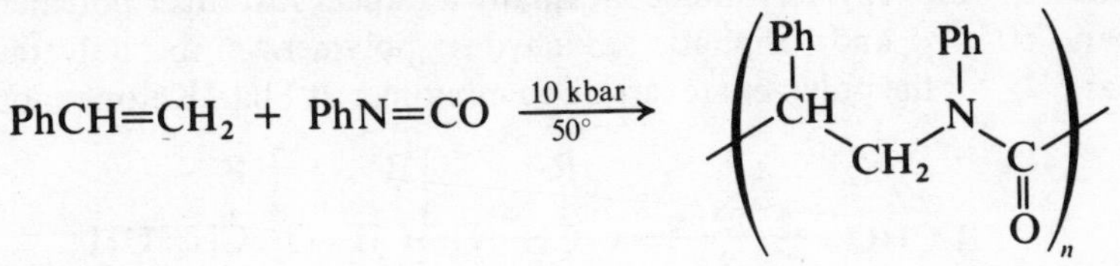

$$PhCH{=}CH_2 + PhN{=}CO \xrightarrow[50°]{10\,\text{kbar}} \left(\begin{array}{c} \overset{Ph}{\underset{|}{}} \quad \overset{Ph}{\underset{|}{}} \\ CH{-}\underset{CH_2}{} {-} N{-}C \\ \underset{O}{\overset{\|}{}} \end{array}\right)_n$$

to give polyimides,

$$nRCN \xrightarrow[>30°\,\text{kbar}]{25°} ..\overset{R}{\underset{|}{C}}{=}N{-}\overset{R}{\underset{|}{C}}{=}N{-}\overset{R}{\underset{|}{C}}{=}N{-}$$

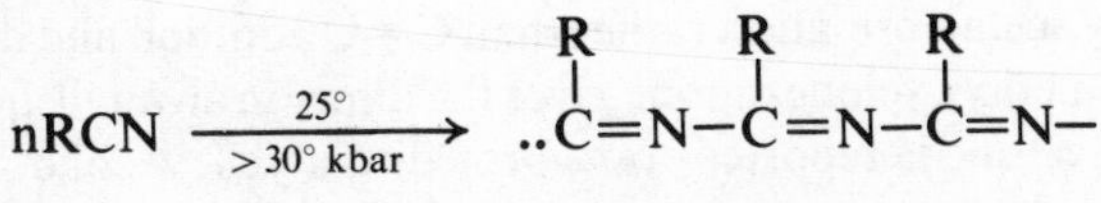

and so even will CS_2 and 2-butyne when sufficiently forced:[411]

$$CS_2 \xrightarrow[110°]{24\text{ kbar}} ..C-S-\overset{\overset{\displaystyle S}{\|}}{C}-S-\overset{\overset{\displaystyle S}{\|}}{C}-S..$$

$$MeC\equiv CMe \xrightarrow[\text{kbar}]{24}$$

Allylic monomers, usually resistant to radical chain polymerization, will do so, e.g. allyl acetate[412] or phthalate[417]

$$CH_2=CH-CH_2OAc \xrightarrow[80°]{8\text{ kbar}} ..\overset{\overset{\displaystyle CH_2OAc}{|}}{CH}\diagdown_{CH_2}\diagup\overset{\overset{\displaystyle CH_2OAc}{|}}{CH}\diagdown_{CH_2}..$$

while α-methylstyrene can be successfully polymerized at 170° under 65 kbar[413] and the polymer which normally reverts to monomer above 60° is stable. Of

$$\underset{Me}{\overset{Ph}{\diagdown}}C=CH_2 \underset{1\text{bar}}{\overset{6\text{ kbar}}{\rightleftharpoons}} ..\diagup\overset{\overset{\displaystyle Ph}{|}}{\underset{|}{C}}\underset{CH_2}{\diagdown}\overset{Me}{}\overset{Ph}{\underset{|}{C}}\underset{CH_2}{\diagdown}\overset{Me}{}..$$

course, the major example of polymerization at high pressure on a large scale is of ethylene and propylene. In the radical chain processes, moderate pressures, 1.5–3 kbar, are used at 150–200°. Since these are reactions of gases, the effect of pressure is mainly to increase the effective concentration of monomer hence these examples will not be further discussed. Among other monomers which yield high molecular weight polymers only under pressure are cinnamate esters, $PhCH=CHCOOR$,[414] and fumarate esters ($trans$-$ROOC\cdot C=CHCOOR$), 1,2-dichloroethylene,[415] indene,[416] acenaphthylene, maleimide[454] and copolymers of maleic anhydride and alkenes[328]. On the other hand, cyclohexene, 1,2-dibromoethylene, diethyl maleate, and stilbene are reported as not polymerizing under pressure so the outcome of a high pressure experiment is difficult to forecast. N,N-diphenylacrylamide, normally a radical scavenger polymerizes at 15 kbar, 60°[207] and aliphatic aldehydes polymerize to polyethers at 10 kbar,[418-420] the polymers reverting to monomer at 1 bar. Ketones appear to

$$n\text{R.CHO} \xrightarrow[-60\text{ to }+25°]{8-10\text{ kbar}} \overset{\overset{\displaystyle R}{|}}{CH}-O\left[\overset{\overset{\displaystyle R}{|}}{CH}-O\right]\overset{\overset{\displaystyle R}{|}}{CH}-OH$$

$$\Delta\overline{V}^{\ddagger} \sim -15\text{ cm}^3\text{ mol}^{-1}$$

polymerize by acetal formation rather than $C=O$ addition and dehydration may follow. Thus, methylcyclohexanone gives the dimer in an anvil apparatus and the trimer also. Acetone is reported transformed to a yellow solid at 50 kbar, 320°, which may be of the same type,[422] though polymeric. Solid Lewis acid catalysts may be advantageously used for these reactions but have been little explored.[423] Polymerization of aldehydes shows none of the characteristics of a radical chain reaction but appears to be catalysed by acid[458,459] and by phenols. The polymer may be stabilized by acylation of the hemiacetal end group.

4.3.6 Inorganic Preparations

The induction of phase change by pressure has been previously mentioned, section 2.3.3. Unusual crystalline modifications of elements and compounds may often be prepared by entering the appropriate region of the phase diagram then cooling so that reversion from the now metastable state is sufficiently slow. Black phosphorus, diamond, dense forms of ice, etc. may be prepared in this way. Numerous successful inorganic reactions require high pressure, such as the transformations of normally inert halides, SF_6, PF_3.

$$SF_6 + 2H_2O \xrightarrow[4\,kbar]{500°} SO_2F_2 + 4HF$$

$$PF_3 \begin{cases} +CO_2 \rightarrow POF_3 + CO \\ +SO_2 \rightarrow POF_3 + S \quad\;\; 300\text{–}500° \\ +CS_2 \rightarrow PSF_3 + C \quad\;\; 0.5 \text{ to } 2\,kbar \\ +O_2 \rightarrow POF_3 \end{cases}$$

Carbon–silicon bonds may be cleaved,

$$Me_4Si + N_2O \longrightarrow Me_3Si\text{–}O\text{–}SiMe_3 + N_2 + C_2H_6$$

and metal carbonyls formed from the oxides and CO;

$$MoO_3 + 9CO \xrightarrow[4\,kbar]{60°} Mo(CO)_6 + 3CO_3$$

On the other hand, all these examples are of processes in which at least one component is gaseous and therefore the effect of pressure may be simply to increase its concentration. The hydrolyses of some metal-silicon bonds take place homogeneously and proceed in good yield with pressure promoting a reaction at temperatures below the decomposition points of the products,[427]

$$Cl_3SiCo(CO)_4 + HBr \xrightarrow[4\,kbar]{90°} Cl_3SiBr + HCo(CO)_4$$

Many compounds of type ABO_3 cannot be prepared at atmospheric pressure since cation sizes do not fit the lattice. They may be synthesized under pressure when constrained to fit some other crystalline form, and are then stable at ordinary pressure.

$$CaO + CrO_2 \xrightarrow[700°]{65\,kbar} CaCrO_3$$

High oxygen pressure (supplied by CrO_3 or H_2O in separate compartments) is necessary for the synthesis of calcium ferrite from the oxygen-deficient brownmillerite; numerous other solid state syntheses have been reviewed.[450] A further

$$Ca_2Fe_2O_5 + (O) \longrightarrow 2CaFeO_3$$

aspect of high pressure synthesis is in increased solubility achieved permitting reagents to combine or to crystallize. This is the basis of the hydrothermal growth of various industrial crystals, particularly quartz which forms large crystals from amorphous material in aqueous alkali at 400°, 1.5 kbar. A great many very sparingly soluble oxides sulphides and salts may be similarly crystallized[451] from appropriate media.

4.3.7 Explosions

Compression of many materials at pressures of tens or hundreds of kbar has been found to result in explosions. This may be the result of a runaway exothermic reaction, frequently polymerization, or it may be the result of an exothermic compound undergoing a spontaneous change. Thus sugar[428] and nitromethane[429] have been shown to detonate on compression. It is likely that the work of compression, $\int P\,dV$, can become sufficiently high and be available in the decomposition mode as to lead to reaction, heat evolution then building up an explosive process. Detonation is more likely to occur if the material is subjected to shear as well as compressive stress. Bridgeman[430] reported the following to detonate on compression in a special anvil apparatus under shear forces; nitro-cellulose, CHI_3, PbO_2, Ag_2O, $KMnO_4$, $(NH_4)_2Cr_2O_7$, $AgNO_3$, $Sr(NO_3)_2$, and even $Al_2(SO_4)_3$.

REFERENCES

1. C. H. Bamford and C. F. H. Tipper (Eds.), *Comprehensive Chemical Kinetics*, Vol. 1, Van Nostrand, New York (1969).
2. H. Eyring and J. Polanyi, *Z. phys. Chem.*, **12B**, 279 (1931).
3. H. Eyring, *J. Chem. Phys.*, **3**, 107 (1934).

4. K. Weale, *Chemical Reactions at High Pressures*, Spon, London (1967).
5. M. G. Gonikberg, *Chemical Equilibria and Reaction Rates at High Pressure*, English trans., N. S. F., Washington (1963).
6. S. D. Hamann, *Physico-Chemical Aspects of Pressure*, Chapter 9, Butterworth, London (1957).
7. C. Eckert, *Rep. Prog. Phys. Chem.*, **23**, 239 (1972).
8. S. D. Hamann, *High Pressure Physics and Chemistry*, R. S. Bradley (Ed.), Academic Press, London (1963).
9. E. Whalley, *Adv. Phys. Org. Chem.*, **2**, 93 (1964).
10. U. Gaarz and H-D. Ludemann, *Ber. Bunsenges. Phys. Chem.*, **80**, 607 (1976).
11. H-D. Ludemann, R. Rauchschwalbe, and E. Lang, *Angew. Chem. Int. Ed.*, **16**, 331 (1977).
12. H. Pleininger and H-O. Schnelle, *Tetrahedron*, **33**, 1197 (1977).
13. F. H. Westheimer, *J. Chem. Phys.*, **15**, 252 (1947).
14. R. Donald, D. R. McKelvey, and K. R. Brower, *J. Phys. Chem.*, **64**, 1958 (1960).
15. K. R. Brower and F. L. Wu, *J. Amer. Chem. Soc.*, **92**, 5303 (1970).
16. C. C. McCune, F. W. Cagle, and S. S. Kistler, *J. Phys. Chem.*, **64**, 1773 (1960).
17. C. Walling and G. Metzger, *J. Amer. Chem. Soc.*, **81**, 5365 (1959).
18. A. E. Nicholson and R. G. W. Norrish, *Disc. Farad. Soc.*, **22**, 97 (1956).
19. C. Walling and J. Pellon, *J. Amer. Chem. Soc.*, **79**, 4786 (1957).
20. R. C. Neuman and J. V. Behar, *J. Amer. Chem. Soc.*, **91**, 6024 (1969).
21. R. C. Neuman and R. P. Pankratz, *J. Amer. Chem. Soc.*, **95**, 8372 (1973).
22. R. C. Neuman and J. V. Behar, *J. Org. Chem.*, **36**, 654 (1971).
23. R. C. Neuman and J. V. Behar, *J. Amer. Chem. Soc.*, **89**, 4549 (1961).
24. R. C. Neuman and G. D. Holmes, *J. Amer. Chem. Soc.*, **93**, 4242 (1971).
25. R. C. Neuman and R. Wolfe, *J. Org. Chem.*, **40**, 3147 (1975).
26. R. C. Neuman, J. D. Lockyer, and M. J. Amrich, *Tet. Lett.*, 1221, (1972).
27. A. H. Ewald, *Disc. Farad Soc.*, **22**, 138 (1956).
28. R. C. Neuman and R. J. Bussey, *Tet. Lett.*, 5859 (1968).
29. R. C. Neuman and R. J. Bussey, *J. Amer. Chem. Soc.*, **92**, 2440 (1970).
30. R. C. Neuman and E. W. Ertley, *J. Amer. Chem. Soc.*, **97**, 3130 (1975).
31. R. C. Neuman and M. J. Amrich, *J. Amer. Chem. Soc.*, **94**, 2730 (1972).
32. R. C. Neuman and E. W. Ertley, *Tet. Lett.*, 1225 (1972).
33. R. C. Neuman, *Acc. Chem. Res.*, **5**, 138 (1972).
34. C. Walling and H. P. Waits, *J. Phys. Chem.*, **71**, 2361 (1967).
35. E. Ishihara, Y. Ogo, and T. Imoto, *Proc. 4th AIRAPT Conf.*, *Kyoto* (1973).
36. R. C. Neuman and J. V. Behar, *Tet. Lett.*, 3281 (1968).
37. C. Walling and J. Peisach, *J. Amer. Chem. Soc.*, **80**, 5819 (1958).
38. J. R. McCabe and C. A. Eckert, *Ind. Eng. Chem. Fundam.*, **13**, 168 (1973).
39. R. A. Grieger and C. A. Eckert, *Trans. Farad. Soc.*, **66**, 2579 (1970).
40. K. Seguchi, A. Sera, and K. Murayama, *Bull. Chem. Soc. Jap.*, **47**, 2242 (1974).
41. G. Swieton and H. Kelm, *J. Chem. Soc.*, *Perkin II*, 769 (1976).
42. C. Brun and G. Jenner, *Tetrahedron*, **28**, 3113 (1972).
43. C. Walling and H. Shugar, *J. Amer. Chem. Soc.*, **85**, 607 (1963).
44. B. Raistrick, R. Sapiro, and D. Newitt, *J. Chem. Soc.*, 1761, (1939).
45. R. B. Woodward and R. Hoffmann, *The Conservation of Orbital Symmetry*, Verlag-Chemie, Weinheim (1970).
46. R. B. Woodward and T. Katz, *Tetrahedron*, **5**, 70 (1959).
47. M. C. Koetzel, *Org. Reactions*, **4**, 1 (1948).
48. J. B. Lambert and J. D. Roberts, *Tet. Lett.*, 1457 (1965).
49. M. Charton, *J. Org. Chem.*, **31**, 3745 (1966).
50. S. Seltger, *J. Amer. Chem. Soc.*, **87**, 1534 (1965).
51. J. Sauer, H. Weist, and A. Mielert, *Chem. Ber.*, **97**, 3183 (1964).
52. P. Brown, *Thesis*, University of Southampton (1964).

344

53. J. Berson, Z. Hamlet, and W. A. Mueller, *J. Amer. Chem. Soc.*, **84**, 297 (1962).
54. K. Seguchi, A. Sera, and K. Maruyama, *Tet. Lett.*, 1585 (1973).
55. R. A. Grieger and C. A. Eckert, *J. Amer. Chem. Soc.*, **92**, 2918 (1970).
56. R. A. Grieger and C. A. Eckert, *Ind. Eng. Chem. Fundam.*, **10**, 369 (1971).
57. J. Rimmelin and G. Jenner, *Tetrahedron*, **30**, 308 (1974).
58. R. A. Grieger and C. A. Eckert, *J. Amer. Chem. Soc.*, **92**, 7149 (1970).
59. C. Reichardt, *Angew. Chem. Int. Ed.*, **4**, 29 (1962).
60. G. Hammond, *J. Amer. Chem. Soc.*, **77**, 334 (1955).
61. J. Rimmelin, G. Jenner, and H. Abdi-Osouki, *Bull. Soc. Chim. France*, 341 (1977).
62. W. J. LeNoble and B. A. Ojosipe, *J. Amer. Chem. Soc.*, **97**, 5939 (1975).
63. R. R. Schmidt, *Angew. Chem. Int. Ed.*, **12**, 212 (1973).
64. A. Padwa, J. Masoraccia, C. L. Osborne, and D. J. Trecker, *J. Amer. Chem. Soc.*, **93**, 3633 (1971).
65. R. Huisgen and G. Steiner, *Tet. Lett.*, 3763, 3769(1973); *J. Amer. Chem. Soc.*, **95**, 4728 (1973).
66. T. L. Jacobs and O. J. Muscio, *Tet. Lett.*, 4829 (1970).
67. R. B. Woodward and R. Hoffmann, *The Conservation of Orbital Symmetry*, Academic Press, Weinheim (1970).
68. G. Sweiton, J. von Jouanne, and H. Kelm, *Proc. 4th AIRAPT Conf. (Kyoto)*, 652 (1975).
69. F. K. Fleischmann and H. Kelm, *Tet. Lett.*, 3773 (1973).
70. N. S. Isaacs and E. Rannala, *J. Chem. Soc., Perkin II*, 1555 (1975).
71. H. J. Friedrich, *Tet. Lett.*, 2981 (1971).
72. J. E. Baldwin and J. A. Kopecki, *J. Amer. Chem. Soc.*, **92**, 4868 (1970).
73. J. E. Baldwin and J. A. Kopecki, *J. Amer. Chem. Soc.*, **91**, 3106 (1969).
74. S-H. Dai and W. R. Dolbier, *J. Amer. Chem. Soc.*, **94**, 3946 (1972).
75. E. K. v. Gustorf, D. V. Wuth, J. Leitich, and D. Henneberg, *Tet. Lett.*, 3113 (1969).
76. R. Montaigne and L. Ghosez, *Angew. Chem. Int. Ed.*, **7**, 221 (1968).
77. N. S. Isaacs and P. Stanbury, *Chem. Comm.*, 1071 (1970).
78. N. S. Isaacs and P. Stanbury, *J. Chem. Soc., Perkin II*, 166 (1973).
79. R. Huisgen and M. Otto, *J. Amer. Chem. Soc.*, **90**, 5922 (1969).
80. G. B. Gill and M. R. Willis, *Pericyclic Reactions*, Chapman and Hall, London (1974).
81. R. Huisgen, *Angew. Chem. Int. Ed.*, **2**, 565 (1963).
82. N. V. Sidgwick and H. D. Springall, *J. Chem. Soc.*, 1532 (1936).
83. G. A. Lawrence and D. R. Stranks, *Inorg. Chem.*, **17**, 1804 (1978).
84. K. Seguchi, A. Sera, and K. Maruyama, *Tet. Lett.*, 1585 (1973).
85. B. S. El Yanov, E. I. Klabunovskii, M. G. Gonikberg, G. M. Porfenova, and L. F. Godunova, *Bull. Acad. Sci. SSSR Ser. Khim.*, 557 (1971).
86. C. Walling and M. Naiman, *J. Amer. Chem. Soc.*, **84**, 2628 (1962).
87. K. R. Brower, *J. Amer. Chem. Soc.*, **83**, 4370 (1961).
88. W. J. LeNoble, *J. Phys. Chem.*, **67**, 2451 (1963).
89. S. Patai (Ed.), *Chemistry of the Ether Linkage*, Academic Press, New York, 635 (1967).
90. D. S. Tarbell, *Org. Rns.*, **2**, 1 (1938); *Chem. Rev.*, **27**, 495 (1940).
91. S. J. Rhoades, *Molecular Rearrangements*, P. deMayo (Ed.), Wiley–Interscience, New York (1969).
92. W. v. E. Doering and W. R. Roth, *Angew. Chem. Int. Ed.*, **2**, 115 (1963).
93. A. Gagneux, S. Winstein, and W. G. Young, *J. Amer. Chem. Soc.*, 5956 (1960).
94. R. P. Bell, *The Proton in Chemistry*, Chapman and Hall, London (1973).
95. K. G. Liphard and A. Jost, *Ber. Bunsenges. Phys. Chem.*, **80**, 125 (1976).
96. C. D. Hubbard, C. J. Wilson, and E. F. Caldin, *J. Amer. Chem. Soc.*, **98**, 1870 (1976).
97. R. Rauchschwalbe, G. Völkel, E. Lang, and H-D. Lüdemann, *J. Chem. Res.*, (S) 448; (M) 5325 (1978).
98. J. C. Turgeon and V. K. LaMer, *J. Amer. Chem. Soc.*, **74**, 9588 (1952).
99. D. T. Y. Chen and K. J. Laidler, *Can. J. Chem.*, **37**, 599 (1959).

100. N. S. Isaacs, unpublished data.
101. R. P. Bell, *Tunnel Effect in Chemistry*, Chapman and Hall, London (1980).
102. R. A. More O'Ferrall, *Adv. Phys. Org. Chem.*, **5**, 331 (1967).
103. N. S. Isaacs and E. Rannala, *Tet. Lett.*, 2039 (1977).
104. C. K. Ingold, *Structure and Mechanism in Organic Chemistry*, Bell, London (1969).
105. S. Winstein, E. Clippling, A. H. Fainberg, R. Heck, and G. C. Robinson, *J. Amer. Chem. Soc.*, **78**, 328 (1956).
106. R. A. Sneen, *Acc. Chem. Res.*, **6**, 46 (1973).
107. M. Szwarc (Ed.), *Ions and Ion-Pairs in Organic Reactions*, Vol. 2, Wiley–Interscience, New York (1974).
108. C. Yamagami, A. Sera, and K. Maruyama, *Bull. Chem. Soc. Jap.*, **47**, 881 (1974).
109. A. Sera, C. Yamagami, and K. Marutama, *Bull. Chem. Soc. Jap.* **46**, 3864 (1973).
110. D. L. Gay and E. Whalley, *Can. J. Chem.*, **48**, 2021 (1970).
111. M. J. Mackinnon, A. B. Lateef, and J. B. Hyne, *Can. J. Chem.*, **48**, 2025 (1970).
112. S. J. Dickinson and J. B. Hyne, *Can. J. Chem.*, **49**, 2394 (1971).
113. D. D. Macdonald and J. B. Hyne, *Can. J. Chem.*, **48**, 2494 (1970).
114. D. Buttner and H. Heydtmann, *Ber. Bunsenges. Phys. Chem.*, **48**, 2025 (1970).
115. A. Sera, T. Miyazawa, T. Madsuda, Y. Togawa, and K. Marayama, *Bull. Chem. Soc. Jap.*, **46**, 3490 (1973).
116. H. C. Brown and Y. Okamoto, *J. Amer. Chem. Soc.*, **80**, 4979 (1958).
117. G. Hills and C. A. N. Viana, *Nature*, **229**, 194 (1971).
118. J. Osugi, M. Sasaki, *et al.*, *Proc. 6th AIRAPT Conf.* (Boulder), 651, Plenum, New York (1979).
119. G. A. Olah and P. v. R. Schleyer, *Carbonium Ions*, Wiley–Interscience (1968–70).
120. K. Kohnstam, *Chem. Comm.*, 1032 (1971).
121. W. J. LeNoble, H. Guggisberg, T. Asano, L. Cho, and G. Grob., *J. Amer. Chem. Soc.*, **98**, 920 (1976).
122. A. Sera, N. Tachikawa, and K. Maruyama, *Proc. 4th AIRAPT Conf.* (*Kyoto*), **648** (1974).
123. B. Capon (Ed.), *Neighbouring Group Participation*, Plenum, London (1976).
124. W. J. LeNoble and B. L. Yates, *J. Amer. Chem. Soc.*, **87**, 3515 (1965).
125. W. J. LeNoble, B. L. Yates, and A. W. Scaplehorn, *J. Amer. Chem. Soc.*, **89**, 3761 (1967).
126. C. J. Lancelot, J. J. Hepner, and P. v. R. Schleyer, *J. Amer. Chem. Soc.*, **91**, 4294 (1969).
127. A. Sera, C. Yamagami, and K. Maruyama, *Bull. Chem. Soc. Jap.*, **47**, 704 (1974).
128. W. J. LeNoble and B. Gabrielson, *Tet. Lett.*, 45 (1970).
129. W. J. LeNoble and A. Shurpik, *J. Org. Chem.*, **35**, 3588 (1970).
130. G. A. Olah and P. v R. Schleyer, *Carbonium Ions*, Interscience, (1972).
131. N. Takisawa, M. Sasaki, and F. Amita, *Chem. Lett.*, 671 (1979).
132. A. H. Ewald and D. J. Ottley, *Austr. J. Chem.*, **20**, 1335 (1967).
133. B. T. Baliga and E. Whalley, *J. Phys. Chem.*, **73**, 654 (1969).
134. B. T. Baliga and E. Whalley, *Can. J. Chem.*, **48**, 528 (1970).
135. C. S. Davis and J. B. Hyne, *Can. J. Chem.*, **51**, 1687 (1973).
136. A. B. Lateef and J. B. Hyne, *Can. J. Chem.*, **47**, 1369 (1969).
137. H. S. Golinkin, I. Lee, and J. B. Hyne, *J. Amer. Chem. Soc.*, **89**, 1307 (1967).
138. J. K. Laidler and R. Martin, *Int. J. Chem. Kinetics*, **1**, 113 (1969).
139. S. Hariya and S. Terasawa, *Nippon Kagaku Zasshi*, **90**, 765 (1969).
140. W. J. LeNoble and B. L. Yates, *Abs. 150th Amer. Chem. Soc. Meeting*, Atlantic City, N. J., September (1965).
141. W. J. LeNoble and B. L. Yates, *J. Amer. Chem. Soc.*, **87**, 3515 (1965).
142. G. J. Hills and C. A. Viani, *Hydrogen-Bonded Solvent Systems*, A. K. Covington and P. Jones (Eds.), Taylor and Francis, London (1968).
143. R. K. Williams, J. J. Loveday, and A. K. Colter, *Can. J. Chem.*, **50**, 1303 (1972).
144. C. S. Davis and J. B. Hyne, *Can. J. Chem.*, **50**, 2270 (1972).

145. K. E. Newman, F. Meyer, and A. Merbach, *J. Amer. Chem. Soc.*, **101**, 1470 (1979).
146. E. Grunwald and S. Winstein, *J. Amer. Chem. Soc.*, **70**, 846 (1948).
147. W. J. LeNoble and R. Daka, *J. Amer. Chem. Soc.*, **100**, 5962 (1978).
148. M. R. Ward, *Acc. Chem. Res.*, **5**, 18 (1972); R. G. Lawler, *Acc. Chem. Res.*, **5**, 25 (1972).
149. K. R. Brower and D. Hughes, *J. Amer. Chem. Soc.*, **100**, 7591 (1978).
150. W. J. LeNoble, *J. Amer. Chem. Soc.*, **87**, 1945 (1961).
151. D. Cram, *Fundamentals of Carbanion Chemistry*, Academic Press, New York (1972).
152. Z. Margolin and F. A. Long, *J. Amer. Chem. Soc.*, **95**, 2757 (1973).
153. N. S. Isaacs and K. Javaid, *J. Chem. Soc., Perkin II*, 1392 (1975).
154. E. S. Lewis *et al.*, *J. Amer. Chem. Soc.*, **89**, 2322, 4337 (1967); R. P. Bell and D. M. Goodall, *Proc. Roy. Soc.*, **294A**, 273 (1966).
155. C. G. Swain, *J. Amer. Chem. Soc.*, **83**, 1945 (1961).
156. N. S. Isaacs, K. Javaid, and E. Rannala, *J. Chem. Soc., Perkin II*, 709 (1978).
157. R. O. Gibson, E. W. Fawcett, and M. W. Perrini, *Proc. Roy. Soc.*, **A150**, 223 (1935).
158. M. G. Gonikberg and B. S. El'yanov, *Izv. Akad. Nauk SSSR Otdel Khim. Nauk.*, 384 (1953).
159. K. E. Weale, *J. Chem. Soc.*, 2959 (1954); *Disc. Farad. Soc.*, **22**, 122 (1956).
160. K. R. Brower, *J. Amer. Chem. Soc.*, **85**, 1401 (1963).
161. E. S. Amis, *Solvent Effects on Reaction Rates and Mechanisms*, Academic Press, New York (1966).
162. J. G. Kirkwood, *J. Chem. Phys.*, **2**, 351 (1934).
163. L. Onsager, *J. Amer. Chem. Soc.*, **58**, 1486 (1936).
164. C. Lassau and J. C. Jungers, *Bull. Soc. Chim. France*, 2678 (1968).
165. Y. Drougard and D. DeCroocq, *Bull. Soc. Chim. France*, 2972 (1969).
166. H. Heydtmann, A. P. Schmidt, and H. Hartmann, *Ber. Bunsenges Phys. Chem.*, **70**, 444 (1966).
167. H. Heydtmann, *Z. für phys. Chem. N.F.*, **54**, 237 (1967).
168. S. D. Hamann, *Austr. J. Chem.*, **28**, 693 (1975).
169. M. G. Gonikberg *et al.*, *Zh. Fiz. Khim.*, **30**, 784 (1956).
170. M. G. Gonikberg *et al.*, *Ivz. Akad. Nauk. SSSR Otdel Khim, Nauk.*, 2103 (1960).
171. M. G. Gonikberg *et al.*, *Doklady Akad. Nauk. SSSR*, **128**, 759 (1959).
172. H. Hartmann, H. D. Brauer, H. Kelm, and G. Rinck, *Z. phys. Chem. N.F.*, **61**, 53 (1968).
173. Y. Kondo, M. Uchida, and N. Tokura, *Bull. Chem. Soc., Jap.*, **41**, 992 (1968).
174. M. Gielen and J. Nasielski, *Rec. Trav. Chim.*, **82**, 228 (1963); *J. Organomet. Chem.*, **1**, 173 (1964).
175. Y. Kondo, M. Ohnishi, and N. Tokura, *Bull. Chem. Soc. Jap.*, **45**, 3579 (1972).
176. W. J. LeNoble and Y. Ogo, *Tetrahedron*, **26**, 4119 (1970).
177. B. S. El'yanov and M. G. Gonikberg, *Bull. Acad. Sci. USSR, Chem. Ser.* (Eng. trans.), 1007 (1967).
178. S. D. Hamann, *High Pressure Physics and Chemistry*, Vol. 2, R. S. Bradley (Ed.), Academic Press, New York (1963).
179. K. E. Weale, *Chemical Reactions at High Pressure*, Chapter 6, Spon, London (1967).
180. M. Okamoto, M. Sasaki, and J. Osugi, *Rev. Phys. Chem. Jap.*, **47**, 33 (1977).
181. H. Hartmann, H-D. Brauer, and G. Rinck, *Z. physik. Chem. N.F.*, **61**, 47 (1968).
182. Y. Kondo, H. Tojima, and N. Tokura, *Bull. Chem. Soc., Jap.*, **40**, 1408 (1967).
183. H-D. Brauer and H. Kelm, *Z. physik. Chem. (Frankfurt)*, **79**, 98 (1972).
184. M. L. Tonnet and A. N. Hambley, *Austr. J. Chem.*, **23**, 2435 (1970).
185. H. Tiltscher and Y. K. Wang, *Z. Physik. Chem. (Frankfurt)*, **90**, 299 (1976).
186. H. Heydtmann and D. Buttner, *Z. Physik. Chem. (Frankfurt)*, **63**, 79 (1969).
187. Y. Kondo, M. Onishi, and N. Tokura, *Bull. Chem. Soc., Jap.*, **45**, 3579 (1972).
188. S. Arakawa, H. Itsuki, and S. Terasawa, *Koatsu Gasu*, **11**, 632 (1974).
189. S. Arakawa, S. Hariya, H. Itsuki, and S. Terasawa, *Nippon Kagaku Zasshi*, 1170 (1974).

347

190. A. H. Ewald and D. J. Otley, *Austr. J. Chem.*, **20**, 1335 (1967).
191. J. M. Stewart and K. E. Weale, *J. Chem. Soc.*, 2849, 2854 (1965).
192. K. R. Brower, *J. Amer. Chem. Soc.*, **85**, 1401 (1963).
193. J. T. Edward, *J. Chem. Ed.*, **47**, 261 (1970).
194. K. R. Brower and J. S. Chen, *J. Amer. Chem. Soc.*, **87**, 3396 (1965).
195. S. L. Johnson, *Adv. Phys. Org. Chem.*, **5**, 237 (1967).
196. K. J. Laidler and D. Chen, *Trans. Farad. Soc.*, **54**, 1026 (1958).
197. B. Anderson, F. Grønland, and J. Olsen, *Acta. Chem. Scand.*, **23**, 2458 (1969).
198. M. L. Tonnet and E. Whalley, *Can. J. Chem.*, **53**, 3414 (1975).
199. H. Heydtmann and H. Stieger, *Ber. Bunsenges. Phys. Chem.*, **70**, 1095 (1966).
200. D. Büttner and H. Heydtmann, *Ber. Bunsenges. Phys. Chem.*, **73**, 640 (1969).
201. C. A. Eckert, *Proc. 6th AIRAPT Conf.* (Boulder), Plenum, New York (1979).
202. C. S. Davis and J. B. Hyne, *Can. J. Chem.*, **51**, 1687 (1973).
203. E. G. Williams, M. W. Perrin, and R. O. Gibson, *Proc. Roy. Soc.*, **154A**, 684 (1936).
204. A. L. T. Moesvelt, *Z. physik. Chem. (Leipzig)*, **105**, 455 (1923).
205. A. L. T. Moesvelt and W. A. T. de Meester, *Z. physik. Chem. (Leipzig)*, **138**, 169 (1928).
206. G. DiSabato, W. P. Jencks, and E. Whalley, *Can. J. Chem.*, **40**, 1220 (1966).
207. Y. Tanaka *et al.*, *Proc. 6th AIRAPT Conf.* (Boulder), 684, Plenum, New York (1979).
208. D. W. Coillet and S. D. Hamann, *Trans. Farad. Soc.*, **57**, 2231 (1961).
209. D. W. Coillet, S. D. Hamann, and E. T. McCoy, *Austr. J. Chem.*, **18**, 1911 (1965).
210. H. Cohn, E. D. Hughes, M. H. Jones, and M. G. Peeling, *Nature*, **169**, 291 (1952).
211. A. J. Ellis, W. S. Fyfe, R. Rutherford, A. Fischer, and J. Vaughan, *J. Chem. Phys.*, **31**, 176 (1959).
212. T. Asano, *Bull. Chem. Soc. Jap.*, **42**, 2005 (1969).
213. T. Asano, R. Goto, and A. Sera, *Bull. Chem. Soc., Jap.*, **40**, 2208 (1967).
214. D. W. Coillet and S. D. Hamann, *Nature*, **200**, 166 (1963).
215. T. Asano, A. Sera, and R. Goto, *Tet. Lett.*, 4777 (1968).
216. G. Jenner, *Nouv. J. Chim.*, **3**, 309 (1979).
217. M. G. Gonikberg and N. I. Prokhova, *Bull. Acad. Sci., USSR, Chem. Ser.*, **1154** (1964).
218. K. R. Brower, *J. Amer. Chem. Soc.*, **81**, 3504 (1959).
219. N. S. Isaacs, *Reactive Intermediates in Organic Chemistry*, John Wiley, Chichester (1974).
220. S. D. Hamann, *Ann. Rev. Phys. Chem.*, **15**, 349 (1964).
221. W. J. LeNoble, *Prog. Phys. Org. Chem.*, **5**, 207 (1967).
222. G. Kohnstam, *Prog. Reaction Kinetics*, **5**, 335 (1970).
223. T. Asano and W. J. LeNoble, *Chem. Rev.*, **407** (1978).
224. Y. Kondo, M. Uchida, and N. Yokura, *Bull. Chem. Soc. Jap.*, **41**, 992 (1968).
225. S. W. Benson and J. A. Berson, *J. Amer. Chem. Soc.*, **86**, 259 (1964).
226. H. S. Golinkin, W. G. Laidlaw, and J. B. Hyne, *Can. J. Chem.*, **44**, 2193 (1966).
227. M. G. Evans and M. Polanyi, *Trans. Farad. Soc.*, 875 (1935); 448 (1937).
228. C. Walling and H. J. Shugar, *J. Amer. Chem. Soc.*, **85**, 607 (1965).
229. W. J. LeNoble, *J. Chem. Ed.*, **44**, 729 (1967).
230. E. Whalley, *Trans. Farad. Soc.*, **58**, 2144 (1962).
231. G. Jenner, *Angew. Chem. Int. Ed.*, **14**, 137 (1975).
232. M. Born, *J. Chem. Phys.*, **7**, 591 (1939).
233. P. Mukherjee, *J. Phys. Chem.*, **65**, 740, 744 (1961).
234. K. J. Laidler and H. Eyring, *Ann. N.Y. Acad. Sc.*, **39**, 303 (1940).
235. E. Whalley, *J. Chem. Phys.*, **38**, 1400 (1963).
236. I. A. Koppel and V. A. Palm, *Adv. Linear Free Energy Relationships*, **1**, N. B. Chapman and J. Shorter, (Eds.), Plenum, New York (1972).
237. J. R. McCabe and C. A. Eckert, *Ind. Eng. Chem. Fundam.*, **13**, 168 (1974).
238. Y. Kondo, H. Tojuma, and Tokuru, *Bull. Chem. Soc. Jap.*, **40**, 1408 (1967).
239. N. S. Isaacs and K. Javaid, *Tet. Lett.*, 3073 (1977).
240. R. P. Bell, *The Proton in Chemistry*, Chapman and Hall, London (1973).

241. M. L. Bender, *J. Amer. Chem. Soc.*, **73**, 1626 (1951).
242. H. Itsuki, B. Matsuda, and S. Terasawa, *Nippon Kagaku Zassi*, **90**, 1016 (1969).
243. G. B. Purohit and S. K. Bhattacharyya, *Ind. J. Chem.*, **8**, 602 (1970).
244. J. Koskikallio and U. Turpinen, *Acta Chem. Scand.*, **25**, 3360 (1971).
245. J. Koskikallio and E. Whalley, *Trans. Farad. Soc.*, **55**, 815 (1959).
246. L. Pyy and J. Koskikallio, *Suom. Kemist.*, **B 40**, 134 (1967).
247. J. G. Pritchard and F. A. Long, *J. Amer. Chem. Soc.*, **55**, 815 (1959).
248. J. Koskikallio and E. Whalley, *Trans. Farad. Soc.*, **55**, 815 (1959).
249. J. E. Early, C. E. O'Rourke, L. B. Clapp, J. O. Edwards, and B. C. Laws, *J. Amer. Chem. Soc.*, **80**, 3458 (1958).
250. E. H. Cordes and H. G. Bull, *Chem. Rev.*, **74**, 581 (1974).
251. J. M. O'Gorman and H. V. Lucas, *J. Amer. Chem. Soc.*, **72**, 5489 (1950).
252. F. Stasink, W. A. Sheppard, and A. N. Bourns, *Can. J. Chem.*, **34**, 123 (1956).
253. B. Capon, *Chem. Rev.*, 417 (1969).
254. N. S. Isaacs, unpublished data.
255. P. J. C. Fierens and P. Vershelten, *Bull. Chem. Soc. Belges*, **61**, 427, 609, (1952).
256. K. Javaid, *Ph.D. Thesis*, University of Reading (1979).
257. B. E. C. Banks, Y. Meinwald, A. J. Rhiud-Tutt, I. Sheft, and C. A. Vernon, *J. Chem. Soc.*, 3240 (1961).
258. A. R. Osborne and E. Whalley, *Z. physik. Chem.* (*Leipzig*), **39**, 597 (1961).
259. E. Cohen and R. B. de Boer, *Z. physik. Chem.* (*Leipzig*), **84**, 41 (1913).
260. V. Rothmund, *Z. physik. Chem.* (*Leipzig*), **20**, 168 (1896).
261. A. R. Osborne and E. Whalley, *Can. J. Chem.*, **39**, 1094 (1961).
262. H. Itsuki and S. Terasawa, *Koatsu Gasu*, **5**, 427 (1968).
263. R. J. Withen, J. E. McAldulf, and E. Whalley, *Can. J. Chem.*, **37**, 1360 (1959).
264. J. Koskikallio, D. Pauli, and E. Whalley, *Can. J. Chem.*, **37**, 1360 (1959).
265. H. Itsuki and S. Terasawa, *Nippon Kagaku Zasshi*, **90**, 1119 (1969).
266. J. Koskikallio and E. Whalley, *Can. J. Chem.*, **37**, 788 (1959).
267. W. J. LeNoble and M. Duffy, *J. Phys. Chem.*, **68**, 619 (1964).
268. P. O. I. Virtanen and T. Kuokkanen, *Suom. Kemist. B*, **46**, 267 (1973).
269. R. E. Parker and N. S. Isaacs, *Chem. Rev.*, **59**, 737 (1959).
270. P. O. I. Virtanen and T. Kuokkanen, *Finn. Chem. Lett.*, 177 (1974).
271. J. Koskikkalio and E. Whalley, *Trans. Farad. Soc.*, **55**, 809 (1959).
272. B. T. Baliga, A. K. Rantamaa, and E. Whalley, *J. Phys. Chem.*, **69**, 1751 (1965).
273. R. J. Withey and E. Whalley, *Trans. Farad. Soc.*, **59**, 901 (1963).
274. E. Whalley, *Trans. Farad. Soc.*, **55**, 798 (1959).
275. R. J. Withey and E. Whalley, *Can. J. Chem.*, **41**, 849 (1963).
276. A. R. Osborne and E. Whalley, *Can. J. Chem.*, **39**, 597 (1961).
277. D. M. Newitt, R. P. Linstead, R. H. Spiro, and E. J. Boarman, *J. Chem. Soc.*, 876 (1937).
278. Peng Shin-Lin, R. H. Sapiro, R. P. Linstead, and D. M. Newitt, *J. Chem. Soc.*, 784 (1938).
279. M. Linton, *Proc. 4th AIRAPT Conf.* (*Kyoto*), 671 (1974).
280. S. K. Bhattacharyya and C. K. Das, *J. Amer. Chem. Soc.*, **91**, 6715 (1969).
281. T. Moriyoshi, *Bull. Chem. Soc. Jap.*, **44**, 2582 (1971).
282. S. K. Bhattacharyya and G. B. Purohit, *J. Phys. Chem.*, **73**, 3278 (1969).
283. B. T. Baliga and E. Whalley, *Can. J. Chem.*, **42**, 1019 (1964).
284. B. T. Baliga and E. Whalley, *Can. J. Chem.*, **43**, 2453 (1965).
285. S. K. Bhattacharyya, F. N. I. Purohit, and G. B. Purohit, *Proc. Ind. Nat. Acad.*, **36A**, 154 (1970).
286. H. Tokaya, N. Tudo, T. Hosoya, and T. Minegishi, *Bull. Chem. Soc., Jap.*, **44**, 1175 (1971).
287. J. J. Scott and K. R. Brower, *J. Amer. Chem. Soc.*, **89**, 2682 (1967).
288. J. F. Bunnett, *J. Amer. Chem. Soc.*, **83**, 4956 (1961).

289. Y. Pocker, *Proc. Chem. Soc.*, 17 (1960).
290. K. Tamura and T. Moriyoshi, *Bull. Chem. Soc. Jap.*, **47**, 2942 (1974).
291. M. H. Abraham, *Comprehensive Chemical Kinetics*, C. H. Bamford and C. F. H. Tipper (Eds.), Vol. 12, Elsevier, Amsterdam (1972).
292. N. S. Isaacs and K. Javaid, *Tet. Lett.*, 3073 (1977).
293. W. H. Saunders and A. F. Cockerill, *Elimination Reactions*, Wiley–Interscience, New York (1973).
294. C. J. M. Stirling, *Essays in Chemistry*, **5**, 123 (1973).
295. F. G. Bordwell, *Acc. Chem. Res.*, **5**, 374 (1972).
296. D. J. McLennon, *Tetrahedron*, **31**, 2999 (1975).
297. Y. Okomoto and T. Yama, *Tet. Lett.*, 493 (1972).
298. K. R. Brower, M. Muhsin, and H. E. Brower, *J. Amer. Chem. Soc.*, **98**, 779 (1976).
299. W. H. Saunders and A. F. Cockerill, *Mechanisms of Elimination Reactions*, Wiley–Interscience, New York (1973).
300. F. Voegtle and P. Neumann, *Chem. Ztg.*, **97**, 600 (1973).
301. K. R. Brower, *J. Amer. Chem. Soc.*, **82**, 4535 (1960).
302. Reference 221; see reference 159 therein.
303. K. R. Brower, B. Gay, and T. L. Konkol, *J. Amer. Chem. Soc.*, **88**, 1681 (1966).
304. E. M. Kosower, *Molecular Biochemistry*, McGraw-Hill, New York (1962).
305. F. H. Westheimer and W. A. Jones, *J. Amer. Chem. Soc.*, **63**, 3283 (1964); G. Frankel, R. L. Bedford, and P. E. Yankwich, *J. Amer. Chem. Soc.*, **76**, 15 (1954).
306. S. D. Hamann and D. R. Teplitsky, *Disc. Farad. Soc.*, **22**, 114 (1956).
307. N. Kornblum, P. Berrigan, and W. J. LeNoble, *J. Amer. Chem. Soc.*, **85**, 1141 (1963).
308. K. R. Brower, R. L. Ernst, and J. S. Chen, *J. Phys. Chem.*, **68**, 3814 (1964).
309. W. J. LeNoble, *J. Amer. Chem. Soc.*, **85**, 1470 (1963).
310. C. Walling, *Free Radicals in Solution*, John Wiley, New York (1957).
311. M. S. Matheson, E. E. Auer, E. B. Bevilaqua, and E. J. Hart, *J. Amer. Chem. Soc.*, **71**, 497 (1949).
312. Y. Ogo, M. Yokawa, and T. Imoto, *Makromol. Chem.*, **171**, 123 (1973).
313. M. Yokawa, Y. Ogo, and T. Imoto, *Makromol. Chem.*, **175**, 179 (1974).
314. M. Yokawa, Y. Ogo, and T. Imoto, *Makromol. Chem.*, **175**, 2913 (1974).
315. M. Yokawa, Y. Ogo, and T. Imoto, *Makromol. Chem.*, **175**, 2903 (1974).
316. M. Yokawa and Y. Ogo, *Makromol. Chem.*, **178**, 443 (1977).
317. C. Walling and G. Metzge, *J. Amer. Chem. Soc.*, **81**, 5365 (1959).
318. G. B. Guarise, *Polymer*, 7, 497 (1966).
319. A. E. Nicholson and R. G. W. Norrish, *Disc. Farad. Soc.*, **22**, 104 (1956).
320. M. Yokawa, Y. Ogo, and T. Imoto, *Proc. 4th AIRAPT Conf. (Kyoto)*, 685 (1974).
321. T. Alfrey and G. Goldfinger, *J. Chem. Phys.*, **12**, 205, 322 (1944).
322. F. de Kok and D. Heikens, *Proc. 4th AIRAPT Conf. (Kyoto)*, 677 (1974).
323. E. Ishihara, Y. Ogo, and T. Imoto, *Proc. 4th AIRAPT. Conf. (Kyoto)*, 681 (1974).
324. C. Walling and J. Pellon, *J. Amer. Chem. Soc.*, **79**, 4776 (1957).
325. V. M. Zholin, M. G. Gonikberg, and V. N. Zagorhinina, *Proc. Acad. Sci. USSR, Chem. Ser.*, **163**, 627 (1965).
326. A. H. Ewald, *Trans. Farad. Soc.*, **55**, 752 (1959).
327. A. C. Toohey and K. E. Weale, *Trans. Farad. Soc.*, **58**, 2446 (1962).
328. M. Kellon and G. Jenner, *Makromol. Chem.*, **180**, 1687 (1979).
329. J. Osugi, M. Sasaki, and I. Onishi, *Rev. Phys. Chem. Jap.*, **36**, 100 (1966).
330. J. Osugi, M. Sasaki, and I. Onishi, *Rev. Phys. Chem. Jap.*, **39**, 57 (1969).
331. J. Osugi, M. Sasaki, and I. Onishi, *Rev. Phys. Chem. Jap.*, **40**, 39 (1970).
332. J. Osugi, M. Sasaki, and I. Onishi, *Rev. Phys. Chem. Jap.*, **41**, 32 (1971).
333. J. Osugi, M. Sasaki, and I. Onishi, *Rev. Phys. Chem. Jap.*, **41**, 42 (1971).
334. D. V. Banthorpe, A. Cooper, and C. K. Ingold, *Nature*, **216**, 232 (1967).
335. N. J. van Hoboken and H. Steinberg, *Rec. Trav. Chim.*, **91**, 153 (1972).
336. N. J. van Hoboken, P. G. Weiring, and H. Steinberg, *Rec. Trav. Chim.*, **94**, 243 (1975).

337. R. Harris and K. E. Weale, *J. Chem. Soc.*, 953 (1956).
338. C. D. Hubbard, C. J. Wilson, and E. F. Caldin, *J. Amer. Chem. Soc.*, **98**, 1870 (1976).
339. N. S. Isaacs and K. Javaid, *J. Chem. Soc., Perkin II*, 3628 (1979).
340. S. D. Hamann and M. Linton, *Austr. J. Chem.*, **30**, 1883 (1977).
341. T. Imoto and K. Aotani, *Nippon Kagaku Zasshi*, **89**, 240 (1968).
342. T. Moriyoshi and K. Mikami, *Rev. Phys. Chem Jap.*, **50**, 38 (1968).
343. P. O. I. Virtanen, *Suomen Kemist*, **B40**, 178 (1967).
344. F. Grønlund and B. Andersen, *Acta. Chem. Scand*, **33 A**, 329 (1979).
345. T. Moriyoshi, *Rev. Phys. Chem. Jap.*, **40**, 102 (1970).
346. T. Moriyoshi, *Rev. Phys. Chem. Jap.*, **41**, 22 (1971).
347. T. Moriyoshi, *Bull. Chem. Soc. Jap.*, **44**, 2582 (1971).
348. C. K. Ingold, reference 104, p. 1029.
349. T. Moriyoshi and M. Hirata, *Rev. Phys. Chem. Jap.*, **40**, 59 (1970).
350. S. J. Cristol and W. P. Norris, *J. Amer. Chem. Soc.*, **76**, 3005 (1954).
351. K. R. Brower and D. Hughes, *J. Amer. Chem. Soc.*, **100**, 7591 (1978).
352. C. G. Swain, *J. Amer. Chem. Soc.*, **83**, 1945 (1961).
353. S. D. Hamann, *Ann. Rev. Phys. Chem.*, **15**, 350 (1964).
354. N. S. Isaacs, K. Javaid, and E. Rannala, *J. Chem. Soc., Perkin II*, 709 (1978).
355. T. Fujii, *Rev. Phys. Chem. Jap.*, **44**, 38 (1974).
356. D. A. Palmer and H. Kelm, *Austr. J. Chem.*, **30**, 1229 (1977).
357. V. M. Zhulin and B. I. Rubinstein, *Bull. Acad. Sci. USSR, Chem. Ser.*, 2055 (1976).
358. D. R. Stranks, *Pure and Appl. Chem.*, **38**, 303 (1974).
359. D. A. Palmer and H. Kelm, *High Pressure Chemistry*, Reidel, (1979).
360. C. H. Longford and H. B Gray, *Ligand Substitution Processes*, Benjamin, New York (1965).
361. T. W. Swaddle and D. R. Stranks, *J. Amer. Chem. Soc.*, **94**, 8357 (1972).
362. S. B. Tong and T. W. Swaddle, *Inorg. Chem.*, **13**, 1538 (1974).
363. D. R. Stranks and T. W. Swaddle, *J. Amer. Chem. Soc.*, **93**, 2783 (1971); M. C. Weekes and T. W. Swaddle, *Can. J. Chem.*, **53**, 3697 (1975).
364. D. R. Stranks and N. Vanderhoek, *Inorg. Chem.*, **15**, 2639 (1976).
365. D. A. Palmer and H. Kelm, *Inorg. Chim. Acta*, **19**, 117 (1976).
366. M. G. Adamson and D. R. Stranks, *J. Chem. Soc., Chem. Comm.*, 648 (1967).
367. W. E. Jones and T. W. Swaddle, *Chem. Comm.*, 998 (1969).
368. W. E. Jones, L. R. Carey, and T. W. Swaddle, *Can. J. Chem.*, **50**, 2739 (1972).
369. N. Ise, M. Ishikawa, Y. Taniguchi, and K. Suzuki, *J. Polym. Sci., Polym. Letts.*, **14**, 667 (1976).
370. D. L. Gay and R. Nalepa, *Can. J. Chem.*, **48**, 910 (1970).
371. H. Lentz and S. O. Oh, *High Temp. High Pres.*, **7**, 91 (1975).
372. G. Guestalla and T. W. Swaddle, *Can. J. Chem.*, **51**, 821 (1973).
373. C. T. Burrin and K. J. Laidler, *Trans. Farad. Soc.*, **51**, 1497 (1955).
374. L. R. Carey, W. E. Jones, and T. W. Swaddle, *Inorg. Chem.*, **10**, 1566 (1971).
375. C. Schenk and H. Kelm, *J. Coord. Chem.*, **2**, 71 (1972).
376. J-M. Lucie, D. R. Stranks, and J. Burgess, *J. Chem. Soc.*, Dalton, 245 (1975).
377. K. R. Brower and T. S. Chen, *Inorg. Chem.*, **12**, 2198 (1973).
378. D. A. Palmer and H. Kelm, *Inorg. Chim. Acta*, **19**, 117 (1976).
379. D. A. Palmer and H. Kelm, *Proc. 4th AIRAPT Conf. (Kyoto)*, 657 (1974).
380. H. Steiger and H. Kelm, *J. Phys. Chem.*, **77**, 290 (1973).
381. B. B. Hasinoff, *Can. J. Chem.*, **54**, 1820 (1976).
382. E. F. Caldin, M. W. Grant, and B. B. Hasinoff, *J. Chem. Soc. Faraday I*, **68**, 2247 (1972).
383. M. W. Grant and C. J. Wilson, *J. Chem. Soc., Faraday I*, **71**, 1362 (1975).
384. E. F. Caldin and M. W. Grant, *J. Chem. Soc., Faraday I*, **69**, 1648 (1973).
385. M. W. Grant, *J. Chem. Soc., Faraday I*, **69**, 560 (1973).
386. W. G. LeNoble, *High Pressure Chemistry*, H. Kelm (Ed.), Reidel, Amsterdam (1978).

387. W. J. LeNoble, R. Goiteu, and A. Shurpik, *Tet. Letts.*, 895 (1969).
388. C. A. Stewart, *J. Amer. Chem. Soc.*, **94**, 635 (1972).
389. W. J. LeNoble and R. Mukhtar, *J. Amer. Chem. Soc.*, **97**, 5938 (1975).
390. G. Jenner, *High Pressure Chemistry*, D. Palmer and H. Kelm (Eds.), Reidel, Amsterdam (1978).
391. A. Khalilpour, G. Jenner, and A. Deluzard, *Bull. Soc. Chim. France*, 583 (1976).
392. R. Mundnich and H. Pleininger, *Tetrahedron*, **32**, 2335 (1976).
393. R. Mundnich, H. Pleininger, and H. Vogler, *Tetrahedron*, **33**, 2661 (1977).
394. C. Brun, G. Jenner, and A. Deluzard, *Bull. Soc. Chim. France*, 2332 (1972).
395. S. K. Shakova and B. S. El'yanov, *Bull. Acad. Sci. USSR, Chem. Div.*, 1461 (1973).
396. K. Seguchi, A. Sera, and K. Maruyama, *Tet. Lett.*, 1585 (1973).
397. S. D. Hamann, *Trans. Farad. Soc.*, 507 (1958).
398. E. F. Caldin and M. W. Grant, *J. Chem. Soc., Faraday I*, **69**, 1648 (1973).
399. J. H. Ridd, *Adv. Phys. Org. Chem.*, **16**, 1 (1978).
400. F. M. Merrett and R. G. W. Norrish, *Proc. Roy. Soc.*, **206 A**, 309 (1951).
401. A. E. Nicholson and R. G. W. Norrish, *Farad. Soc. Disc.*, **22**, 104 (1956).
402. W. G. Dauben and H. O. Krabbenhoft, *J. Amer. Chem. Soc.*, **98**, 1993 (1976).
403. W. Jarre, D. Bieniek, and F. Korte, *Angew. Chem. Int. Ed.*, **14**, 181 (1975).
404. J. Roemer-Moehler, D. Bieniek, and F. Korte, *Z. Naturforsch.*, **30B**, 290 (1975).
405. M. Linton, *Mol. Rate Processes Symp.* (1975); *Chem. Abs.*, **88**, 104330.
406. M. Kurabayashi, K. Yanagita, and M. Yasumoto, *Proc. 4th AIRAPT Conf. (Kyoto)*, 663 (1974).
407. G. I. Nikishin, S. S. Spectov, G. P. Shakovskii, V. G. Glukhovtsev, and V. M. Zhulin, *Izv. Akad. Nauk, SSSR*, 1664 (1976).
408. H. Tojata, *Proc. 4th AIRAPT Conf. Kyoto)*, 718 (1974).
409. R. K. Lyon, *J. Catal.*, **30**, 21 (1973).
410. N. S. Isaacs, unpublished data.
411. *Nat. Phys. Lab. (U.K.) Reports*, 1966–1970 (1972).
412. C. Walling and J. Pellon, *J. Amer. Chem. Soc.*, **79**, 4782 (1957).
413. J. G. Kilroe and K. E. Weale, *J. Chem. Soc.*, 3849 (1960).
414. C. S. Marvell and G. H. McCain, *J. Amer. Chem. Soc.*, **75**, 3272 (1953).
415. K. Weale, *J. Chem. Soc.*, 2223 (1952).
416. W. A. Holmes-Walker and K. E. Weale, *J. Chem. Soc.*, 2295 (1955).
417. I. Tatsuya and N. Takao, *Chem. Abs.*, **68**, 520 (1966).
418. M. Okamoto, *Rev. Phys. Chem. Jap.*, **44**, 77 (1974).
419. J. Osugi and T. Mizukami, *Nippon Kagaku Zasshi*, **87**, 1157 (1966).
420. A. Novak and E. Whalley, *Trans. Farad. Soc.*, **55**, 1490 (1959); *Can. J. Chem.*, **37**, 1710, 1718 (1959).
421. T. Takahashi, K. Hara, and J. Osugi, *Rev. Phys. Chem. Jap.*, **44**, 93 (1974).
422. I. S. Bengelsdorf, reference 6, p. 187.
423. R. Palland and G. V. Ansterweil, *Compt. Rend.*, **242**, 506 (1956).
424. K. E. Weale, *Quart. Rev.*, **16**, 267 (1962).
425. H. G. Drickamer and C. W. Franck, *Electronic Transitions and the High Pressure Chemistry and Physics of Solids*. Chapman and Hall (1973).
426. M. I. Kuhlman and H. G. Drickamer, *J. Amer. Chem. Soc.*, **94**, 2867 (1972).
427. A. P. Hagen, *J. Chem. Ed.*, **55**, 620 (1978).
428. E. Teller, *J. Chem. Phys.*, **36**, 901 (1962).
429. D. W. Coillet and S. D. Hamann, *Trans. Farad. Soc.*, **57**, 2231 (1961).
430. P. W. Bridgman, *Rev. Phys.*, **48**, 825 (1935).
431. Yu. E. Raifel'd, B. S. El'yanov, and S. M. Makin, *Izv. Akad. Nauk. SSSR*, **1090** (1976).
432. B. S. El'yanov, S. M. Makin, and Yu. E. Raifel'd, *Ivz. Akad. Nauk. SSSR*, 831, 836 (1976).
433. S. M. Makin and Yu. E. Raifel'd, and B. S. El'yanov, *Izv. Akad. Nauk. SSSR*, 1094 (1976).

352

434. A. V. Kamernitzky, I. S. Levina, E. I. Martikova, V. M. Shitkin, and B. S. El'yanov, *Tetrahedron*, **33**, 2135 (1977).
435. J. A. Gladysz, S. J. Lee, J. A. V. Tomasello, and Y. S. Yu, *J. Org. Chem.*, **42**, 4170 (1977).
436. H. M. R. Hoffman, *Angew. Chem.* (Int. Ed.), **8**, 556 (1969).
437. J. A. Gladysz and Y. S. Yu, *J. Chem. Soc. Chem. Comm.*, 599 (1978).
438. W. J. LeNoble, personal communication.
439. G. Jenner and S. Aieche, *J. Polym. Sci., Polym. Chem.*, **16**, 1017 (1978).
440. S. Aieche and G. Jenner, *Polymer*, **19**, 1236 (1978).
441. J. W. Mitchell, R. C. de Vries, and R. W. Roberts, *Reactivity of Solids*, J. Wiley, New York (1969).
442. K. R. Brower, *J. Amer. Chem. Soc.*, **94**, 5747 (1972).
443. A. Kivinen and A. Viitala, SUOM Chemist. B41372 (1968).
444. R. Rauchschwalbe, G. Vokel, E. Lang, and H-D. Ludemann, *J. Chem. Res. (S)*, 448 (1978).
445. E. Lang, R. Rauchschwalbe, and H-D. Ludemann, *High Temp. High Pres.*, **9**, 519 (1977).
446. M. Kurabayashi, *19th Ann. Meeting of H. P. Sendai*, October (1978).
447. M. Kurabayashi, *Abs. Chem. Soc. Jap., 38th Meeting*, Osaka (1978).
448. V. M. Zhulin, *Izv. Akad. Nauk. SSSR, ser. Khim.*, 2361 (1971).
449. V. M. Zhulin, M. Yu. Botnikov, and I. Kh. Milyarskaya, *Izv. Akad. Nauk. SSSR*, 1131 (1975).
450. J. B. Goodenough, J. A. Kafalos, and J. M. Longo, *Preparative Methods in Solid State Chemistry*, P. Hagenmuller (Ed.), Academic Press, New York (1972).
451. C. J. M. Rooymans, *Preparative Methods in Solid State Chemistry*, P. Hagenmuller (Ed.), Academic Press, New York (1972).
452. W. J. LeNoble, personal communication.
453. V. L. Antonovskii, L. D. Bezborodin, and M. E. Yasel'man, *Zhur. Fiz. Khim.*, **43**, 2286 (1969).
454. T. Bartnik and B. Baranowski, *Pol. J. Chem.*, **53**, 741 (1979).
455. L. G. Savedoff, *J. Amer. Chem. Soc.*, **88**, 664 (1966).
456. A. J. Ellis, W. S. Fyfe, R. I. Rutherford, A. Fischer, and J. Vaughan, *J. Chem. Phys.*, **31**, 176 (1959).
457. T. Asano, A. Sera, and R. Goto, *Tet. Lett.*, **46**, 4777 (1968).
458. C. Walling and T. A. Angurt, *J. Amer. Chem. Soc.*, **88**, 4163 (1966).
459. Y. Obsuka and C. Walling, *J. Amer. Chem. Soc.*, **88**, 4167 (1966).
460. G. A. Lawrence and D. R. Stranks, *Acc. Chem. Res.*, 2173 (1967).
461. N. Serpone and D. G. Bickley, *Progr. Inorg. Chem.*, **17**, 392 (1972).
462. F. Basolo, J. C. Hayes, and H. M. Neumann, *J. Amer. Chem. Soc.*, **75**, 5102 (1978).
463. R. G. Wilkins and M. J. G. Williams, *J. Chem. Soc.*, 1763 (1957).
464. G. A. Lawrence and D. R. Stranks, *Inorg. Chem.*, **16**, 929 (1977).
465. D. A. Palmer and H. Kelm, *Inorg. Chem.*, **16**, 3139 (1977).
466. M. V. Twigg, *Inorg. Chem. Acta*, **24**, 184 (1977).
467. G. A. Lawrence, D. R. Stranks, and S. Suvachattanont, *Inorg. Chem.*, **18**, 83 (1979).
468. R. D. Gillard, *Coord. Chem. Rev.*, **16**, 67 (1975).
469. J. P. Candlin and J. Halpern, *Inorg. Chem.*, **4**, 1086 (1965).
470. G. A. Lawrence and D. R. Stranks, *Inorg. Chem.*, **17**, 1804 (1978).
471. P. L. Kendall and G. A. Lawrence, *Inorg. Chem.*, **17**, 1166 (1978).
472. D. R. Stranks and N. Vanderhoek, *Inorg. Chem.*, **15**, 2639 (1976).
473. S. Funahashi, K. Ishihara, and M. Tanaka, *Inorg. Chem. Acta*, **35**, L351 (1979).
474. H. Vanni and A. E. Merbach, *Inorg. Chem.*, **18**, 2758 (1979).
475. F. K. Meyer, W. L. Earl, and A. E. Merbach, *Inorg. Chem.*, **18**, 888 (1979).
476. F. K. Meyer, K. E. Newman, and A. E. Merbach, *Inorg. Chem.*, **18**, 2142 (1979).
477. Y. Ducommun, W. L. Earl, and A. E. Merbach, *Inorg. Chem.*, **18**, 2754 (1979).
478. W. L. Earl, F. K. Meyer, and A. E. Merbach, *Inorgan. Chim. Acta*, **25**, L91 (1977).

479. K. G. Liphard, *Rev. Sci. Instr.*, **50**, 1089 (1979).

480. B. Andersen and F. Grønlund, *Acta. Chem. Scand.*, **A33**, 275 (1979).

481. W. J. LeNoble and Y. S. Chang, *J. Amer. Chem. Soc.*, **94**, 5402 (1972); *J. Chem. Ed.*, **50**, 418 (1973).

482. T. Asano and W. J. LeNoble, *Chem. Rev.*, 407 (1978).

483. W. J. LeNoble and R. Mukhtar, *J. Amer. Chem. Soc.*, **96**, 6191 (1974).

484. C-M. Backman, S. Claesson, and M. Szwarc, *Trans. Farad. Soc.*, **74**, 3061 (1970).

485. A. Jost, *Proc. 6th AIRAPT Conf.* (Boulder), Plenum, London (1979) 668.

486. T. W. Swaddle, *Proc. 6th AIRAPT Conf.* (Boulder), 631, Plenum, London (1979).

487. J. v. Jouanne, H. Kelm, and R. Huisgen, *J. Amer. Chem. Soc.*, **101**, 151 (1979).

488. K. Tamura and T. Imoto, *Bull. Chem. Soc. Japan*, **48**, 369 (1975).

489. J. v. Jouanne, D. A. Palmer, and H. Kelm, *Bull. Chem. Soc. Japan*, **51**, 463 (1978).

490. C. T. Burris and K. J. Laidler, *Trans. Farad. Soc.*, **51**, 1497 (1955).

491. K. R. Brower, *J. Amer. Chem. Soc.*, **94**, 5747 (1972).

492. H. Plieninger and H. P. Kraemer, *Angew. Chem. Int. Ed.*, **15**, 243 (1976).

493. W. J. LeNoble and S. K. Palit, *Tet. Lett.*, 493 (1972).

494. B. B. Hasinoff, *Can. J. Chem.*, **57**, 77 (1979).

495. M. Buback and H. Lendle, *Z. Naturforsch.*, **34A**, 1482 (1979).

496. F. Grønlund and B. Andersen, *Acta Chem. Scand.*, **A33**, 329 (1979).

497. A. Marani and G. Talamini, *High Temp. High Pres.*, **4**, 183 (1972).

498. G. Brunton, D. Griller, L. R. C. Barclay, and K. U. Ingold, *J. Amer. Chem. Soc.*, **98**, 6803, (1976); **100**, 4197 (1978).

499. P. R. Marriott and D. Griller, private communication.

Chapter 5

Effects of Pressure on Biochemical Systems

Life on land has evolved in a pressure environment close to 1 bar but in the sea organisms have had to cope with additional hydrostatic pressure of approximately 1 bar for each 10 m depth. At the greatest depths of the ocean therefore, such as at the bottom of the Marianas Trench (12 km), the pressure reaches 1200 bar and water is compressed by some 4%. Nonetheless, bacteria live in bottom deposits and fish, crustacea, and other creatures penetrate to great depths and may be able to adapt to a wide range of pressures. It is clearly of interest to enquire how pressure affects biological molecules—protein, nucleic acid, cell walls, and membranes—as a means of investigating biochemical reaction mechanisms and also to determine whether the chemistry of abyssal creatures is especially adapted. A great deal of data has been accumulated describing the effects of pressure on biological reactions, on the function of organs, and on whole organisms. Most of these data cannot be interpreted in any detail on account of the complexity of the systems and it is only recently that some degree of understanding is beginning to appear following improved knowledge of molecular biology. High pressure techniques in biology have been discussed by Suzuki[40] and by Hawley[63] and Heremans.[41]

5.1   PRESSURE EFFECTS ON BIOPOLYMERS

Since biochemical reactions are inevitably mediated by enzymes which depend for their activity on retaining an intricate conformation, it is appropriate to begin with an examination of their stability under pressure. Enzymes consist of protein, poly-α-aminocarboxylic acids, which are capable of existing in a helical conformation maintained by internal hydrogen-bonding. One such form, the α-

helix, is shown in fig. 5.1(a). Additionally, the double helical chain may be further convoluted and this tertiary structure is secured by both covalent bonding—disulphide bridges—and a variety of non-bonded attractive forces. Oligomerization of several such protein subunits (quaternary structure) may be a feature of certain species. The result of this ordering is to produce a structure of the exact conformation which is required for biological activity, the 'native' form, fig. 5.1(b). The native form of many enzymes is fragile and may be disrupted by heat, pressure, ionic strength, or by hydrogen-bonding solutes such as urea. This process is called 'denaturation' and is manifest in a loss of biological activity due to a conformational change ranging from a small component of tertiary structure to almost complete uncoiling of the α-helix. Denaturation may be irreversible or, usually only for quite small proteins, reversible and may be accompanied by spectroscopic changes or even precipitation from solution. In general, pressure brings about protein denaturation[19] although the conditions necessary vary quite widely. Many enzymes lose their activity fairly sharply at pressures between 4 and 10 kbar depending on the species and also temperature, etc., and denaturation, as judged by spectroscopic properties, occurs similarly with non-enzymic proteins, figs. 5.2, 5.3. Metmyoglobin undergoes a reversible denaturation, well-studied under numerous variables.[64,15] A complex interplay of pressure, temperature, and pH* is evident from the curves fig. 5.4. The process $N \rightarrow D$ is characterized by a volume change which is negative (ca. $-100 \, cm^3 \, mol^{-1}$) though pH- and temperature-dependent. It is known that denaturation of metmyoglobin is accompanied by protonation of imidazole rings present as histidine residues. However, for this, $\Delta \bar{V}$ is only around $+1 \, cm^3 \, mol^{-1}$ if protonated by H_3O^+ or -23 if by H_2O so the experimental volume change may represent also an inherent change in the molecule as it unfolds. $\Delta \bar{V}$ for denaturation in alkaline conditions is less negative, ca. $-60 \, cm^3 \, mol^{-1}$, and now ionization of phenolic groups in tyrosine residues is implicated although to what extent they contribute to the experimental $\Delta \bar{V}$ is not known. Other proteins which exhibit reversible changes, apparently between just two conformations, are ribonuclease and chymotrypsinogen. Ribonuclease is a very small enzyme whose molecular structure is known. The volume change, $\Delta \bar{V}$, is strongly dependent on pH (-46 at pH 2 and -5 at pH 4) evidently due to differing extents of protonation. Though loss of activity occurs with a minor conformational change, further ionization and volume contraction continues as the pH is lowered so that some 12 protons are absorbed in the limit.[33] Chymotrypsinogen shows a large negative volume change and also a negative entropy change ($\Delta \bar{V} = -143 \, cm^3 \, mol^{-1}$, $\Delta S = -227$ eu). The approach to equilibrium is slow and analysis of the two forms can be achieved by high pressure electrophoresis.[45,46] The larger proteolytic enzymes, trypsin, and chymotrypsin are partially denatured above 4 kbar but retain a fraction of their activity, the amount depending on pH and temperature, but which remains above 8 kbar, fig. 5.3. The processes are irreversible and associated with activation volumes -20

---

* For the effect of pressure on buffer pH, see Chapter 3.

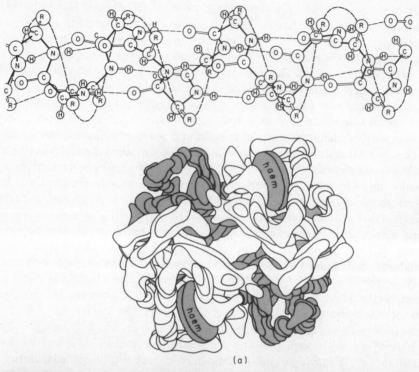

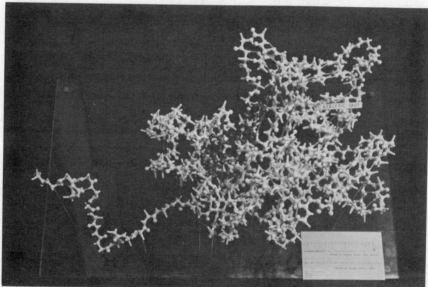

Fig. 5.1. (a) Secondary structure (α-helix) of a section of protein showing hydrogen bonding
(—H ———O—); side chains which are characteristic of individual aminoacids are
denoted R. (b) Model of haemoglobin showing tertiary structure.[2]
(Reproduced by permission of B. H. Nicholson)

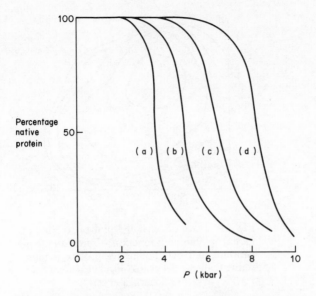

Fig. 5.2. Denaturation of proteins by 5 minutes exposure to high pressure and spectrophotometric estimation of residual native protein; (a) β-lactoglobulin; (b) γ-globulin; (c) ovalbumin; (d) α-amylase. (After Susuki and Taniguchi[1])

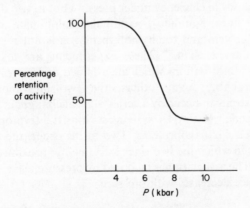

Fig. 5.3. Denaturation of trypsin by 5 minutes exposure to high pressure and measurement of activity at 1 bar[38]

and $-40\,\text{cm}^3\,\text{mol}^{-1}$ (pH 7.6, 25°) respectively.[38,39] This behaviour suggests that two sites of activity are present, one pressure-sensitive and one not, although an alternative view that the denatured form of these rather unspecific enzymes possesses a low order of catalytic activity, seems tenable. Changes of fluorescence provide a sensitive index of conformational change. The fluor may be tryptophan residues in the protein or substrates which are capable of binding. The UV fluorescence of *riboflavin-binding protein* shows a reversible shift under pressure

358

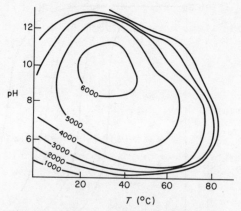

Fig. 5.4. Three-dimensional plot of stability of
metmyoglobin as a function of pressure, temperature, and pH. The native protein is more
stable than the denatured inside each contour.
(After Scheraga et al.[14])

from 342 to 354 nm, the latter being identical with that of free aqueous
tryptophan. It is inferred that denaturation is accompanied by transference of the
tryptophan from an organic to an aqueous environment.[47] Fluorescence of
lysozyme decreases in efficiency under pressure but in two distinct steps and the
binding of the fluor, 8-anilino-1-naphthalenesulphonate, which occurs above
6 kbar to this protein and to chymotrypsinogen is not related to fluorescence
change of the tryptophan.[48] These experiments are taken to indicate the
multiplicity of conformations which are attained under pressure so that in these
cases, and probably many others, the two-state model is inadequate.
Denaturation seems to occur by a series of local changes in the protein. Studies
have been extended to simpler systems such as the tryptophanylflavin derivatives, (1), models for flavoproteins. Two forms occur, the open form and the
associated one in which the two rings exist in close proximity and which is non-
fluorescent.[49] The association is favoured by pressure, $\Delta \bar{V} = -1.8 \text{ cm}^3 \text{ mol}^{-1}$, so
that fluorescence decreases with pressure.

$$(CH_2)_n-CO-NH-CH \overset{CH_2}{\underset{COOCH_3}{|}}$$

H$_3$C

H$_3$C

$n = 3,5.$

(1)

The opposite effect is found in flavodoxins, simple flavinyl proteins.[50] Fluorescence intensities increase with pressure indicating the release of flavin, $\Delta \bar{V} - +70 \, \text{cm}^3 \, \text{mol}^{-1}$. Evidently in these systems large sections of the protein molecule are involved in the pressure-induced change. The forces which are affected by pressure in denaturation processes are evidently hydrogen-bonding, both internal and to solvent water; hydrophobic interactions; and electrostatic forces.[1]

Hydrogen bonding has been previously discussed (section 3.2.4), and simple systems shown to increase this type of interaction under pressure, $\Delta \bar{V} - 3$ to $-7 \, \text{cm}^3 \, \text{mol}^{-1}$ per bond. However, most of these estimates were made in non-aqueous media. In water, hydrogen-bonding of solvent to solute is made at the expense of solvent–solvent bonds so that these effects would be very small as has been recently expressed by Schneider.[51] Therefore, the breaking of the hydrogen-bonded helix and formation of solvated NH and C=O groups would only have a small pressure-dependence. Hydrophobic interactions imply a thermodynamic preference of a non-polar component of structure such as an alkyl chain or aromatic ring, to be juxtaposed with similar structural units, rather than be exposed to solvation by water. It is often found, for example, that 'fatty' side chains such as those of leucine or phenylalanine tend to be concentrated in the interior of native protein and resist conformational changes which will bring them into an aqueous environment or the proximity of ionic groups. The thermodynamics of hydrophobic interactions have been studied.[14,52] A model system is the association of long-chain alkyl cations and anions (studied by conductance) such as $R(CH_2)_n \overset{+}{N}Me_3 \; \bar{O}_2C(CH_2)_mCH_3$ and in which the ionic interactions can be allowed for, so as to isolate alkyl chain association. The free energy increases with temperature as a result of the decrease in entropy of association. Volumes of hydrophobic association here are positive; values amount to ca. $+30 \, \text{cm}^3 \, \text{mol}^{-1}$ for $n, m = 10$ but are less for shorter chains. It is estimated that the volume increase amounts to some $10.2 \, \text{cm}^3 \, \text{mol}^{-1}$ for each additional pair of interacting methylene groups, which is held to be due to the increasing promotion of ordered water structure in the vicinity of the hydro-carbon moiety. This is a consequence of the lack of strong attractive forces between solute and solvent and the attempt to reduce the free energy of the system by increasing water–water forces. The formation of micelles (section 3.2.8) is an example of hydrophobic forces at work and, in a similar manner, is disfavoured by pressure. Other models which are relevent are solubilities of hydrocarbons in water (sections 4.2.9, 4.2.10); since the volume change for association of hydrocarbon molecules is positive, they are more soluble in water (i.e. as unassociated species) at high pressure. It seems likely furthermore[53] that a distinction must be made between volume changes for association of aliphatic and for aromatic species. Aromatic association (e.g. (1)) will generally show a negative volume change and charge-transfer may be important.[55-57] All these effects will become compensating when a large protein molecule uncoils and can explain the range of values found for the volume changes. Electrostatic interactions include ion-pairing, the association of positively and negatively charged groups (section 3.2.3) also ion–dipole and dipole–dipole attractions.

Ion-pair formation tends to reduce ion-solvent interaction and is therefore associated with a positive volume change, $\Delta \bar{V} + 7$ to $+25 \, cm^3 \, mol^{-1}$ so that pressure would tend to induce dissociation of the ions. Further, a protein contains many ionizable groups, mainly —COOH and —NH$_2$ whose ionic dissociation will greatly affect conformational stability as a change in pH is known to do. Volumes of ionization of synthetic protein such as poly(glutamic acid) or poly(lysine) are similar to those of the monomeric species and these model systems undergo uncoiling of the helical conformation when the degree of ionization is around 25%.[6] An increase in pressure of 5 kbar will only bring about an increase in dissociation from 2.8 to 5.4% ($\Delta \bar{V} = -12.7 \, cm^3 \, mol^{-1}$) but a native protein molecule is likely to be much more sensitive to this effect than are the models which only exist as open helical structures or random coils. The transformation between these types has also been observed by NMR line-width changes.[28,29]

Many proteins are known to exist as oligomers of smaller subunits, dimers, trimers, or tetramers. Haemoglobin, fig. 5.1(b), is an example of a tetramer. The same types of weak interaction maintain the integrity of the oligomer which usually tends to dissociate to inactive monomer under the action of pressure.[26] Light scattering (which is an indication of molecular weight) has been used to monitor the dissociation of glutamate dehydrogenase[3] and fibrin[4] to their monomers. β-Casein is initially dissociated but above 2.5 kbar undergoes association, presumably into a different form. Also, glutamate dehydrogenase forms rod-like aggregates of molecules for whose association phosphate is involved and $\Delta \bar{V} + 350 \, cm^3 \, mol^{-1}$.[30] From these examples it will be clear that individual proteins differ widely in their tolerance of pressure and, although the principles are becoming better understood, it remains extremely difficult to interpret the effects of pressure on a whole protein molecule since so many interactions are affected. As a final example, consider the effects of pressure on the two related model polymers, poly(methacrylic acid) and poly(acrylic acid),

$$\left(\underset{H_2C}{\overset{H_3C}{\diagdown}} C \overset{COOH}{\diagup} \right)_n \qquad \left(\underset{H_2C}{\overset{H}{\diagdown}} C \overset{COOH}{\diagup} \right)_n$$

Dissolution of polyacrylic acid is accompanied by a decrease in volume so that the polymer is more soluble in water under pressure. The opposite is true of poly(methacrylic acid) in which, presumably, there are additional hydrophobic interactions[1] offsetting hydrogen bonding or ionic dissociation effects.

The other important biopolymer, nucleic acid, behaves quite differently to protein and is conformationally stable towards pressure.[7] The double helical form is the most stable, providing that the base pairs match, i.e. the strands are

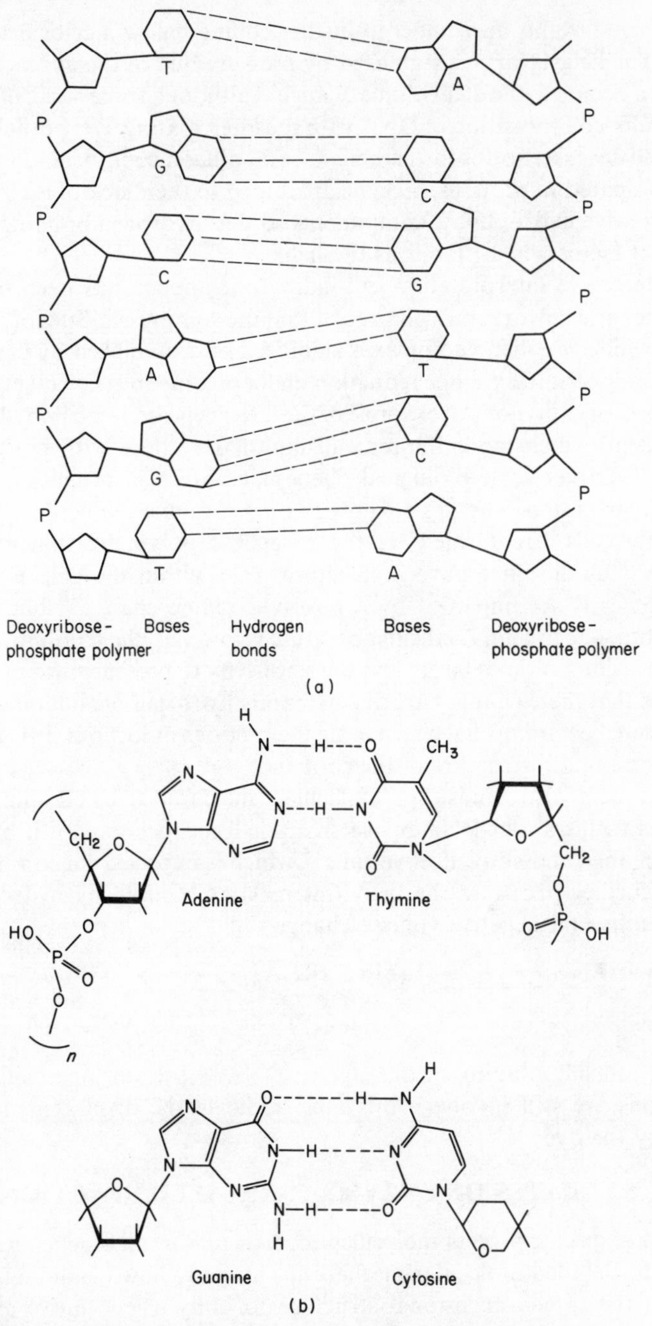

Fig. 5.5. (a) Complementarity of two strands of DNA by pairing of the bases thymine (T), with adenine (A) and guanine (G) with cytosine (C). (b) Detail of the hydrogen bonding between base pairs in DNA

complementary and may undergo further coiling below a critical temperature, fig. 5.5. The helical form is stabilized by pressure but, in the absence of salt, the transition occurs without a volume change[53] although some small effect is found for the helix-coil transition of DNA with the intercalating dye, proflavin, which is associated by insertion into the helix.[54] The difference in behaviour of nucleic acids, compared to proteins, must be attributed to their lack of ionizable groups or of extensive hydrophobic components so that hydrogen bonding is the most important factor which stabilizes the helix.

An alternative interpretation of binding phenomena has been proposed by Drickamer and coworkers, who suggest that the compressibilities of binding sites can determine whether a complex is stabilized or destabilized by pressure. 'Soft' binding sites permit a volume reduction under pressure by skeletal bond rotation while 'hard' ones do not. An example of the latter type is the carbohydrate poly(β-cyclodextrin) which binds fluorescent naphthalene dyes with positive volume changes.[58] Larger scale biological phenomena are also sometimes a result of protein aggregation. The fibres known as microtubules, which are found in all eukaryotic cells, are formed by the reversible polymerization of a protein monomer, tubulin, and have a structural role within the cell. Formation of microtubules is accompanied by a positive volume change[59] but this is very temperature dependent. As the temperature is lowered, the equilibrium constant for aggregation becomes larger and the sensitivity to pressure also increases. It is suggested that these changes are due to mainly hydrophobic interactions at high temperature becoming changes in ionic interactions at low but little is known as yet concerning the structure of the protein.

Nucleic acid also binds dye molecules; the ionic dye, ethidium bromide, associates with transfer-RNA by electrostatic forces accompanied by a positive volume change and activation volume[15] which is expected for ion-pairing. The dye can further penetrate the helix (intercalate), binding by hydrogen bonds, which results in a negative volume change

$$\text{Eth.Br} + t\text{RNA} \underset{\longleftarrow}{\xrightarrow{\hspace{1.5cm}}} \underset{\substack{\text{surface} \\ \Delta\bar{V} = +17}}{[\text{EthBr}\ldots t\text{RNA}]} \underset{\longleftarrow}{\xrightarrow{\hspace{1.5cm}}} \underset{\substack{\text{intercalated} \\ \Delta\bar{V} = -20\,\text{cm}^3\,\text{mol}^{-1}}}{[\text{EthBr}-t\text{RNA}]}$$

However, outside binding of this dye to DNA is pressure-insensitive. If ionic interactions are still involved, this must indicate the displacement of bound cations by the dye.[61,41]

5.2   RATES OF ENZYME-CATALYSED REACTIONS

Enzymes, specific protein molecules, mediate almost all biochemical reactions though the detailed molecular mechanisms are only now beginning to come to light and the three-dimensional structures of only a few native enzymes are known from crystallographic studies.[9] One of the best understood examples is the hydrolysis of amides and esters of aromatic amino acids by the digestive enzyme, α-chymotrypsin (fig. 5.6).[10] The substrate fits into a cleft in the enzyme molecule (the active site) with the aryl group in a hydrophobic 'pocket' and the

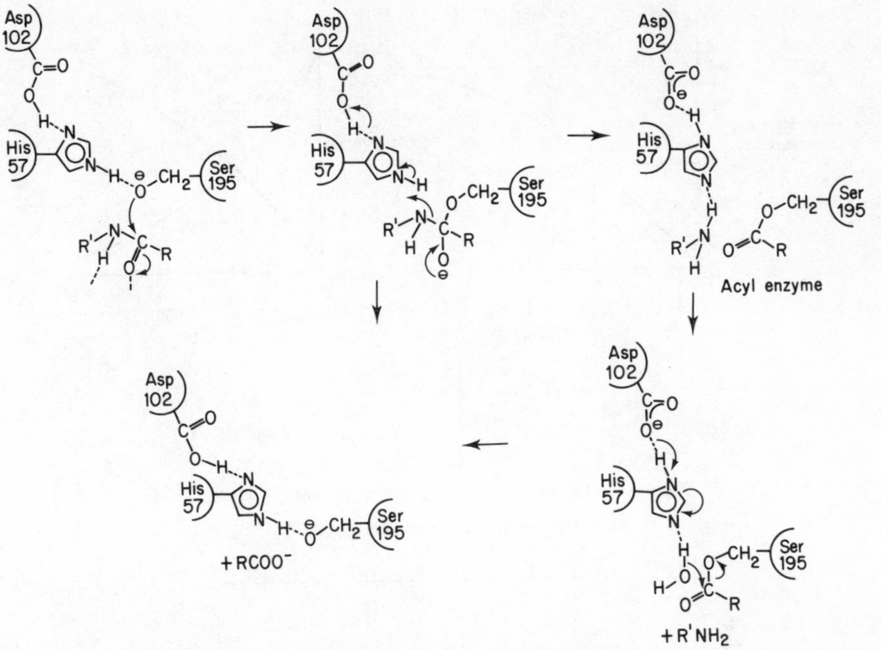

Fig. 5.6. Proposed mechanism of amide hydrolysis catalysed by α-chymotrypsin[10] (there is probably a further water molecule relaying the proton between ser-195 and bis-57)

cooperative effects of a suitably-placed histidine and serine side chains permit hydrolysis to occur with a great lowering of free energy of activation by what is essentially neighbouring group participation. Kinetically, the following scheme may be taken as general (Michaelis–Menten kinetics); the substrate, S, is reversibly associated with the enzyme, E, to form a complex, ES, which decomposes to products and enzyme via an intermediate identified as the acylated enzyme, E-ac:

$$E + S \xrightleftharpoons[k_{-1}]{k_1} ES \xrightarrow{k_2} \text{E-ac} \longrightarrow \text{Products}$$

The slow step is k_2 but the steady-state approximation does not hold as the concentration of the complex is not constant but builds up to a maximum and falls again (fig. 5.7), i.e. $k_1 < k_{-1} \gg k_2$. Now,

$$\text{rate} = k_2[ES]$$

and

$$K_s = \frac{k_1}{k_{-1}} = \frac{[E][S]}{[ES]} = \frac{([E_0] - [ES])[S]}{[ES]}$$

where $[E_0]$ is the amount of enzyme initially added. Then the reaction velocity,

$$\text{rate}, v = \frac{k_2[E_0]}{K_s/[S] + 1}$$

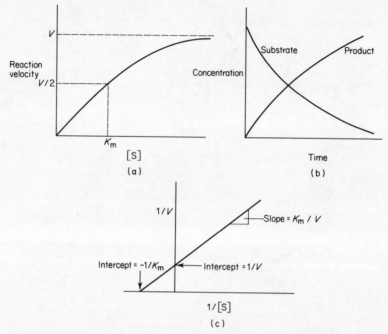

Fig. 5.7. Some kinetic plots; (a) reaction velocity as a function of substrate concentration; (b) Concentrations of substrate and product as a function of time; (c) Lineweaver–Burk plot

v increases with [S] but non-linearly, reaching a maximum value, v_{max}, which corresponds to saturation of all active sites when

$$v_{max} = k_2[E_0]$$

and hence

$$v = \frac{v_{max}}{K_s/[S] + 1}$$

further, if $K_s = [S]$ then $v = v_{max}/2$. The substrate concentration which leads to a reaction velocity $v_{max}/2$ is known as the Michaelis constant, K_m (fig. 5.7(b)) and

$$K_m = \frac{k_{-1} + k_2}{k_1} = \frac{k_{-1}}{k_1}$$

if k_2 is $\ll k_1, k_{-1}$. The parameters v_{max} and K_m are those normally used to characterize enzymic reaction rates and are obtained from a linear plot such as the Lineweaver–Burk relationship, fig. 5.7(c), obtained from

$$1/v = 1/[S] \cdot (K_m/v_{max}) + 1/v_{max}$$

In order to understand the effects of pressure upon enzymic reactions one should determine the changes in both these parameters over a range of pressure insufficient to cause denaturation. Few studies as detailed as this have been attempted. The hydrolysis of some nitrophenyl esters catalysed by the synthetic 'enzyme' have been examined by Suzuki[11]

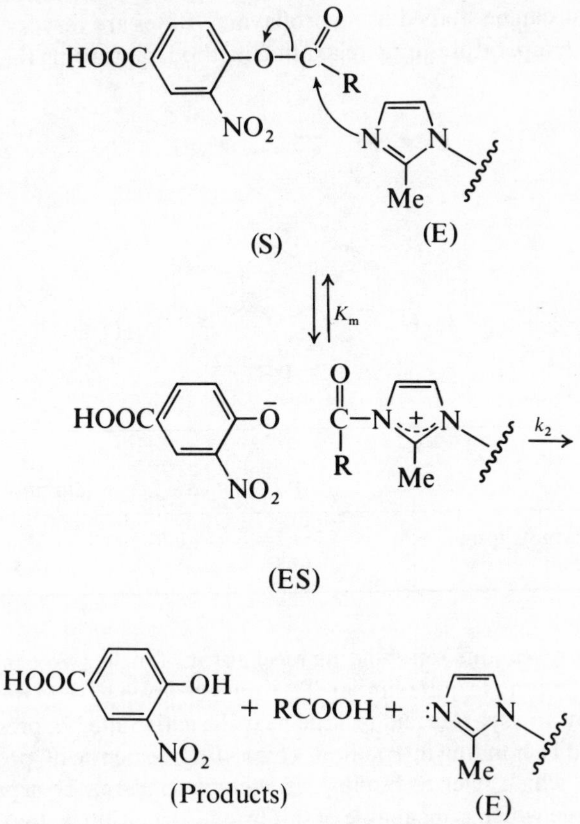

HOOC—⟨ ⟩—O—C(=O)—R
        NO$_2$ N—N
                      |
                     Me

(S)   (E)

$\Updownarrow K_m$

HOOC—⟨ ⟩—Ō   C(=O)—N—(+)N
        NO$_2$ R   Me $\xrightarrow{k_2}$

(ES)

HOOC—⟨ ⟩—OH + RCOOH + :N—N
        NO$_2$ |
                                Me
(Products)   (E)

Michaelis–Menton kinetics are observed and both K_m and k_2 increase with pressure:

$$\Delta \bar{V}^m = -RT \partial \ln K_m / \partial P = -11 \, \text{cm}^3 \, \text{mol}^{-1} \quad \text{and}$$
$$\Delta V_2^{\ddagger} = -RT \partial \ln k_2 / \partial P = -12 \text{ to } -20 \, \text{cm}^3 \, \text{mol}^{-1}$$

depending somewhat on pH. Also negative but smaller values are obtained for a similar reaction mediated by chymotrypsin.[12] $\Delta V_{obs}^{\ddagger}$ (for the overall reaction rate) $= -2$ to $-6 \, \text{cm}^3 \, \text{mol}^{-1}$, fig. 5.6. It is inferred that under the conditions used $k_{obs} = k_2$ since similar volumes of activation characterize the deacylation steps (k_2) of isolated acylenzymes[13]

$$(\alpha\text{-Chy})\text{-OCOR} \xrightarrow[\Delta V^{\ddagger}]{k_2} (\alpha\text{-Chy}) \xrightarrow[\text{fast}]{(\leftarrow)K_{PF}} (\alpha\text{-Chy}).\text{PF}$$
$$(\text{E-ac}) \qquad\qquad +\text{RCOOH}$$
$$\Delta V^{\ddagger} = -5 \, (\text{R} = \text{iso Pr}), \quad -2 \, (\text{R} = t\text{Bu}) \, \text{cm}^3 \, \text{mol}^{-1}$$

The reaction was followed by the spectral change associated with the uptake of proflavin, PF, an orange dye which occupies the active site as soon as it is vacated.

The reversible binding of a substrate to α-chymotrypsin, uncomplicated by further reaction, can be studied using proflavin.[15] Rates are very fast and need to be studied by a temperature-jump relaxation method, perturbing the equilibrium K_{PF}:

$$E + PF \underset{k_{-1}}{\overset{k_1}{\rightleftarrows}} E.PF$$

PF

	$\Delta V_1^{\ddagger}$	$\Delta V_{-1}^{\ddagger}$	$\Delta \bar{V}$ (cm^3 mol^{-1})
E = α-Chymotrypsin	8.7	9.5	0
= Trypsin	17.5	1	16.5

These figures reveal different binding mechanisms for the two enzymes whose mechanisms of action are quite similar. Trypsin is believed to bind proflavin in its cationic form by an aspartic acid residue near the active site by, presumably, an electrostatic and H-bonding interaction. This is in agreement with positive values of $\Delta V_1^{\ddagger}$ and $\Delta \bar{V}$ which refer to binding. In α-chymotrypsin, the aspartic acid is replaced by serine which is incapable of salt bridge formation so that some other mode of binding is probably adopted. Further complications need to be considered, however. Chymotrypsin exists in at least two conformations in equilibrium, the proportions of each of which depend upon pH, temperature, pressure, ionic strength, etc. The two conformers differ in one possessing a salt bridge, an attractive interaction between protonated isoleucine-16 and aspartate-194 residues.[43] The active form possesses the salt bridge and its breaking is accompanied by loss of enzymatic activity, fig. 5.8. The preliminary stages of the reaction, represented by the binding of proflavin, PF, may be represented as:

$$E_i \overset{\Delta \bar{V}_a}{\underset{}{\rightleftarrows}} E_a \overset{PF}{\longrightarrow} E.PF$$

where E_i and E_a are the inactive and active forms of the enzyme. Breaking a salt bridge is favoured by pressure since it increases individual solvation of the ions, the volume change for activation (measured by stopped-flow kinetics) amounting to $+23$ cm^3 mol^{-1}.[44] It is likely that pressure effects on other enzymes depend on dissociation of salt bridges.

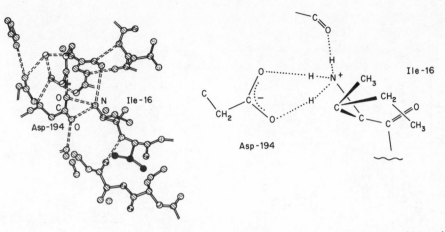

Fig. 5.8. Salt bridge in active form of α-chymotrypsin, between aspartic acid-194 and isoleucine-16 residues[43]

The activity of the enzyme ATP-ase diminishes with pressure but there is an inflection point in the plot of ln v–P plot which suggests a structural change.[60] The pressure at which this occurs varies from 100 to 400 bar as the temperature is raised from 20° to 30° giving a pressure-dependence of the transition temperature, $dT/dP = 28$ K kbar^{-1}. This is of the same order as phase transitions ('melting') in many phospholipids and a parallel transformation within the enzyme by pressure is suggested. Another example of this is provided by nitrogenase.[68]

Table 5.1 gives values of the volume change for substrate binding, $\Delta \bar{V}_m$, and volumes of activation for the decomposition of the enzyme-substrate complex, $\Delta V_2^{\ddagger}$. The nature of the reactions is all so different that discussion of each individually is required. It seems clear that the hydrolysis of methyl hydrocinnamate, $PhCH_2CH_2COOMe$, by α-chymotrypsin is as well understood as any enzymic reaction. Enzyme-substrate binding is weak (*ca.* 14 kJ mol^{-1}) and does not involve ionizable groups: the Michaelis–Menten scheme is applicable. Presumably these considerations apply to other substrates such as L-tyrosine ethyl ester and the negative volume change indicates binding by at least two hydrogen bonds, evidently different forces than the binding of proflavin.[18] Further volume constriction occurs in the breakdown of the enzyme-substrate complex which may reflect covalency formation or conformational changes but, overall, the reaction is accelerated by pressure,[16] some 30% at 500 bar. This conclusion is evident for the majority of examples in Table 5.1. Myosin with $\Delta \bar{V}_m$ positive is believed to bind its substrate ATP with charge neutralization, inhibited by pressure. The volume of activation may be quite sensitive to the presence of ions or other low molecular weight substrates which can interact with the enzyme.[20] This is held to support the view that the activation process, k_2, involves changes in the volume of the protein itself due to conformational changes, and also in hydration. In fact, the sign of $\Delta V_2^{\ddagger}$ can be reversed by the

Table 5.1. Volumes of binding, $\Delta \bar{V}_m$, and volumes of activation, $\Delta V_2^{\ddagger}$ for some enzyme-catalysed reactions[a]

Enzyme	Substrate	$\Delta V_2^{\ddagger}$ (cm³ mol⁻¹)	Reference
Chymotrypsinogen			
α-Chymotrypsin	L-Tyrosine ethyl ester	21	62
Trypsin	Benzoyl-L-argininamide	−13.5	16
	Benzoyl-L-arginine isoPr ester	0	17
	L-Arginine-Me ester	0	
	Serum albumin	−5.5	
Sucrase	Sucrose	−13.7	21
Amylase	Starch	−8	22
	ATP	−28	
Alcohol dehydrogenase (see text)		ca. −30	32
Malate dehydrogenase	Malate (+200 mM KF)	−27	20
	K₂SO₄	−13	
	KCl	−2	
	KBr	0	
	KI	20	
	KSCN	23	
Lactate dehydrogenase	Lactate (+100 mM KF)	−11	
	K₂SO₄	−2	
	KCl	6	
	KBr	13	
	KI	16	
	KSCN	20	
Phosphokinase	Phosphoenolpyruvate (+200 mM K₂SO₄)	7	
	KCl	14	
	KBr	25	
	KI	37	
	KSCN	54	

Table 5.1 (*cont.*)

Enzyme	Substrate	$\Delta V_2^{\ddagger}$ (cm³ mol⁻¹)	Reference
Isocitrate dehydrogenase	Isocitrate	ca. 30	
Poly (Vinyl-2-methyl-imidazole)	3-Nitrophenyl-butyrate	−21	11

Enzyme	Substrate	$\Delta \bar{V}_m$	$\Delta V_2^{\ddagger}$	Reference
Trypsin	β-Lactoglobulin	−36		17
Chymotrypsin	Casein	−13.8	9	16
Pepsin	Gelatin	+22		22, 62
Lipase	Tributyrin	+13		23
Myosin	ATP	+8 to +23		11
Poly(vinyl-2-methyl-imidazole)	3-Nitrophenyl-butyrate	−11		
Glyceraldehyde-3-phosphate dehydrogenase	D-Glyceraldehyde-3-phosphate	+60		42
Lactic dehydrogenase	Lactic acid	0		42
Invertase	Sucrose 30°		−8	66
	40°		−69	
Luciferase			12	67
ATP-ase (ex. *Antimora rostrata*) (ex. *Onchorhyncus kisutch*)			−25 47	68

[a] These figures are not to be taken too seriously. They most probably represent true values only over a narrow pressure range near 1 bar, at a particular temperature and pH. Moreover, they may be species dependent. A large number of enzymic activation volumes (indicating their origins) is given by Morild.[65]

addition of tightly binding anions to malate dehydrogenase (Table 5.1) while isocitrate dehydrogenase in unaffected. It is clear that great care must be exercised when attempting to interpret high pressure data since so many variables can contribute to the effects. Lactic and glyceraldehyde-3-phosphate dehydrogenases appear to show very different pressure effects. These enzymes, however, are very sensitive to denaturation which is complete at 2 kbar so it is difficult to carry out pressure measurements in the reversible region.[42] Volumes of activation for the deacylation step of chymotrypsin, with p-nitrophenyl esters as substrates, have been determined both from Michalis–Menten treatment[25] and by direct determination of reactions of the preformed ES-complex[24] and found to be small and negative, values comparable to those of base-catalysed hydrolyses:

$$\Delta V_2^{\neq} = -3 \text{ to } -6 \text{ cm}^3 \text{ mol}^{-1}$$

Lysozyme, which causes lysis of bacterial cells, is isolable as the crystalline enzyme and its activity has been studied to 500 bar.[34] The volume of activation for the overall reaction is negative at low ionic strength ($-10 \text{ cm}^3 \text{ mol}^{-1}$ at 0.004 M NcCl) but is sensitive to salt concentration ($\Delta V^{\neq} = +1.5 \text{ cm}^3 \text{ mol}^{-1}$ at 0.125 M NaCl). It is suggested that the presence of these ions produces enzyme species with different numbers of salt bridges and associated ions, each species reacting at its own rate.

Alcohol dehydrogenase catalyses the redox reaction between ethanol and acetaldehyde by nicotinamide adenine dinucleotide (NAD),

$$\begin{array}{cc} CH_3CH_2OH + NAD^+ & \xrightarrow{E} \quad CH_3CHO + NADH + H^+ \\ (S_2) \qquad (S_1) & \qquad (P_2) \qquad (P_1) \end{array}$$

Since the two substrates, S_1 and S_2, must be combined and the two products released from the enzyme, a more complicated scheme must be drawn.[31]

$$E + S_1 \underset{k_{-1}}{\overset{k_1}{\rightleftharpoons}} ES_1$$

$$ES_1 + S_2 \underset{k_{-2}}{\overset{k_2}{\rightleftharpoons}} EP_1 + H^+ + P_2$$

$$EP_1 \underset{k_{-3}}{\overset{k_3}{\rightleftharpoons}} E + P_1$$

There is a further complication in that inhibition of reaction is produced by ethanol, itself a substrate, and this is interpreted as due to the formation of a complex with enzyme which is not capable of proceeding to products but merely reduces the enzyme concentration $E + S_2 - K_i \rightarrow ES'_2$. The remarkable observation is that, while rates of reaction rise with pressure, this inhibition is also removed.[32] The interpretation follows from the individual volume changes summarised below:

$\Delta V_1^{\ddagger}$	$\Delta V_{-1}^{\ddagger}$	$\Delta V_2^{\ddagger}$	$\Delta V_3^{\ddagger}$	$\Delta \bar{V}_i \, (\text{cm}^3 \, \text{mol}^{-1})$
$+30$	$+80$	$+70$	-30	$+50$

Since k_3 is rate determining, overall acceleration is observed since $\Delta V_3^{\ddagger}$ is negative. A positive volume change is associated with both ES_1 and ES_2 complex formation, the latter being responsible for the removal of substrate inhibition at high pressure. The complex volume profile for the process is shown in fig. 5.9. The majority of activation volumes shown in Table 5.1, for reactions under more or less biological conditions, are negative though much of the older literature

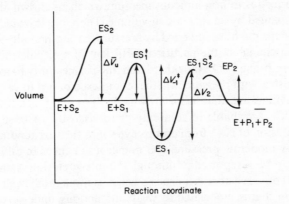

Fig. 5.9. Schematic volume profile for the reaction between ethanol (S_2) and NAD (S_1) mediated by alcohol dehydrogenase. (After Morild[32]

records a fairly general inactivation of enzymes by pressure. When pressures above 1 kbar are used, these results are more than likely to be due to changes in the enzyme structure rather than a reflection of the mechanism of the reaction under investigation.[27]

5.3   MEMBRANE PROPERTIES

Functional biological molecules such as enzymes and their substrates are often separated by membranes which exert vital effects such as maintaining the correct ionic strength in appropriate parts of the cell but also necessitate a transport step in bringing together reagents. While many types of membrane exist, a common form is an oriented bilayer of phospholipid, long-chain alkanes with polar end groups at the water interface, fig. 5.9. This stable arrangement is maintained by hydrogen-bonding between water and polar groups and by the hydrophobic interactions between chains and between layers in much the same way as a micelle exists. Protein molecules with biological function may be attached to either surface or embedded in the membrane and a certain amount of fluidity exists in that phospholipid molecules may migrate within the membrane. The best packing arrangement of the lipid chains is the all-trans conformation, but transformations introducing one or more gauche links are possible. This has the effect of increasing the area of the membrane and of decreasing its thickness, in all increasing the volume by 1.4%. This can be seen as a change from crystalline state to liquid crystal with accompanying changes in light scattering.[41] Increasing pressure increases the temperature of the transition. Compression of a membrane should render the whole structure more rigid both by inhibiting conformational change and by increasing the effective 'viscosity'. Diminished motion of the alkane groups under pressure is evident from esr spectra using a spin-labelled probe.[35] Diffusion of non-polar molecules involves similar physical changes to transfer from aqueous to non-aqueous medium which, as shown above (section 4.1.6), is accompanied by an increase in volume. Therefore it would be expected that such diffusions might be impeded by pressure. Experimentally such an effect is observed as a change in the partition coefficient of molecules between water and isolated membrane.[36] On the other hand, the permeability to ions, Na^+ and K^+, also diminishes with pressure, whether complexed or not to valinomycin (an antibiotic behaving somewhat like a 'crown ether') so that it is not clear to what extent the effect is ascribable to solubility or to viscosity, or to more complex phenomena. Effusion of Na^+ from erythrocytes into the surrounding medium is also retarded by moderate pressure, and numerous changes in diffusion of ions from tissues have been reported[36] although the interpretation is uncertain. The action potential of nerve fibre is a process which is moderately well understood. Adjacent protein chains, associated by weak interactions, undergo displacement of divalent cations such as Ca^{++} by monovalent ions when stimulated causing conformational changes in the chains. Isolated nerves have been shown to change in excitability under pressure, both to increase and to decrease subsequently. Though the resting potential was largely unaffected, the shape of the action current pulses changed, the duration of the fall being increased some 5-fold at 600 bar.[37] This may relate to nervous response to pressure of whole organisms, the tremors and convulsions induced in many species from mammals to crustacea. On the whole, pressure effects on membrane properties are greater than those on proteins or nucleic acids. Membrane-bound enzymes are likely to be very sensitive to pressure.

REFERENCES

1. K. Susuki and Y. Taniguchi, *Soc. Exptl. Biol. Symposium*, **XXVI**, 103 (1972).
2. P. Doty, 'Proteins', *Scientific American*, September 1957.

3. K. A. H. Heremans, *Proc. 4th AIRAPT Conf.*, Kyoto, 627 (1974).
4. D. Collen, G. Vandereycken, and L. DeMayer, *Nature*, **228**, 669 (1974).
5. T. A. J. Payens and K. A. H. Heremans, *Biopolymers*, **8**, 335 (1969).
6. K. Susuki and Y. Taniguchi, *Biopolymers*, **6**, 215 (1968).
7. K. Susuki, Y. Miyasawa, and Y. Taniguchi, *J. Biochem. Tokyo*, **69**, 595 (1971).
8. K. Susuki and M. Tsuchiya, *Bull. Inst. Chem. Res. Kyoto Univ.*, **47**, 270 (1969).
9. M. Dixon and E. C. Webb, *Enzymes*, Longmans, London (1964).
10. J. J. Birktoft, *Phil. Trans. Roy. Soc.*, **B257**, 67 (1970).
11. K. Susuki, Y. Taniguchi, K. Shimokowa, and H. Hisatomi, *Proc. 4th AIRAPT Conf.*, Kyoto, **632**, (1974).
12. R. C. Neuman and G. Lockyer, *Proc. 4th AIRAPT Conf.*, Kyoto, 635 (1974).
13. R. C. Neuman, D. Owen, and G. D. Lockyer, *J. Amer. Chem. Soc.*, **98**, 2982 (1967).
14. G. Nemethy and H. A. Scheraga, *J. Phys. Chem.*, **66**, 1773 (1962).
15. K. A. H. Heremans, J. Snaewaert, H. R. Vandersypen, and Y. Van Nuland, *Proc. 4th AIRAPT Conf.* Kyoto, 623 (1974).
16. H. Werbin and A. D. McLaren, *Arch. Biochim.*, **31**, 285 (1951).
17. H. Werbin and A. D. McLaren, *Arch. Biochim. Biophys.*, **32**, 325 (1951).
18. K. J. Laidler, *Disc. Farad. Soc.*, **20**, 83 (1954).
19. W. Kauzmann, *Proc. 4th AIRAPT. Conf.*, Kyoto, 619 (1974).
20. P. S. Low and G. N. Somero, *Proc. Nat. Acad. Sci. USA*, **72**, 3014 (1975).
21. P. Talwar, D. Macheboeuf, and H. Basset, *J. Colloid. Sci.*, Suppl. 1, 14 (1954).
22. K. J. Laidler, *Arch. Biochem.*, **30**, 226 (1951).
23. L. Onellet, K. J. Laidler, and N. Morales, *Arch. Biochim. Biophys.*, **39**, 37 (1952).
24. R. C. Neuman, D. Owen, and G. D. Lockyer, *J. Amer. Chem. Soc.*, **98**, 2982 (1976).
25. G. D. Lockyer, D. Owen, D. Crew, and R. C. Neuman, *J. Amer. Chem. Soc.*, **96**, 7303 (1974).
26. J. T. Penniston, *Arch. Biochem. Biophys.*, **142**, 322 (1971).
27. K. Susuki, *Methods in Enzymol.*, **26**(C), 424 (1972).
28. R. K. Williams, C. A. Fyfe, R. Epand, and D. Bruck, *Biochem.*, **17**, 1506 (1978).
29. R. K. Williams, C. A. Fyfe, D. Bruck, and L. VanVeen, *Biopolym.* **18**, 89 (1979).
30. H. H. Halvarsen, *Biochem.*, **18**, 2480 (1979).
31. H. Theorell, *Act. Chem. Scand.*, **5**, 1127 (1951).
32. E. Morild, *J. Phys. Chem.*, **81**, 1162 (1977); *Biophys. Chem.*, **16**, 351 (1977).
33. J. F. Brandts, R. J. Oliveira, and G. Weber, *Biochem.*, **9**, 1038 (1970).
34. W. M. Neville and H. Eyring, *Proc. Nat. Acad. Sci. USA*, **69**, 2417 (1972).
35. J. R. Trudell, W. L. Hubbell, and E. N. Cohen, *Biochim. Biophys. Acta*, **291**, 328 (1973).
36. A. G. Macdonald and K. W. Miller, *Biochem. Biophys. Perspect. Marine Biol.*, **3**, 117 (1976).
37. C. S. Spyropoulos, *J. Gen. Physiol.*, **40**, 849 (1957); *Am. J. Physiol.*, **189**, 214 (1957).
38. K. Miyagawa and K. Susuki, *Rev. Phys. Chem. Jap.*, **32**, 43 (1963).
39. K. Miyagawa and K. Susuki, *Rev. Phys. Chem. Jap.*, **32**, 51 (1963).
40. K. Susuki, *Methods in Enzymol.*, **26**, 424 (1972).
41. K. A. H. Heremans, *High Pressure Chemistry*, H. Kelm (Ed.), Reidell, (1978).
42. G. Schmidt, H.-D. Lüdemann, and R. Jaenicke, *Biophys. Chem.*, **3**, 90 (1973).
43. A. R. Fersht, *J. Mol. Biol.* **64**, 497 (1972).
44. H. A. H. Heremans, F. Certerick, J. Snaewaert, and J. Wautes, *Techniques and Applications of Fast Reactions in Solution*, W. J. Gettins and E. Wynn-Jones (Eds.), 429 (1979).
45. S. A. Hawley, *Biochem.*, **10**, 2436 (1971).
46. S. A. Hawley and R. M. Mitchell, *Biochem.*, **14**, 3257 (1975).
47. T. M. Li, J. W. Hook, H. G. Drickamer, and G. Weber, *Biochem.*, **15**, 3205 (1976).
48. T. M. Li, J. W. Hook, H. G. Drickamer, and G. Weber, *Biochem.*, **15**, 5571 (1976).
49. A. J. W. G. Visser, T. M. Li, H. G. Drickamer, and G. Weber, *Biochem.*, **16**, 4883 (1977).
50. A. J. W. G. Visser, T. M. Li, H. G. Drickamer, and G. Weber, *Biochem.*, **16**, 4879 (1977).

374

51. G. M. Schnieder, *Proc. 7th AIRAPT Conf.* (Le Creusot), 1978; in preparation.
52. D. Oakenfell and D. E. Fenwick, *Aust. J. Chem.*, **30**, 741 (1977).
53. S. A. Hawley and R. M. Macleod, *Biopolym.* **13**, 1417 (1974).
54. K. A. H. Heremans, *High Pressure Science and Technology*, Vol. I, K. D. Timmermans and M. S. Barber (Eds.), Plenum, (1979).
55. G. Weber and L. B. Young, *J. Biol. Chem.*, **239**, 1415 (1964).
56. G. Weber and F. J. Farris, *Biochem.*, **18**, 307 (1979).
57. D. D. Kasarda, *Biochim. Biophys. Acta*, **217**, 535 (1976).
59. Y. Engelborghs, K. A. H. Heremans, and L. C. M. DeMeyer, *Nature*, **259**, 686 (1976).
60. H. DeSmedt, R. Borghgraef, F. Ceuterick, and K. A. H. Heremans, *Biochim. et Biophys. Acta*, **556**, 479 (1979).
61. K. A. H. Heremans, J. Snaewaerts, H. Vandersypen, and Y. Van Nuland, *Proc. 4th AIRAPT Conf.*, Kyoto, 829 (1974).
62. A. L. Curl and E. F. Jansen, *J. Biol. Chem.*, **185**, 713 (1950).
63. S. A. Hawley, *Methods in Enzymol.*, **49**, 14 (1978).
64. A. Zipp and W. Kauzmann, *Biochem.*, **12**, 4217 (1973).
65. E. Morild, to be published. *Adv. Protein Chem.* (1981).
66. H. Eyring, F. H. Johnson, and R. L. Gensler, *J. Phys. Chem.*, **50**, 453 (1946).
67. K. J. Laidler, *Arch. biochem.*, **30**, 226 (1951).
68. F. Ceuterick, J. Peeters, K. A. H. Heremans, H. De Smedt, and H. Olbrechts, *Eur. J. Biochem.*, **87**, 401 (1978).

Chapter 6

Effects of Pressure on Spectroscopic Properties and on Photochemistry

Spectroscopy in its broadest sense provides means of probing energy level differences between quantum states of an atom or molecule by means of the emission or absorption of radiation such that $\Delta E = h\nu$. In the liquid state and in solution, vibrational transitions may be examined by infrared absorption or by the Raman effect while in the ultraviolet and visible regions, electronic levels are probed. Magnetic environments of protons and other magnetic nuclei may be studied by NMR spectroscopy and of unpaired electrons by ESR, each in the radiofrequency region. All these measurements are possible under pressure although the limits for each technique may be very different—about 2 kbar for NMR and ESR but up to 1 Mbar for infrared. Two types of pressure dependence may be revealed; spectra may change in response to changes in the species present as a result of the pressure; either chemical or phase changes may become evident and the spectroscopic tool is used in an analytical function. There may in addition be changes brought about by pressure in the energy levels being probed which can therefore reveal fundamental properties of the molecules.

6.1 INFRARED AND RAMAN SPECTRA

Techniques for obtaining infrared and Raman spectra of solutions using window cells or of solids using the diamond anvil have been previously mentioned (section 1.2.3).[1] In the liquid phase, bond properties are not much

376

affected by moderate pressure so that vibrational frequencies should be affected principally by changes in the environment, solvation, and the intramolecular potential. Drickamer has shown that a number of observed spectral shifts can be ascribed to the solute–solvent interaction terms of the potential function,[2] eq. (6.1):

$$\Delta v \sim \frac{1}{8\pi c^2 m_r \bar{v}} - \left(\kappa \frac{\partial E}{\partial r} + \frac{\partial^2 E}{\partial r^2} \right) \tag{6.1}$$

where Δv is the pressure-induced shift for an oscillator of reduced mass m_r, wave number $\bar{v}$; the derivatives express the changes in solvation energy, E, with the internuclear distance, r, of the oscillator. It is observed that most infrared stretching bands undergo a pressure shift though it is often very small, Table 6.1.

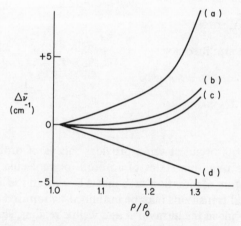

Fig. 6.1. Changes in infrared CH stretching frequencies of CH_2Cl_2; antisymmetric stretch, (a) in CS_2, (b) in $CFCl_3$; symmetric stretch, (b) in CS_2, (d) in $CFCl_3$. (After Drickamer et al.[2])

A red-shift (i.e. towards longer wavelength) is found for O–H and N–H stretching bands and also for C–H stretching in a symmetric mode (in CH_2 and CH_3 groups) but a blue shift may occur for the asymmetric C–H modes (fig. 6.1). These effects are solvent-dependent; for the series of X–H bonds, the red-shift is greater the more polarizable the solvent[3] and also the more electronegative is X. These factors will both contribute to the degree of interaction between solute and solvent under Franck–Condon conditions, i.e. in the short time necessary for absorption of the quantum of radiation which permits no relaxation of the geometry of the system. Multiple bonds, —CN, C=O show blue shifts in their stretching frequencies with pressure.[4] Inductive polarization of the solvent by the solute is held to be responsible for these effects.

The causes of these spectroscopic changes must also include changes in the dielectric constant, ε, of the medium under pressure (section 2.1.12) for which the relationship, eq. (6.2):

$$\Delta v \sim \frac{1}{r^3}\frac{(\varepsilon-1)}{(2\varepsilon+1)} \tag{6.2}$$

has been proposed[5] where r is the radius of the species. The same effects may show up in the vibrational fine-structure of electronic transitions. For instance, the skeletal vibration bands of naphthalene shift by $200\,cm^{-1}$ over a few kbar, observable in the ultraviolet spectrum. Raman spectroscopy gives essentially similar information to infrared but often complementary because of the difference in selection rules which apply.[9] The Raman spectrum of water has been measured under pressure.[7] Interpretation is easier if a very little deuterium is introduced, the stretching frequency of the O–D bond in an environment of H_2O is then measured and found to decrease at the rate $2.3\,cm^{-1}\,kbar^{-1}$. This is a function of the O–O internuclear distance in the hydrogen bond, (1),

(1)

which from compressibility measurements is known to change by $-0.0030\,\text{Å}\,kbar^{-1}$.

Table 6.1. Pressure-induced infrared shifts[1a]

Vibration	Solvent	$\Delta\bar{v}\,(cm^{-1}\,kbar^{-1})$			
		None	CS_2	$CFCl_3$	Reference
C–H stretch					
$CHCl_3$			−1.72	+0.19	
CH_2Cl_2 Symm			−0.48	+0.25	64
Asymm			+0.20	+0.92	
CHI_3			−1.22	—	
C–F stretch; $CFCl_3$		−1.77			
O–H stretch; BuO–H			−1.1		3
PhO–H	i			−1.46	2
HO–D	(in H_2O)	−2.3			7
N–H stretch; $PhNH_2$, Symm				−0.08	2
Asymm				0	
S–H stretch; nPrS–H			−0.56		

[a] Values are averages over 6–10 kbar and are not linear.

	Mode				
Solids	$\bar{v}_3$ (cm^{-1})	$\Delta \bar{v}_3/\Delta P$ (cm^{-1} kbar^{-1})	$\bar{v}_4$ (cm^{-1})	$\Delta \bar{v}_4/\Delta P$ (cm^{-1} kbar^{-1})	Reference
Rb_2TeCl_6	262	0.8	148	1.1	6
Rb_2TeBr_6	201	0.55	107	0.8	
K_2TeI_6	160	0.25	98	1.3	
K_2SeBr_6	321	0.4	132	1.0	
$(NH_4)_2SeBe_6$	227	0.35	133	2.5	
K_2PtCl_6	345	0.4	187	0.8	
Tl_2PtCl_6	335	0.65	181	0.5	

A value is obtained of -770 cm^{-1} Å^{-1} for the spectroscopic shift as a function of O–O distance, a relationship which holds for the various ice phases and is useful in their characterization.[8] The shift in O–H stretching frequency of n-butanol has been attributed to aggregation by hydrogen-bonding.[3]

The diamond anvil cell enables small samples to be compressed in excess of 500 kbar and is suitable for infrared and Raman spectroscopy. All samples will, of course be solid under these conditions.[10, 11] Crystal lattice vibrations of simple cubic crystals such as NaCl have longitudinal and transverse modes (phonon frequencies) owing to the fact that mutual polarization of anions and cations reduces the charge on each to a value less than unity—that is, there is a degree of covalency. The splitting of these modes is dependent on the effective charge, z^*, on the ions and is found to diminish with pressure. Values of this effect for some ionic crystals are:[12]

	dz^*/dP (10^3 kbar^{-1})
LiF	-0.9
NaF	-1.35
KCl	-1.87
CsBr	-8.2
ZnS	-1.5

The pressure at which the effective charge becomes zero may be estimated; for CsBr the change from $z^* = 1$ to $z^* = 0$ should take place by $(1/8.2 \times 10^{-3}) = 120$ kbar above which the compound should be covalent, other effects being ignored. The Raman spectrum of diamond shows a lattice vibration at 1334.0 cm^{-1} which increases linearly with pressure to at least 20 kbar at the rate, $d\bar{v}/dP = +0.296$ cm^{-1} kbar^{-1}.[9] The change is brought about by reduction in the equilibrium distance between the appropriate crystal planes and the change in force constant which this entails. There would be no such effect if the crystal were perfectly harmonic so this value provides a measure of the

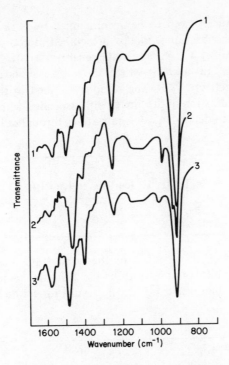

Fig. 6.2. Infrared spectra of methylammonium iodide; 1, phase I at 1 bar: 2,
phase II at 6 kbar: 3, phase III at
15 kbar.[13a] (Reproduced by permission of
Pion Limited)

anharmonicity of the potential function for the lattice,[1] and in turn leads to an
estimation of the compressibility of diamond.

Infrared spectroscopy frequently shows up phase changes as pressure is
increased, for example, methylammonium iodide, fig. 6.2,[13] although X-ray
crystallography would be necessary to determine the new structure. The
interpretation of complex spectra in terms of crystal symmetry is, however,

(2) v_3 (3) v_4

difficult and such studies need to be accompanied by X-ray crystallographic information. Certain vibrational modes of octahedral complexes of the type M_2Xhal_6 where X = Se, Te, Pt, are quite sensitive to pressure, in particular the modes v_3 and v_4 (2) and (3) which show blue shifts of about 0.5 and $2\,cm^{-1}\,kbar^{-1}$ respectively.[15-17] The reasons ascribed to this behaviour are changes in the hybridization of the 'inert pair' associated with the central atom permitting a pressure-dependent delocalization throughout the lattice. For example:

Cs_2TeCl_6:	$\bar{v}_3$ (cm^{-1})	$\bar{v}_4$ (cm^{-1})
1 bar	256	135
40 kbar	265	164

Of organic compounds investigated in the solid state, many disubstituted benzenes show blue shifts of almost all bands although some are much more sensitive to pressure than others.[13c, 19] The results for 1,4-dichlorobenzene are given below:

Cl⟨⟩Cl

Mode	v_2 C–H s str	v_{20} C–H ass str	v_{19a}, v_{19b} C–C str	v_{14} C–C str	v_{12} ring i.p. def
$\bar{v}$	3035	3093	1473, 1391	1265	1084
$d\bar{v}/dP$ (cm^{-1} kbar^{-1}	+1	+1.7	+0.1, +0.45	0	0.22

Mode	v_{17a} C–H oop	v_{18a} C–H ip	v_{17b} CH oop	v_{20a} C–Cl str	$v_{16a, 16b}$ ring def
$\bar{v}$	950	1013	820	542	408, 485
$d\bar{v}/dP$	+0.75	+0.275	+0.45	+0.525	+0.05, 0.725

A series of bicyclic compounds including adamantane, (4), and some bicyclo-[2, 2, 2]-octanes, (5), have been shown by infrared investigations at high pressure to undergo phase transitions; large entropy changes suggest these changes are evidently of the disorder–order type[20]

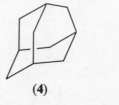

(4)   (5)

Information concerning molecular motion in the liquid phase is available from Raman spectra, in particular from the anisotropic component of the scattered rotation–vibration bands. From an analysis of the band shape a correlation function, R_f, is obtained having the physical significance of a rate at which a molecule reorients itself after displacement, a measure of the rotational mobility of the molecule. This technique has been applied under pressure to methyl iodide, as pure liquid,[33] with results shown in fig. 6.3. Correlation functions fall with pressure whether under conditions of constant temperature or constant volume.

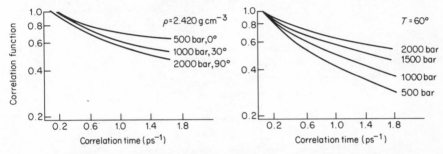

Fig. 6.3. Reorientation correlation function for CH_3I: (a) at constant density, (b) at constant pressure. (After Jonas et al.[33])

The latter is smaller showing the importance of density on rotational rates. By treating molecular rotation as an activation process, energies and volumes of activation can be obtained. The model on which this treatment is based regards molecular motion as resulting from a molecule possessing sufficient kinetic energy to displace surrounding molecules in an activation step.

6.2 ELECTRONIC TRANSITIONS

It is frequently found that molecules, both organic and inorganic, possess electronically excited states which differ in energy from their ground states by $150–600\,kJ\,mol^{-1}$ corresponding to transitions in the visible and near ultraviolet regions, 800–200 nm wavelength. With few exceptions, the ground state will contain all paired electrons and is called a singlet state whereas the excited states may be singlets or may be triplets possessing two unpaired electrons. Transitions between various states may proceed as expressed in fig. 6.4. Absorption of radiation taking the molecule from S_0, the ground state to S_1 the first excited singlet, requires some 10^{-16} s and is thereby much faster than molecular vibrations. The excitation takes place therefore in the absence of geometrical changes of the molecule (the Franck–Condon principle). If the excited molecule has an equilibrium geometry which differs from that of the ground state in some respect, the most probable 'vertical' transition will result in a vibrationally as well as an electronically excited state figs. 6.5, 6.6. The probability of the transition therefore depends upon the vertical juxtapositioning of vibronic energy levels

382

expressed by the Franck–Condon factor, F;

$$F = \int \psi'_v \psi''_v \, dr \qquad (6.3)$$

where ψ'_v, ψ''_v are vibrational wavefunctions for the excited and ground states and dr the change in normal coordinate for the vibration considered between upper and lower states. The fate of the initially-produced excited molecule may lie along

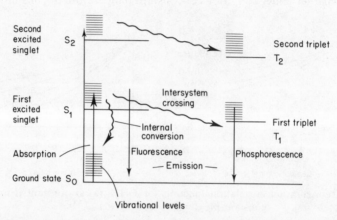

Fig. 6.4. Modified Jablonsky diagram showing relationship of schematic energy levels and transitions between them

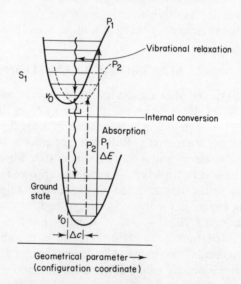

Fig. 6.5. Relationship between potential curves of two electronic states and transitions between them

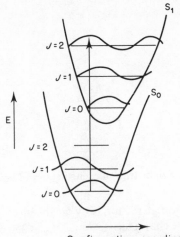

Fig. 6.6. Vibronic energy levels of two states and the effect of a vertical transition

one of several paths. Simple molecules may dissociate as a result of the large amount of vibrational energy concentrated in a few modes. Larger molecules may equilibrate thermally in the excited state, their geometry relaxing to the most stable one and vibrational energy being passed to the surroundings (time scale *ca.* 10^{-12} s). The emission of a quantum of radiation results in return to the ground state, a process known as fluorescence (*ca.* 10^{-11} s), otherwise internal conversion to an excited vibrational level of the ground state may occur followed by thermal equilibration. If a triplet state lies near the excited singlet, intersystem crossing may take place. S–T interconversions are formally forbidden and hence are of low probability which is why the T_1 state is not normally accessible directly from the ground state. The triplet then becomes deactivated by emission of radiation (phosphorescence)—a relatively slow process (10^{-4}–1 s) since it is also forbidden.

Absorption processes   In general it takes rather high pressures to affect the energies of molecular orbitals but changes in excitation energies do occur, particularly in transitions which involve rather weakly bound electrons, π-orbitals or the d–d transitions of metal complexes. Polycyclic aromatic molecules show spectra in the near ultraviolet whose absorption occurs at wavelengths becoming progressively longer the more conjugated the system. Ground states are all singlets and each molecule has several excited singlet states accessible, each with its appropriate energy and differing in distribution of π-electrons from each other. In benzene there are three accessible excited states, fig. 6.7, with approximate π-electron distributions as shown. Compared to the ground state all are polar, the extent of charge redistribution (transition moment) being proportional to the intensity of the absorption or probability of interaction with

384

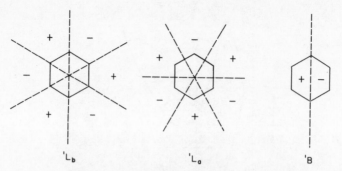

Fig. 6.7. Symmetries of excited states of benzene

the photon. The ^{1}B transition at 180 nm is by far the most intense as the excited state is highly dipolar.[21] The characteristic band ^{1}L$_b$ at 250 nm is proportionately weak. Interactions between excited state and solvent molecules (presumably polarization) increase with polarity of the former and solvatochromic behaviour (change of excitation energy with solvent) also increases in the order: ^{1}L$_b$ < ^{1}L$_a$ < ^{1}B. The effect of pressure may be expected to show a similar sensitivity, so far as spectral shifts brought about by the environment are concerned. The variation of excitation energy by pressure should depend upon the type of transition. The limited data available bear out this conclusion. Fig. 6.8 shows the red shifts produced by pressure on the spectra of chrysene, a polycyclic aromatic with similar types of excited state to benzene. The shift in wavelength of

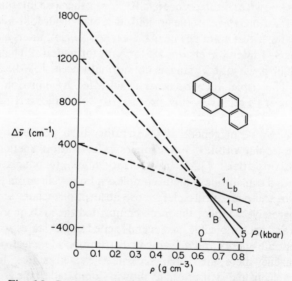

Fig. 6.8. Comparative shifts of three UV transitions of chrysene in pentane by pressure, as a function of the density of the solution. (After Robertson et al.[22])

Table 6.2. Pressure-induced electronic spectral shifts

	Transition	Solvent	$\dfrac{d\bar{\nu}}{dP}$ (cm^{-1} kbar^{-1})		Reference
Aromatic hydrocarbons					
Benzene	^1L_b	Pentane	-100		22, 23
Naphthalene	^1L_b	Pentane	-70		
Phenanthrene	^1L_b	Pentane	-96		
Naphthalene	^1L_b	Ethanol	-68		
Azulene	^1L_b	Ethanol	$+42$		
Naphthalene	^1L_a	Pentane	-325		
Phenanthrene	^1L_b	Pentane	-310		
Anthracene	^1L_b	Pentane	-203		
Tetracene	^1B_b	Pentane	-465		
Azulene	^1B_b	Ethanol	-135		
Tetracene	^1B_b	Ethanol	-220		
Pentacene	^1L_a	Ethanol	-83		
Anilines	$n-\pi^*$				
2-Nitro		Water	-75		25
3-Nitro			-95		
4-Nitro		Water	-125		
		Ethanol	-160		
		Hexane	-280		
Phenols, phenolate ions $n-\pi^*$			ArOH	ArO$^-$	
2-Nitro		Water	-50	$+30$	
3-Nitro			-90	$+35$	
4-Nitro			-125	$+40$	
2,6-Dinitro			-95	-25	
2,4,6-Trinitro			-65	-50	
1-Naphthanol			-10	$+25$	
2-Naphthol			-30	0	

Table 6.2 (cont.)

Transition	Solvent	$\dfrac{d\bar{\nu}}{dP}$ (cm^{-1} kbar^{-1})	Reference
Miscellaneous n$-\pi^*$			
1,4-Benzoquinone	Water	-20	25
Phenol red			
(acid form)		-90	
(alkaline form)		-5	
Crystal violet		-35	
Malachite green		-80	
Phenanthroline $\pi-\pi^*$		-32	26
2,4,6-Triphenylpyridinium-N-(3,5-diphenylphenol) betaine	CHCl$_3$	$+270$	60
	CH$_2$Cl$_2$	230	
	Me$_2$CO	200	
	MeOH	200	
	EtOH	160	
	MeCN	160	
	DMF	125	
	DMSO	125	
Emission spectra			
Azulene (S$-$S$_0$)	Polymethylmethacrylate	-15	27
2-Chloroazulene (S$_2-$S$_0$)	Polymethylmethacrylate	-18	
1,3-Dicarbethoxy-2-chloro-6-bromoazulene (S$_1-$S$_0$)	Polymethylmethacrylate	$+13$	
1,3-Dicarbethoxy-2-amino-6-bromoazulene (T$_1-$S$_0$)	Polymethylmethacrylate	$+10$	

the bands is linearly related to the dielectric constant of the medium (here, pentane) which changes under pressure according to the density (section 2.1.12). Table 6.2 includes further data for polycyclic aromatics. The solvatochromic shift of aromatic bands which is observed is interpretable in terms of the dielectric constant and refractive index of the solvent,[24] eq. (6.4),

$$\Delta v = (AL+B)\frac{n^2-1}{2n^2+1}+C\frac{\varepsilon-1}{\varepsilon+2}\frac{n^2-1}{n^2+2}+F\frac{\varepsilon-1}{\varepsilon+2}\frac{n^2-1}{n^2+2} \qquad (6.4)$$

A, B, C, and F are constants for a given transition, L, a polarizability term. It seems that pressure effects are directly comparable with those produced by solvent change.[23] The pressure effect is itself solvent dependent; red shifts are generally greater in more polar solvents such as ethanol than in alkanes. Azulene is an exception to these generalizations. In the ground state there is considerable polarity which can be expressed in valence bond terms,

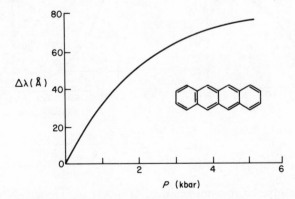

etc.

The excited state ^1L_b is less polar than the ground state so there is a blue shift both in more polar solvents and under pressure; the intense ^1B_b state is also less polar than is the case for the other hydrocarbons and the pressure sensitivity is less than that for, say, tetracene (fig. 6.9). Red shifts of $30-100\,cm^{-1}kbar^{-1}$ are also observed for phenols in their longest wavelength absorption (n–π^* band), for anilines and for quinones.[25] Phenolate ions may show red or blue shifts. This is despite the fact that n–π^* transitions exhibit generally blue shifts on raising the polarity of the medium and suggests that equating of the pressure and solvent effects, at least with these compounds, is not realistic. There seems little systematic relationship that may be deduced between structure and the pressure dependence of a transition.

Fig. 6.9. Shift of the ^1L_a band of tetracene with pressure.
(After Roberson et al.[22])

The displacement along the configuration coordinates of the excited states as a result of pressure has consequences upon the emission of fluorescence. The relaxation of the excited (S_1) state to the ground state can take place by either fluorescence or by internal conversion—the transference of energy from an excited electronic mode to a highly vibrationally excited mode of the ground state or into lattice vibrations, followed by collisional loss of vibrational energy as heat. The relative probabilities of these processes depends upon (a) the energy difference between S_0 and S_1, ΔE, and (b) the displacement, Δc, between the two states along the configuration coordinate. The probability of emission increases as ΔE increases and as Δc decreases.

This can be seen as a result of the Franck–Condon principle operating on Δc and the need for overlapping vibrational levels of S_0 and S_1 affecting ΔE. Fluorescence of azulene has been examined under pressure[27] by Drickamer. This hydrocarbon has two accessible singlet excited states and emission from both S_1 and S_2 could be studied.

Fluorescence from the S_2 state was associated with a red shift induced by pressure, non-linear, and around -10 to $-30\,\mathrm{cm}^{-1}\,\mathrm{kbar}^{-1}$. Emission from the S_1 state was subject to a blue shift of similar magnitude. Quantum yields also showed opposite trends; a decrease with pressure from S_2 and an increase from S_1 (fig. 6.10). The same trends were observed for heavy atom derivatives such as iodoazulene. Since in the latter case inter-system crossing should be facilitated it was concluded that this mode of deactivation is unimportant. Fluorescence efficiency is strongly dependent upon the energy gap separating upper and lower states, the results indicate that internal conversion competes with emission as

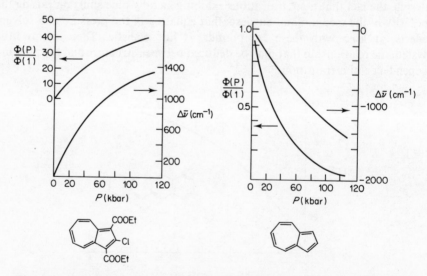

Fig. 6.10. Relative fluorescence emission, $\Phi(P)/\Phi(1)$ and pressure-shifts of their maxima ΔJ from some azulenes. (After Drickamer *et al.*[27]). Medium, polymethylmethacrylate

relaxation processes and the pressure dependence of their relative importance results from changes in the relative energy levels. Phosphorescence $(T_1 \rightarrow S_0)$ may be observed in some azulene derivatives. The maximum shows a blue shift with pressure (fig. 6.11). In all these examples, very high pressures (140 kbar) and a polymeric 'solvent' were used.

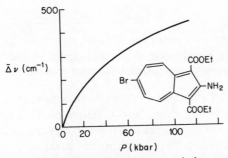

Fig. 6.11. Phosphorescence emission maximum as a function of pressure for an azulene derivative. (After Drickamer *et al.*[27])

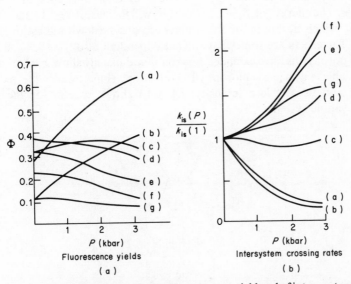

Fig. 6.12. Pressure dependence of fluorescence yield and of inter-system crossing rate for some anthracenes: (a) 9-methylanthracene, (b) 9-chloroanthracene, (c) anthracene, (d) 1-methylanthracene, (e) 2-methylanthracene, (f) 2-chloroanthrocene, (g) 1-chloroanthracene. (After Jonas *et al.*[33])

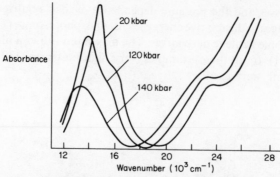

Fig. 6.13. Absorption spectrum of Fe(II) phthalocyanine
at different pressures. (After Drickamer et al.[28])

The decay of excited (S_1) anthracene occurs by fluorescence or phosphorescence from a nearby triplet, T_2, state. Non-radiative decay is negligible. The relative effects of pressure on fluorescence and inter-system crossing can therefore be studied as a function of pressure by measuring the quantum yields for the two emissions, which occur at different wavelengths. The effect depends upon the position of substituents present. For 1- and 2-substituted anthracenes the fluorescence yield diminishes as the rate of intersystem crossing increases with pressure (fig. 6.12) while the opposite is found for 9-substituted analogues. Red shifts in the emissions also occur. It is inferred[33] that pressure lowers the energy of the S_1 state relative to the T_2 on account of their difference in symmetry properties. The energy gap $S_1 - T_2$, however, will increase if $S_1 < T_2$ and decrease if $S_1 > T_2$ which accounts for the differences observed with substituted compounds. The effects are somewhat solvent dependent also.

Metal porphyrins have extended π-systems and usually show $\pi-\pi^*$ transitions in the visible, e.g. haem, chlorophyll. Transitions shift towards the red under pressure and usually lose intensity; fig. 6.13 shows spectra for the related phthalocyanine system.[28]

Transitions in inorganic compounds   A great deal of information is available concerning pressure-induced shifts in transition metal compounds, most of it referring to the solid state.[28-31] Optical spectra are due to transitions between d- or f-orbitals split by the ligand field. The splitting 10 Dq consistently increases

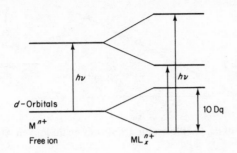

with pressure for d-group elements (fig. 6.14) and transitions shift to higher energy if they depend on this quantity. Molecular orbital calculations suggest[32] that 10 Dq varies as r^{-5} where r is the interatomic distance in the crystal. This appears to be the case; a plot of $\log(10\,Dq)_p/(10\,Dq)_0$ against P has approximately the slope predicted (fig. 6.15). Phase transitions sometimes show up as discontinuities in the spectral shift with pressure, as for example occurs in $Ni(NH_3)_6Cl_2$. The

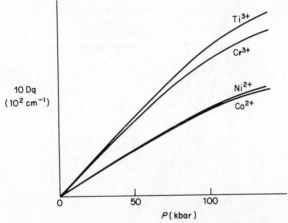

Fig. 6.14. Variation of the *d*-orbital splitting parameter, Dq with pressure for some transition metal ions in MgO. (After Drickamer[29])

complex spectra of lanthanides frequently show four or more transitions in the UV-visible region; the shifts under pressure depend upon their symmetries.[31] Crystal packing can have an effect on observed effects. Whereas a blue shift is observed for octahedral complexes of nickel as mentioned, the dimethylglyoxime complex shows a red shift (fig. 6.16). This is believed to be due to changes in the

Ni–Ni distance within the crystal and mixing of the atomic orbitals of adjacent metal atoms.[31]

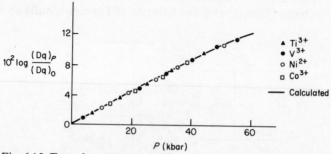

Fig. 6.15. Test of eq. (6.5) for some transition metal ions in Al_2O_3

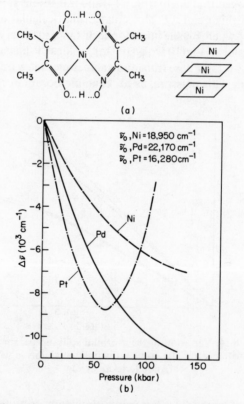

(a)

(b)

Fig. 6.16. (a) Nickel bis(dimethylglyoxime) and its arrangement in the crystal. (b) Effect of pressure on visible transitions of some transition metal glyoxime complexes.[31] (Reproduced by permission of John Wiley and Sons Inc.)

Energy levels in solids are best described in terms of bands—effectively, continua of permitted energy states between limits of energy, and which may be filled, partially filled, or vacant of electrons. The width of the energy gap between a filled and vacant band is obtained from the frequency at which intense absorption sets in—the absorption edge. Pressure reduces the band gap for elements such as iodine, phosphorus, arsenic, silicon, selenium, and sulphur and for covalent compounds such as Hg_2I_2, AgI, CI_4, SnI_4, and TlhaI (fig. 6.17). The effects are quite large, of the order of $-100\,cm^{-1}\,kbar^{-1}$ and accompanied by drastic falls in resistance (section 2.4.1). Pressure can therefore in principle induce semiconductor properties in a material which is an insulator at 1 bar although semiconductors such as germanium and GaAs show an increase in their band gap with pressure attributable to complexities in the band structure. At the extreme of pressure, it is predicted that metallic properties will develop (at pressures in excess of 1 Mbar), a phenomenon known as a Mott transition and which is still being sought.

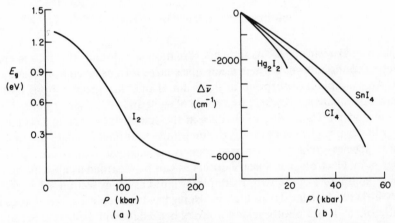

Fig. 6.17. Pressure dependence of the band gap in (a) iodine and (b) some covalent iodides. (After Drickamer[29])

Chemical change may occur, the best documented being the reduction of Fe(III) to Fe(II) by the application of pressure in the solid state. For many compounds, a relationship of the type,

$$\frac{[Fe(II)]}{[Fe(III)]} = K = AP^M$$

where A and M are constants, is obeyed.[30] Changes in the equilibrium constant with pressure are observable from changes in the spectra of the component ions. The reaction involved is 1-electron donation from a ligand, the ease of reduction being correlated with donor properties of the ligands in a series of acetylacetonates. Alkali metal halide crystals which have been irradiated with ionizing radiation develop colours associated with halide ion vacancies into which an electron has been substituted. These behave as isolated 'particles in a box', excitation energies being related to the dimensions of the box—the crystal lattice:[33]

$$(\Delta E/\text{eV})(r/\text{cm}) = 1.76 \times 10^{-19}$$

Since the lattice parameters change with pressure, the transitions would be expected to shift to show a blue shift dependent upon the compressibility of the crystal, which is observed (fig. 6.18).[31]

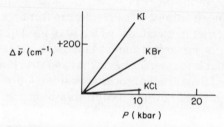

Fig. 6.18. Spectral blue shifts for F-centre absorption in alkali halide crystals[31]

Charge-transfer transitions   Among coordination compounds this means excitation to a configuration in which electronic charge is transferred from ligand to metal. An example is absorption in the red and the blue–green regions by the hexabromo-osmate ion, $OsBr_6^=$. Red shifts of -50 and $-12.5 \, \text{cm}^{-1} \text{kbar}^{-1}$ are observed under pressure due to changes in the spin–orbit coupling.[26] Charge-transfer in organic chemistry means the intermolecular interaction of donor and acceptor species giving rise to new transitions characterized by the partial transfer of an electron. The spectra observed can be regarded as absorption by transient complexes of the two. Examples are discussed in section 4.2.7.

Very little photochemistry at high pressures has been reported. No change in the ratio of 1,2- to 1,3-photocycloaddition of benzene to vinyl ethers is found at 2 kbar.[55] On the other hand, a photo product has been found to form when pyrene is irradiated in the presence of a little oxygen and an alcoholic solvent.[56] The ultraviolet spectrum of this product may be clearly seen after irradiation but disappears when pressure is removed. Its nature is not yet known. The alkaline hydrolysis of photo-excited m-nitroanisole occurs by the following scheme:[61]

The quantum yield for this reaction is reduced by pressure[62] possibly due to an increase in k_{-1} relative to k_1. This would be plausible on account of the very small volume of OH^-.

6.3 NUCLEAR MAGNETIC RESONANCE

A nuclear magnetic resonance signal is sensitive to the average environment of the nucleus under investigation, both the nature of the chemical species and the local environment as defined by the conformation of the molecule in which it resides and the surrounding medium. Pressure may affect the nature of the species present and the technique can therefore be used analytically for studying equilibria, rates, etc. Examples are discussed in section 4.3.1 of conformational changes which may be examined in this way. In this section is discussed the role of NMR spectroscopy in revealing details of molecular motion. A resonance signal will be characterized by the magnetic field at which it occurs (at a fixed radio frequency)—the chemical shift—and by the width of the resonance line. Other features which may be present are hyperfine splittings due to adjacent magnetic nuclei and which are defined by coupling constants.[34, 35] When pulse techniques

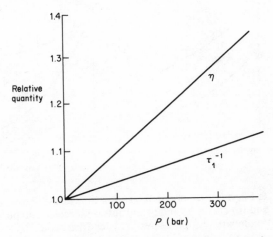

Fig. 6.19. Effect of pressure on relative relaxation time and on relative viscosity for benzene. (After Jonas et al.[37])

are used the decay period of the upper spin state, the spin–lattice relaxation time, T_1, may also be obtained. Relaxation, which is analogous to fluorescence, is induced by the fluctuating electrical and magnetic fields which result from molecular motion. In a crystal they are often very long, of the order of hours, in a liquid, of the order of seconds. Viscosity affects spin–lattice relaxation in a complex manner, fig. 6.19, showing that at least two effects are superposed. At high viscosity there will be more encounters in a given time tending to increase the rate of relaxation. On the other hand, molecular motions will be slowed so that only low-frequency perturbations, of low efficiency in causing relaxation, will be experienced. The behaviour of acetone is shown in fig. 6.20(a).[36] Relaxation is related to diffusion in that both result from molecular translational motion. The simplest model of diffusion views a molecule as moving in a continuous medium of viscosity η, according to the Stokes–Einstein equation:

$$D = \frac{kT}{6\pi\eta r} \tag{6.6}$$

where D is the coefficient of diffusion and r the radius of the molecule. The relaxation time is related to viscosity by this model according to eq. (6.7),[47]*

$$1/\tau_1 = (\alpha + \beta N)\eta/T \tag{6.7}$$

where α, β are constants referring to intra- and inter-molecular processes and N is the number density of the molecules. It follows that $\tau_1 \eta$ should be constant at any temperature but experiment shows that relaxation increases faster than viscosity with pressure,[37] fig. 6.19. It is possible to partition the relaxation time into its

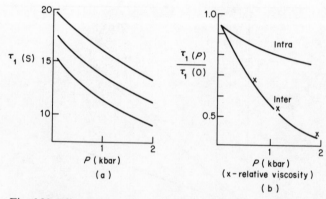

Fig. 6.20. Effect of pressure on relaxation time of acetone: (a) effect on τ_1, and (b) effects on inter- and intra-molecular components, compared to the viscosity. (After Jonas et al.[36])

intra- and inter-molecular components. Their pressure dependence is quite different, fig. 6.20(b), and the intra-molecular contribution is more closely related to viscosity indicating a relationship between the molecular motions for these phenomena. Evidently rotation of the molecule is less affected by pressure than is translation. This conclusion holds also for methyl iodide, benzene, toluene, cyclohexane, trichloroethane, and chloroform.[48] In all these, the value of T_1 falls with pressure except, in the case of cyclohexane, where solidification sets in and T_1 then rises.

Relaxation data are available from 'spin–echo' measurements, the time for a set of precessing nuclei to regain coherence. The self-diffusion coefficients of liquids which contain magnetic nuclei can be obtained in this way (section 2.2). In methanol, for example, self-diffusion falls with pressure[44] and similar results are

---

*Two relaxation times should be distinguished, longitudinal, τ_1 and transverse, τ_2; for freely tumbling molecules, as in a fluid, there is little difference between these quantities.

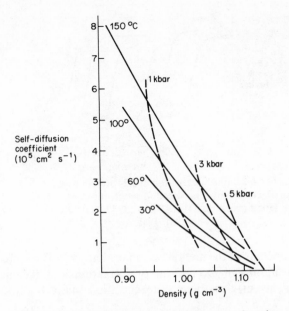

Fig. 6.21. Self-diffusion of liquid pyridine as a function
of pressure. (After Jonas[37])

found for pyridine,[45] fig. 6.21, and perfluorocyclobutane.[40] In gaseous systems, such as CF_4 in argon, spin–lattice relaxation increases under pressure because of the increased collision rate.[39,46] Water, as usual, shows anomalies. At temperatures below 10° the self-diffusion coefficient shows a maximum with pressure confirming that the mobility of water first increases with pressure as the hydrogen-bonding network is distorted. After this it decreases as the free volume of the liquid diminishes.[41,42,49] The chemical shift δ of the OH proton in water also goes through a minimum.[43] Above 30° the liquid behaves normally; T_1 and D fall while δ increases with pressure.

Phase change may show up in an NMR experiment under variable temperature, for example dinitropropane undergoes a transition in the solid phase at $-5°$ above which relatively free rotation is permitted. This is accompanied by a sudden sharpening of the NMR line and the discontinuity is shifted to higher temperatures as the pressure increases, fig. 6.22, corresponding to a volume increase of $7.5 \, cm^3 \, mol^{-1}$[38] in good agreement with density measurements.

Polyisobutene also shows a phase change of this type around $-70°C$.[50] Other solid phase studies include the alkali metals each of which shows line-broadening and shift to lower field at high pressures. These parameters are affected by diffusion and by the paramagnetism of the conduction electrons.[51,52]

In transition metal complexes, resonance of nuclei in a ligand may be affected by exchange processes such as

$$Ni^{++}(MeCN)_6(ClO_4^-)_2$$

$$+ MeCN* \rightleftarrows Ni^{++}(MeCN)_5(MeCN*)^1(ClO_4^-)_2 + MeCN$$

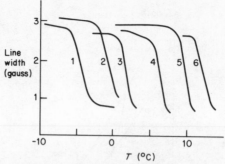

Fig. 6.22. Phase transition in 2,2-dinitropropane as shown by line-width discontinuity, at different pressures: (1) 1; (2) 210; (3) 420; (4) 640; (5) 850; (6) 1060 bar

resulting in broadening of the methyl resonance line.[58, 59] The exchange process contributes an additional mechanism of relaxation and the line-width, Δv (measured at half height) gives directly the relaxation time (actually the transverse relaxation time, T_2),

$$1/T_2 = \pi . \Delta v$$

The exchange rate may be measured from T_2 and the chemical shift, δ, for bound as compared to free ligand by means of eqs. (6.8) and (6.9) due to Swift and Connick[57]

$$\frac{1}{T_2} - \frac{1}{T_2^0} = \frac{x_s}{\tau_c} \frac{[T_2^{c-2} + (T_2^c \tau_c)^{-1} + \delta_c^2]}{[(T_2^{c-1} + \tau_c^{-1})^2 + \delta_c^2]} \tag{6.8}$$

$$\delta = \frac{x_s . \delta_c}{(\tau_c/T_2^c + 1)^2 + \tau_c^2 \delta_c^2} \tag{6.9}$$

where relaxation times, T_2, are: T_2 (experimental), T_2^c (bound ligand in the absence of exchange), T_c^0 (uncomplexed ligand); chemical shifts are: δ (observed), δ_c (in the absence of exchange—e.g. at low temperature); τ_c is the average lifetime of a bound ligand molecule and x_c the mole fraction of bound ligand. The equations may be modified to include complexation in the outer solvation sphere.[58] Pressure brings about a reduction in exchange rate ($\Delta V^{\neq} = 11 \text{ cm}^3 \text{ mol}^{-1}$) and the relaxation time of the bound ligand protons becomes shorter.

6.4 NUCLEAR QUADRUPOLE RESONANCE

Magnetic resonance of a nuclear quadrupole may be observed at radio frequencies since its orientation is also quantized in a magnetic field. Nuclei with a spin number $> \frac{1}{2}$ may have a quadrupole moment; chlorine, ^{35}Cl ($I = \frac{3}{2}$) and nitrogen, ^{14}N ($I = 1$) are examples which are often studied. In a similar way to NMR spectroscopy, nuclear quadrupole resonances give information concerning

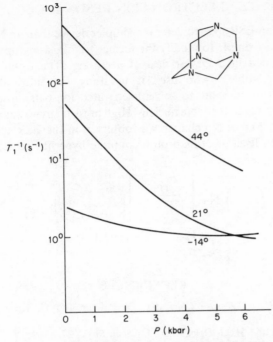

Fig. 6.23. Relaxation times for crystalline hexamine as a function of temperature and pressure. (After Zussman and Rapoport[54])

the magnetic environment of the nucleus and of its time-dependence. Molecular motion, diffusion, and relaxation phenomena may thus be studied by this technique and a few studies under pressure have been reported. Measurements of the electronic field gradients in the vicinity of the nucleus have been made for crystals such as Na$\underline{Cl}O_3$, Na$\underline{N}O_3$ and Na$\underline{Br}$ (resonating nucleus underlined).[53] NQR is appropriate for the study of long relaxation times and has been used to examine molecular reorientation in the spherical molecule hexamethylenetetramine, (6). T_1 increases with pressure, the effect increasing at higher temperature, fig. 6.23.[54] An activation energy of 70 kJ mol^{-1} and activation volume of 24 cm^3 mol^{-1} are obtained from the data which are surprisingly high for such a symmetrical molecule and must point to a high anisotropy of the intermolecular forces and considerable dislocation of the lattice required for rotation about a three-fold axis.

(6)

6.5 ELECTRON-SPIN RESONANCE

The cavity of an ESR spectrometer is a simpler device than an NMR probe so that it is relatively simple to modify the former to take a high pressure sample tube, thermostatted in a Dewar flask if necessary. The technique has been described by LeNoble[63] who used it to study the transition of sodium naphthalenide from intimate to solvent-separated ion-pair, observable in the reduction of coupling to the ^{23}Na nucleus. High pressure pyrex samples tubes can withstand about 1 kbar. No change was observed in the ESR spectrum of the paramagnetic ion itself or of its proton coupling hyperfine structure.

$$\Delta \bar{V} = -15 \text{ cm}^3 \text{ mol}^{-1}$$

REFERENCES

1. E. Whalley, *High Pressure Chemistry*, D. Palmer and H. Kelm (Eds.), Reidel, Amsterdam (1978).
2. A. M. Benson and H. G. Drickamer, *J. Chem. Phys.*, **27**, 1164 (1957).
3. E. Fishman and H. G. Drickamer, *J. Chem. Phys.*, **24**, 548 (1956).
4. R. R. Wiederkier and H. G. Drickamer, *J. Chem. Phys.*, **28**, 311 (1958).
5. E. Bauer and M. Magat, *P. Phys. radium*, **9**, 319 (1938).
6. O. G. Weigang and W. W. Robertson, *J. Chem. Phys.*, **30**, 1413 (1959).
7. G. E. Wolrafer, *J. Sol. Chem.*, **2**, 159 (1973).
8. E. Whalley, *J. Chem. Phys.*, **63**, 5205 (1975).
9. E. Whalley, A. Lavergne, and P. T. T. Wong, *Rev. Sci. Instr.*, **47**, 845 (1976).
10. D. M. Adams and S. T. Payne, *Ann. Rep. Chem. Soc., A*, **69**, 3 (1972).
11. P. T. T. Wong and E. Whalley, *Rev. Sci. Instr.*, **45**, 904 (1974).
12. S. S. Mitra and K. V. Namjoshi, *J. Chem. Phys.*, **55**, 1817 (1971).
13. S. D. Hamann and M. Linton, *High Temp. High Pres.*, **10**, 97 (1978) and earlier papers.
14. D. M. Adams, *Inorganic Solids, An Introduction to Concepts in Solid State Structural Chemistry*, Wiley-Interscience, New York, N.Y. (1974).
15. D. M. Adams and S. J. Payne, *Ann. Rep. Chem. Soc. (A)*, **69**, 3 (1972).
16. D. M. Adams, J. D. Findlay, and M. C. Coles, *J. Chem. Soc. Dalton*, 371 (1976).
17. D. M. Adams and J. S. Payne, *J. Chem. Soc. Dalton*, 407 (1974).
18. D. M. Adams and J. S. Payne, *J. Chem. Soc. Dalton*, 215 (1975).
19. S. D. Hamann, *High Temp. High Pres.*, **10**, 97 (1978).
20. S. D. Hamann, *High Temp. High Pres.*, **10**, 445 (1978).
21. H. H. Jaffe and M. Orchin, *Theory and Applications of Ultraviolet Spectroscopy*, Wiley, New York (1964).
22. W. W. Roberton, O. E. Weiging, and F. A. Matsen, *J. Mol. Spectr.*, **1**, 1 (1957).
23. W. W. Robertson and A. D. King, *J. Chem. Phys.*, **34**, 1511 (1961).
24. E. G. McRae, *J. Phys. Chem.*, **61**, 562 (1957).
25. S. D. Hamann and M. Linton, *Aust. J. Chem.*, **28**, 701 (1975).
26. H. G. Drickamer, *Angew. Chem., Int. Ed.*, **13**, 39 (1974).
27. D. J. Mitchell, H. G. Drickamer, and G. B. Schuster, *J. Amer. Chem. Soc.*, **99**, 7489 (1977).

28. H. G. Drickamer and C. W. Franks, *Electronic Transitions and the High Pressure Chemistry and Physics of Solids*, Chapman and Hall, London (1973).
29. H. G. Drickamer, *Solid State Physics*, **17**, 1 (1965).
30. H. G. Drickamer and C. W. Frank, *Ann. Rev. Phys. Chem.*, **23**, 39 (1972).
31. H. G. Drickamer and J. C. Zahner, *Adv. Chem. Phys.*, **4**, 161 (1962).
32. G. Burns and J. D. Axe, *J. Chem. Phys.*, **45**, 4362 (1966).
33. J. H. Campbell, J. F. Fisher, and J. Jonas, *J. Chem. Phys.*, **61**, 346 (1974).
34. J. Pople, W. G. Schneider, and H. J. Bernstein, *Nuclear Magnetic Resonance*, McGraw-Hill, New York (1959).
35. L. M. Jackman and S. Sternhell, *Applications of NMR Spectroscopy in Organic Chemistry*, Pergamon, London (1969).
36. J. Jonas and J. A. Akai, *J. Chem. Phys.*, **66**, 4946 (1977).
37. M. Fury, G. Munie, and J. Jonas, *J. Chem. Phys.*, **70**, 1260 (1979).
38. R. J. Finn, M. Wolfe, and J. Jonas, *J. Chem. Phys.*, **67**, 4004 (1977).
39. M. Wolfe, E. Arndt, and J. Jonas, *J. Chem. Phys.*, **67**, 4012 (1977).
40. R. J. Finn, M. Fury, and J. Jonas, *J. Chem. Phys.*, **66**, 760 (1977).
41. J. Jonas, T. de Fries, and D. J. Wilbur, *J. Chem. Phys.*, **66**, 582 (1976).
42. D. J. Wilbur, T. de Fries, and J. Jonas, *J. Chem. Phys.*, **65**, 1783 (1976).
43. J. W. Linowski, N.-I. Liu, and J. Jonas, *J. Chem. Phys.*, **65**, 3383 (1976).
44. J. E. Bull and J. Jonas, *J. Chem.*, **52**, 2779 (1970).
45. D. E. Woessner and B. S. Snowden, *J. Chem. Phys.*, **52**, 1621 (1970).
46. J. T. Billings and A. W. Nolle, *J. Chem. Phys.*, **29**, 214 (1958).
47. J. A. S. Smith, *High Pressure Chemistry and Physics*, R. S. Bradley (Ed.), Academic Press, London (1963).
48. A. W. Nolle and P. P. Mahendroo, *J. Chem. Phys.*, **31**, 863 (1960).
49. T. de Fries and J. Jonas, *J. Chem. Phys.*, **66**, 896 (1977).
50. A. W. Nolle and J. T. Billings, *J. Chem. Phys.*, **30**, 84 (1959).
51. T. Kushida and G. B. Benedek, *J. Phys. Chem. Solids*, **5**, 241 (1958).
52. R. G. Barnes, R. D. Engardt, and R. A. Hu, *Rev. Phys. Letters*, **2**, 202 (1959).
53. R. A. Bernheim and H. S. Gutowsky, *J. Chem. Phys.*, **32**, 1072 (1960).
54. A. Zussman and E. Rapoport, *J. Chem. Phys.*, **61**, 5098 (1974).
55. N. S. Isaacs, unpublished data.
56. Y. Tarihashi, A. Itaya, and N. Mataga, *Chem. Letters*, 325 (1973).
57. T. J. Swift and R. E. Connick, *J. Chem. Phys.*, **37**, 307 (1962).
58. K. E. Newman, F. K. Meyer, and A. E. Merbach, *J. Amer. Chem. Soc.*, **101**, 1470 (1979).
59. F. K. Meyer, W. L. Earl, and A. E. Merbach, *Inorg. Chem.*, **18**, 888 (1979).
60. J. v. Jouanne, D. A. Palmer, and H. Kelm, *Bull. Chem. Soc. Jap.*, **51**, 463 (1978).
61. R. O. de Jongh and E. Havinga, *Rec. Trav. Chim.*, **85**, 275 (1966).
62. H. Aomi, M. Sasaki, and J. Osugi, *Rev. Phys. Chem. Jap.*, **47**, 111 (1977).
63. W. J. LeNoble and F. Staub, *J. Organometallic Chem.*, **156**, 26 (1978).
64. M. Buback, E. U. Franck, and H. Lendle, *Z. Naturforsch.*, **34A**, 1489 (1979).

Appendix 1

The Units of Pressure

Pressure, P, is defined as a force, F, acting upon an area $A = l^2$:

$$P = F/A = F/l^2$$

and force is that which imparts uniform acceleration to mass, m; thus

$$F = m.l.t^{-2}$$

and

$$P = m.l^{-1}.t^{-2}$$

in ultimate units of mass, length and time. Also, since energy has the dimensions $E = m.l^2.t^{-2}$ it follows that $P = E.l^{-3}$ or energy per unit volume.

Rational units   Units of force are the dyne and its modern (SI) equivalent, the Newton, N, defined as follows:

> 1 dyne imparts an acceleration of $1 \, cm \, s^{-2}$ to $1 \, g$ mass

> 1 Newton imparts an acceleration of $1 \, m \, s^{-2}$ to $1 \, kg$ mass

thus

$$1 \, N = 10^5 \, dynes$$

Pressure is commonly expressed as $dynes \, cm^{-2}$ or, as recommended today, in $Newtons \, m^{-2}$. The specific name for the latter unit is the Pascal Pa

$$1 \, N \, m^{-2} = 1 \, Pa = 1 \, J \, m^{-3}$$

and this is the SI unit of pressure. The Pascal is quite a small unit (about 10^{-5} of the pressure of the atmosphere) so that a gas at 1 Pa pressure is a pretty good vacuum. In high pressure work the megapascal, MPa, or gigapascal, GPa, are more appropriate

$$1 \, MPa = 10^6 \, Pa, \quad 1 \, GPA = 10^9 \, Pa$$

The older unit of dynes cm^{-2} ('barye') is still in wide use, however, in the form of the *bar* where

$$1\,\text{bar} = 1\,\text{megabarye} = 10^6\,\text{dynes}\,\text{cm}^{-2}$$

In particular the kilobar, $\text{kbar} = 10^3$ bar, is an appropriate and widely used unit in high pressure work and will be used throughout this book. For those who prefer the SI unit, conversion is simple since

$$1\,\text{kbar} = 100\,\text{MPa} = 0.1\,\text{GPa}$$

Gravity based units   One pound mass acting under the earth's gravitational field exerts a force defined as 1 pound force and similarly $1\,\text{kg}$ produces $1\,\text{kg}$ force. Common units of pressure derived from this concept are $\text{lb.}\,\text{in}^{-2}$ (pound per square inch, p.s.i.) and $\text{kg}\,\text{cm}^{-2}$.

For precision, a standard gravitational constant needs to be specified, and the direct usage of standard mass to produce a force necessitates a knowledge of the local value of g. In practice this is so nearly constant it may be ignored.

Another measure relying on gravity is the height of a column of liquid which expresses pressure at its base. Mercury is frequently used in such measurements on account of its high density and low vapour pressure. Temperature must be specified and in principle also, g. Pressure then may be expressed as $\text{cm}\,\text{Hg}$ or $\text{m}\,\text{Hg}$. As a pressure scale it is not strictly linear on account of the compression of the fluid (section 1).

Atmospheric units   Though variable, the pressure of the atmosphere has a significance easily comprehended and placed in context. The standard atmosphere is defined in terms of the bar to which it is so nearly equivalent, as is the $\text{kg}\,\text{cm}^{-2}$.

1 standard atmosphere = $1.01325\,\text{bar} = 1.0332\,\text{kg}\,\text{cm}^{-2}$
   (atmos)

$\qquad\qquad\qquad = 1013.25\,\text{millibar}$ (the meteorological unit)

Atmosphere and bar are frequently assumed to be synonymous though a significant difference between the two exists.

Appendix 2

Bibliography of Books and Reviews on High Pressure Chemistry

1957 S. D. Hamann, *Physico-Chemical Effects of Pressure*, Butterworth, London.

1963 S. D. Hamann, *High Pressure Chemistry and Physics*, R. S. Bradley (Ed.), Academic Press, London.

1963 M. G. Gonikberg, *Chemical Equilibria and Reaction Rates at High Pressure*, trans. Nat. Sci. Found., Washington.

1964 E. Whalley, Use of volumes of activation for determining reaction mechanism, *Adv. Phys. Org. Chem.*, **2**, 93.

1964 S. D. Hamann, High pressure chemistry, *Ann. Rev. Phys. Chem.*, **15**, 349.

1967 K. E. Weale, *Chemical Reactions at High Pressures*, Spon, London.

1967 W. J. LeNoble, Reactions in solution under pressure, *Progress in Physical Organic Chemistry*, **5**, 207.

1967 W. J. LeNoble, Reactions in solution under pressure, *J. Chem. Ed.*, **44**, 729.

1970 G. Kohnstam, The kinetic effects of pressure, *Prog. Reaction Kinetics*, **5**, 335.

1972 R. C. Neumann, Pressure effects as mechanistic probes of organic radical reactions, *Acc. Chem. Res.*, **5**, 381.

1974 D. R. Stranks, Elucidation of inorganic reaction mechanisms by high pressure studies, *Pure App. Chem.*, **38**, 303.

1975 G. Jenner, 'High pressure kinetic investigations in organic and macromolecular chemistry', *Angew. Chem. Int. Ed.*, [14], 137.

1978 T. Asano and W. J. LeNoble, Activation and reaction volumes in solution, *Chem. Rev.*, **407**.

1978 A. P. Hagen, High pressure synthetic chemistry, *J. Chem. Ed.*, **55**, 620.

1978 H. Kelm (Ed.), *High Pressure Chemistry*, Proceedings of NATO Advanced Study Institute, Reidel, Amsterdam.

1980 G. A. Lawrence and D. R. Stranks, The role of activation volume in the elucidation of reaction mechanisms in octahedral coordination complexes, *Acc. Chem. Res.* (in press).

Appendix 3

Firms Supplying High Pressure Equipment

1. American Instrument Co. (Aminco), Silver Spring, Maryland 20910, U.S.A. (UK agents, PSIKA Ltd., Glossop, Derbyshire.)
2. Autoclave Engineers Ltd., 2930 West 22nd Street, Erie, Pennsylvania 16512, U.S.A.
   (U.K. supplier) Cross Lances Road, Hounslow, Middlesex, England.
   (French supplier) Burton Corblin, 78–80 Bd Saint-Marcel, 75224 Paris, Cedex 05, France.
3. Basset–Bretagne–Loire, 8 Rue Troyon, 92310 Sevres, France.
4. Coleraine Instrument Co. Ltd., 82 Killowen Street, Coleraine, N. Ireland.
5. High Pressure Equipment Ltd., 122 Linden Avenue, Erie, Pennsylvania 16505, U.S.A.
6. Pressure Products Industries (U.K.) Ltd., Commercial Avenue, Stanley Green Industrial Estate, Cheadle Hulme, Cheshire SK8 6QH, U.K.
7. Nova Swiss, Nova Werke, AG, Vogelsangstrasse 24, CH-8307, Effretikon Switzerland.
   (U.K. supplier), Olin Energy Systems Ltd., North Hylton Road, Sunderland SR5 3JD, England.
8. PSIKA Pressure Systems Ltd., 19 Royle Avenue, Glossop, Derbyshire SK13 9RD, England.
9. Tem-Pres Research, 1401 South Atherton Street, State College, Pennsylvania 16801, U.S.A.

Ancilliary Equipment

Hydraulic pumps

Tangye EPCO Ltd., Gough Road, Greet, Birmingham BT1 2NH, England.
Charles E. Madan and Co. Ltd., Atlantic Street, Broadheath, Altrincham, Cheshire.

Gauges
Budenberg Gauge Co. Ltd., P.O. Box 5, Altrincham, Cheshire WA14 4ER, U.K.

(N. American supplier), Peacock Bros. Ltd., Box 1040, Montreal, Canada.
Sangamo Weston Controls Ltd., North Bersted, Bognor Regis, Sussex
PO22 9BS, England.
Transducers (CEL) Ltd., 3 Trafford Road, Reading, Berks, England.
Coleraine Instrument Co. (q.v.).

Electrical seals

Specitec, 155 rue de Charonne, 75011 Paris, France.

High strength steels

Sanderson Kayser L td., Sheffield, England.
International Nickel Ltd., Thames House, Millbank, London SW1P 4QF,
England.

Diaphragm pumps

Olin Energy Systems Ltd. (Haskell Pumps)—see above.

List of Tables of Data

408

Index

410

LIST OF SYMBOLS USED

α	thermal expansivity; optical rotation
β_T, β_s	compressibility; isothermal, adiabatic
$\bar{\beta}$	compressibility, partial molar
γ	ratio of principle specific heats; activity coefficient related to ideal dilute solution
γ_v	pressure-temperature coefficient
Δ	difference; heating operation
ε	dielectric constant
η	viscosity
$\Lambda^0, \Lambda^\infty$	molar conductivity
λ	single ion conductivity, wavelength
κ	pressure derivative of $\Delta\bar{V}$; transmission coefficient
σ	stress; radius; substituent constant (Hammett); surface tension
ρ	density; reaction constant (Hammett)
v	stoichiometric numbers; frequency; $\bar{v}$ = wave number
μ	chemical potential
τ	relaxation time
σ	stress; radius
ϕ	pressure coefficient of viscosity
Φ	$1/\varepsilon^2 . \partial\varepsilon/\partial P$; quantum yield
Φ_v	apparent partial molar volume
χ	electric field strength
a	activity
c	concentration (molar)
C	capacitance
C_p, C_v	specific heat
D	coefficient of diffusion
e	electronic charge
E, E_A	energy; Arrhenius activation energy, E.M.F.
f	activity coefficient relating to ideal mixtures
g	acceleration due to gravity
G	Gibbs free energy
H	enthalpy
ΔH_L	latent heat
h	Planck's constant
k	specific rate constant; Boltzmann constant
K	equilibrium constant; radius ratio for a hollow cylinder
K_A	acid dissociation constant
K	secant bulk modulus